PUBLICATIONS OF THE NEWTON INSTITUTE

T0269353

Solar and Planetary Dynamos

Publications of the Newton Institute

Edited by P. Goddard
Deputy Director, Isaac Newton Institute for Mathematical Sciences

The Isaac Newton Institute of Mathematical Sciences of the University of Cambridge exists to stimulate research in all branches of the mathematical sciences, including pure mathematics, statistics, applied mathematics, theoretical physics, theoretical computer science, mathematical biology and economics. The four six-month long research programmes it runs each year bring together leading mathematical scientists from all over the world to exchange ideas through seminars, teaching and informal interaction.

Associated with the programmes are two types of publication. The first contains lecture courses, aimed at making the latest developments accessible to a wider audience and providing an entry to the area. The second contains proceedings of workshops and conferences focusing on the most topical aspects of the subjects.

SOLAR AND PLANETARY DYNAMOS
Proceedings of a NATO Advanced Study Institute
held at the Isaac Newton Institute, Cambridge, September 1992

edited by

M. R. E. Proctor

P. C. Matthews

and

A. M. Rucklidge

University of Cambridge

CAMBRIDGE
UNIVERSITY PRESS

CAMBRIDGE UNIVERSITY PRESS
Cambridge, New York, Melbourne, Madrid, Cape Town, Singapore, São Paulo

Cambridge University Press
The Edinburgh Building, Cambridge CB2 8RU, UK

Published in the United States of America by Cambridge University Press, New York

www.cambridge.org
Information on this title: www.cambridge.org/9780521454704

First published 1993
This digitally printed version 2008

A catalogue record for this publication is available from the British Library

ISBN 978-0-521-45470-4 hardback
ISBN 978-0-521-05415-7 paperback

Contents

Preface

Magnetes Geheimnis, erkläre mir das!
Kein größer Geheimnis als Lieb und Haß.

— Goethe

The secret of magnetism, expound me its history!
Save loving and hating, there's no greater mystery.

This volume contains papers contributed to the NATO Advanced Study Institute 'Theory of Solar and Planetary Dynamos' held at the Isaac Newton Institute for Mathematical Sciences in Cambridge from September 20 to October 2 1992. Its companion volume 'Lectures on Solar and Planetary Dynamos', containing the texts of the invited lectures presented at the meeting, will appear almost contemporaneously. It is a measure of the recent growth of the subject that one volume has proved insufficient to contain all the material presented at the meeting: indeed, dynamo theory now acts as an interface between such diverse areas of mathematical interest as bifurcation theory, Hamiltonian mechanics, turbulence theory, large-scale computational fluid dynamics and asymptotic methods, as well as providing a forum for the interchange of ideas between astrophysicists, geophysicists and those concerned with the industrial applications of magnetohydrodynamics.

The papers included have all been refereed as though for publication in a scientific journal, and the Editors are most grateful to the referees for helping to get all the papers ready in such a short time. They also wish on behalf of the Scientific Organising Committee to record their appreciation of the dedication of the staff of the Isaac Newton Institute, who coped cheerfully with many bureaucratic complexities, and to give special thanks to the Deputy Director, Peter Goddard, for making the whole meeting possible.

The Editors strove to produce an unambiguous classification of the papers by subject; this proving beyond their abilities, the papers have been placed in alphabetical order by author, with a full subject index at the back.

It is just over 20 years since the last NATO Advanced Study Institute to be held in Cambridge on problems connected with dynamo theory (*Magnetohydrodynamic Phenomena in Rotating Fluids*: June–July 1972) and the report of that meeting makes interesting reading (H.K. Moffatt 1973; *J. Fluid Mech.* **57**, 625–649). It is a pleasure to note that all four of the surviving Session Chairmen attended the present meeting. At that time, the discussions of 'mean-field electrodynamics' were unconstrained by comparisons with detailed numerical simulations, and so the complexities that have recently emerged regarding nonlinear effects were below the horizon. Nor was there much discussion of the role of large-scale flows in limiting field amplitudes – a problem which is still not fully resolved. There was no 'fast dynamo theory' or descriptions of galactic fields. But the problem of geomagnetic reversal, whose resolution was the challenge given by Sir Edward Bullard at the beginning of the earlier meeting, was given little attention at this conference. One hopes that the enthusiasm so evident among the younger participants will translate into a proper theory of reversals that can be presented at the next NATO meeting in 2012!

M.R.E. Proctor
P.C. Matthews
A.M. Rucklidge

Cambridge, March 1993

Scientific Organising Committee

A. Brandenburg	(Denmark)	NORDITA, Copenhagen
S. Childress	(USA)	Courant Institute, New York University
H.K. Moffatt	(UK)	DAMTP, University of Cambridge
M.R.E. Proctor (Director)	(UK)	DAMTP, University of Cambridge

Local Organising Committee
A.D. Gilbert
P.C. Matthews
M.R.E. Proctor (Chairman)
A.M. Rucklidge

List of Participants

S.W. Anderson	(USA)	Phillips Laboratory, MA
A.P. Anufriev	(Bulgaria)	Bulgarian Academy of Sciences
C.F. Barenghi	(UK)	University of Newcastle upon Tyne
D.M. Barker	(UK)	University of Manchester
K. Bajer	(Poland)	University of Warsaw
B.J. Bayly	(USA)	University of Arizona
P. Bell	(UK)	University of Newcastle upon Tyne
S.I. Braginsky	(USA)	University of California
A. Brandenburg	(Denmark)	NORDITA, Copenhagen
D.P. Brownjohn	(UK)	University of Cambridge
F.H. Busse	(Germany)	University of Bayreuth
P. Cardin	(France)	Ecole Normale Superieure, Paris
S. Childress	(USA)	New York University
A.Y.K. Chui	(UK)	University of Cambridge
P. Coffey	(UK)	University of Newcastle upon Tyne
P. Collet	(France)	Ecole Polytechnique, Palaiseau
I. Cupal	(Czech Republic)	Czechoslovak Academy of Sciences
S. Drew	(UK)	University of Glasgow
D.R. Fearn	(UK)	University of Glasgow
A. Ferriz-Mas	(Germany)	Kiepenheuer-Institut, Freiburg
M. Foth	(Germany)	Universitäts-Sternwarte, Göttingen
P.A. Fox	(USA)	High Altitude Observatory, Boulder
U. Frisch	(France)	Observatoire de Nice
A. Gailitis	(Latvia)	Latvian Academy of Sciences
B. Galanti	(Japan)	ICFD, Tokyo
D.J. Galloway	(Australia)	University of Sydney
M. Ghizaru	(Romania)	Romanian Academy of Sciences
A.D. Gilbert	(UK)	University of Cambridge
P. Heptinstall	(UK)	University of Newcastle upon Tyne
R. Hollerbach	(UK)	University of Exeter
D.W. Hughes	(UK)	University of Leeds
A. Jackson	(UK)	University of Oxford
D. Jault	(France)	Institut de Physique du Globe, Paris
C.A. Jones	(UK)	University of Exeter
T. Kambe	(Japan)	University of Tokyo
R.R. Kerswell	(UK)	University of Newcastle upon Tyne
Y. Kimura	(USA)	NCAR, Boulder

I. Klapper	(USA)	University of Arizona
E. Knobloch	(USA)	University of California
W. Kuang	(USA)	Harvard University
R.M. Kulsrud	(USA)	Princeton University
C. Lamb	(UK)	University of Glasgow
A. Lazarian	(UK)	University of Cambridge
D. Linardatos	(UK)	University of Cambridge
D.F. Loper	(USA)	Florida State University
W.V.R. Malkus	(USA)	Massachusetts Institute of Technology
P.C. Matthews	(UK)	University of Cambridge
H.K. Moffatt	(UK)	University of Cambridge
D. Moss	(UK)	University of Manchester
M. Núñez	(Spain)	Universidad de Valladolid
A. Ossendrijver	(Netherlands)	Laboratory for Space Research, Utrecht
N. Platt	(USA)	NavSWC, Silver Spring, MD
A. Pouquet	(France)	OCA, Nice
T. Prautzsch	(Germany)	Universitäts-Sternwarte, Göttingen
M.R.E. Proctor	(UK)	University of Cambridge
P.H. Roberts	(USA)	University of California
A.M. Rucklidge	(UK)	University of Cambridge
A.A. Ruzmaikin	(Russia)	IZMIRAN, Troitsk
R. Schlichenmaier	(Germany)	Kiepenheuer-Institut, Freiburg
C. Sellar	(UK)	University of Glasgow
H. Shen	(USA)	Dartmouth College
H. Shimizu	(USA)	Florida State University
A.M. Shukurov	(Russia)	Moscow State University
D.D. Sokoloff	(Russia)	Moscow State University
A.M. Soward	(UK)	University of Newcastle upon Tyne
E.A. Spiegel	(USA)	Columbia University
T.F. Stepinski	(USA)	Lunar and Planetary Institute, Houston
M.G. St. Pierre	(USA)	University of California
P.L. Sulem	(France)	Observatoire de Nice
L. Tao	(USA)	University of Chicago
S.I. Vainshtein	(USA)	University of Chicago
M. Vergassola	(France)	Observatoire de Nice
P. Watson	(UK)	University of Cambridge
N.O. Weiss	(UK)	University of Cambridge
J. Wicht	(Germany)	University of Bayreuth
K. Zhang	(UK)	University of Exeter
O. Zheligovsky	(Russia)	Academy of Sciences USSR
V. Zheligovsky	(Russia)	Academy of Sciences USSR
E.G. Zweibel	(USA)	Astrophysical Sciences, Boulder

Magnetic Noise and the Galactic Dynamo

S.W. ANDERSON* & R.M. KULSRUD

Princeton Plasma Physics Lab
Princeton, NJ 08543 USA

*Current address:
Space Physics Directorate
Phillips Lab, Geophysics Directorate
Hanscom Air Force Base, MA 01731 USA

Galactic magnetic fields are widely thought to be the product of a turbulent mean field dynamo. We find, however, that kinematic mean field theory is inapplicable for galactic parameters because there is no effective way to destroy the small-scale fluctuating magnetic fields. We find that this 'magnetic noise' grows exponentially with a time constant of 10^4 years, while the dynamo grows with a 2×10^8 year time scale. The dynamo field quickly becomes unobservable under such conditions and the kinematic approximation fails before the mean field grows significantly.

1 INTRODUCTION

Our galaxy and others are permeated by magnetic fields. They play an important role in star formation, in the support of molecular clouds against collapse, and in cosmic ray confinement. With a field strength of a few microgauss, they are comparable in in energy density to thermal energy, radiation, and cosmic rays. These fields are widely assumed to be the result of a dynamo operating on an initial seed field.

Dynamos work by folding magnetic field lines back on themselves constructively more often than destructively. Mean field theory assumes that the many folds in the field with no net contribution are destroyed, usually by resistivity. What would happen if these small disordered fields were not destroyed? They would obscure the growing large-scale field and might dominate the total magnetic energy. This is indeed a concern for galactic dynamo theory as magnetic loops 0.1 pc across need 10^{22} years to decay ohmically.

1

M.R.E. Proctor, P.C. Matthews & A.M. Rucklidge (eds.)
Theory of Solar and Planetary Dynamos, 1–7
©1993 Cambridge University Press.

We consider homogeneous, isotropic turbulence with a Kolmogorov spectrum extending from a driving scale Λ_{min}^{-1} to a viscous scale Λ_{max}^{-1}. We take the 100 pc intercloud spacing to be the driving scale and, from the viscosity of neutrals, take 0.1 pc to be the viscous scale. The smallest turbulent eddies then have a turnover time of $\gamma^{-1} = 10^4$ years. We neglect differential rotation and helicity since they act on the dynamo time scale, that is, much more slowly than the growth of small-scale fluctuations. Given these ingredients, how quickly do small-scale magnetic fluctuations grow compared to typical dynamo growth rates of 10^8 years (Ruzmaikin *et al.* 1988)?

2 MAGNETIC ENERGY SPECTRUM

How can we quantify the criticism that the small-scale fields dominate the large-scale dynamo field $\langle \mathbf{B} \rangle$? The easiest way is to divide the magnetic energy into its spectrum $M(k)$;

$$B^2 = \int M(k)\, dk. \tag{1}$$

Following $M(k,t)$ will tell us whether energy builds up more quickly on the small or large scales. The two limiting cases are:

i) Magnetic energy is concentrated on large scales. A mean field dynamo successfully explains galactic magnetic fields.

ii) There is much more energy on small scales than large. The dynamo works, but its field is completely obscured by the growing magnetic fluctuations.

We find unambiguous confirmation of ii). For an initial large-scale field of 10^{-17} gauss we find that the total energy $\langle B^2 \rangle$ grows to an equipartition strength of 10^{-6} gauss before the mean field energy $\langle B \rangle^2$ has grown by even one percent.

Vainshtein (1970) demonstrated that the average field $\langle \mathbf{B} \rangle$ can grow exponentially in a turbulent fluid. We use the same formalism to find an equation for the total magnetic energy $\langle B^2 \rangle$, and then divide it into its spectrum. The starting point is the induction equation

$$\frac{\partial \mathbf{B}}{\partial t} = \nabla \times (\mathbf{v} \times \mathbf{B}) + \eta \nabla^2 \mathbf{B}. \tag{2}$$

We Fourier transform $\mathbf{B}$ and $\mathbf{v}$, and introduce a series solution

$$\tilde{\mathbf{B}} = \tilde{\mathbf{B}}_0 + \tilde{\mathbf{B}}_1 + \tilde{\mathbf{B}}_2 + \ldots \tag{3}$$

which gives the recursion relation

$$\tilde{\mathbf{B}}_{n+1}(\mathbf{k}, t) = \int_0^t \int \mathbf{k} \times \left[\tilde{\mathbf{v}}(\mathbf{k} - \mathbf{k}_0) \times \tilde{\mathbf{B}}_n(\mathbf{k}_0, t') \right] d\mathbf{k}_0 dt'. \tag{4}$$

The magnetic energy $|\tilde{\mathbf{B}}|^2 = (\tilde{\mathbf{B}}_0 + \tilde{\mathbf{B}}_1 + \ldots) \cdot (\tilde{\mathbf{B}}_0^* + \tilde{\mathbf{B}}_1^* + \ldots)$ can then be written as a function of the initial velocity and field.

We move from a particular system of $\mathbf{v}$ and $\mathbf{B}$ to the ensemble average of many similarly prepared systems by replacing every occurrence of the tensor $\tilde{\mathbf{v}}\tilde{\mathbf{v}}$ with the ensemble averaged tensor

$$\langle \tilde{\mathbf{v}}(\mathbf{k},t)\tilde{\mathbf{v}}(\mathbf{k}',t')\rangle = \left\{\left(\mathbf{I} - \hat{\mathbf{k}}\hat{\mathbf{k}}\right)U_1(k) + \mathbf{I} \times \hat{\mathbf{k}}\,U_2(k)\right\}\delta\left(t - t'\right)\delta(\mathbf{k} - \mathbf{k}'), \quad (5)$$

where $U_1(k)$ and $U_2(k)$ are the spectra of the non-helical and helical turbulent motions, respectively, and $\hat{\mathbf{k}}$ is the unit vector in the direction of $\mathbf{k}$. By doing so we find the ensemble average of the magnetic energy spectrum M. This form of $\langle \tilde{\mathbf{v}}\tilde{\mathbf{v}}\rangle$ guarantees incompressibility and isotropy.

The small parameter that justifies the expansion in equation (3) is $\delta t/\tilde{v}k$; we only advance the equation for $\tilde{\mathbf{B}}^2$ a time δt short compared to the evolution time of the total magnetic energy. Using the short correlation time explicit in equation (5) we can then advance another short time step, etc. After advancing by δt we subtract $M(k,0)$ and divide by δt to find (Kulsrud & Anderson 1992)

$$\frac{\partial M(k)}{\partial t} = \int_{k-\Lambda_{max}}^{k+\Lambda_{max}} K(k,k_0)M(k_0)\,dk_0 - 2\left(\eta_T + \eta\right)k^2 M(k) \quad (6)$$

where the kernel K and turbulent resistivity η_T are given by

$$K(k,k_0) = 2\pi k^4 \int \frac{\sin^3\theta}{k_1{}^2}(k^2 + k_0^2 - kk_0\cos\theta)U_1(k_1)\,d\theta \quad (7)$$

$$\eta_T = \frac{1}{3}\int U_1(k)\,dk \quad (8)$$

and $k_1{}^2 = k^2 + k_0^2 - 2kk_0\cos\theta$. This equation was found in different form by Kraichnan & Nagarajan (1967) and by Kazantsev (1968). It does not include the possibility of dynamo growth since we have neglected the anti-symmetric portion of the velocity correlation tensor U_2. However, the dynamo effect has a time scale of 10^8 years, while the production of magnetic noise proceeds with a 10^4 year period. After 10^6 years the total magnetic energy has grown to the point of invalidating the kinematic assumption, long before the alpha effect has had time to act.

These equations yield a characteristic magnetic energy growth rate

$$\gamma = \frac{1}{3}\int k^2 U_1(k)\,dk, \quad (9)$$

which is roughly the turnover rate of the smallest eddies. The smallest eddies rotate the fastest and therefore wrap up the field lines most quickly.

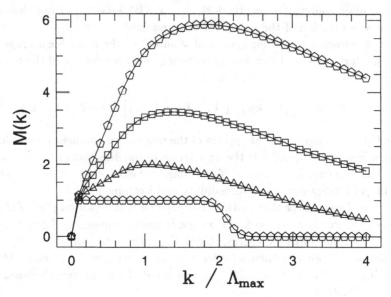

Figure 1. Early evolution of an initially flat spectrum. The lowest curve is the initial spectrum $M(k,0)$, followed by the spectra at times $t = 0.5\gamma^{-1}$, γ^{-1}, and $1.5\gamma^{-1}$.

In the absence of resistivity our integral equation (6) for $\partial M/\partial t$ can be easily integrated over all k to find that the total magnetic energy grows as $E_M \sim e^{2\gamma t}$. The energy increases in two ways: the spectrum cascades to smaller scales and the energy on each scale increases. When resistivity is added the spectrum M can no longer continually cascade to smaller scales, and we find that the magnetic energy growth is reduced to $E_M \sim e^{3\gamma t/4}$. There is no steady state magnetic spectrum in the kinematic limit; the total magnetic energy grows exponentially (even with resistivity) until nonlinear effects set in.

Our strategy is substantially simpler than the EDQNM approach (Pouquet *et al.* 1976) since we are asking the kinematic question of how a weak magnetic field responds to a turbulent fluid instead of the dynamical question of how the velocity and field evolve together. This is appropriate as we are investigating the scope of validity for the widely used kinematic mean field dynamo equation.

3 NUMERICAL RESULTS

3.1 Methods
All length scales are normalized to the viscous scale $L_{vis} = \Lambda_{max}^{-1}$ as we are interested in both the Kolmogorov turbulence on larger scales and in the smaller scales where magnetic energy is created. Time is measured in turnover

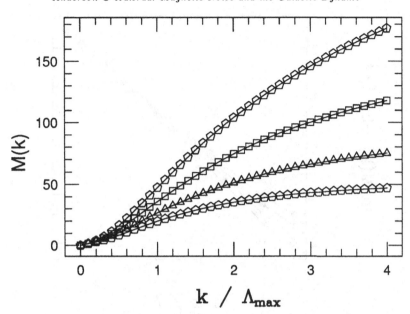

Figure 2. Later evolution of the same initial spectrum. Times $t = 3.5\gamma^{-1}, \ldots, t = 5\gamma^{-1}$ are shown in ascending order.

times of eddies on the viscous scale.

The resistive scale is 10^8 times smaller than the viscous scale, making it impossible to follow $M(k)$ on every scale on which magnetic energy is building up. Instead, we evolve the large-k tail using a Green's function outlined by Kulsrud & Anderson (1992). Energy builds up on scales too small to follow, but does so in a known way whose effects we can include.

The spectrum $M(k, t)$ is advanced in time using a 4th-order Runge–Kutta scheme with adaptive step size control. The integration in equation (6) can be done with the 5th-order Bode's method since the kernel $K(k, k_0)$ is smooth.

3.2 Results

Figure 1 shows how an initially flat magnetic energy spectrum evolves; we see that most of the energy is being created on small scales. The characteristic growth time is on the order of 10^4 years, or about the smallest eddy turnover time. This is much shorter than the 10^8 year mean field dynamo growth time.

Figure 2 shows the same initial spectrum later in time. The peak in the magnetic spectrum is well below the viscous scale after even a few smallest eddy turnover times. The spectrum assumes a definite shape with an overall growth factor. Figure 3 is the same data as Figure 2, but on a log-log scale and with the exponential time factor $e^{3\gamma t/4}$ taken out. We see that the spectrum approaches the form

$$M(k, t) \sim e^{3\gamma t/4} k^{3/2} \tag{10}$$

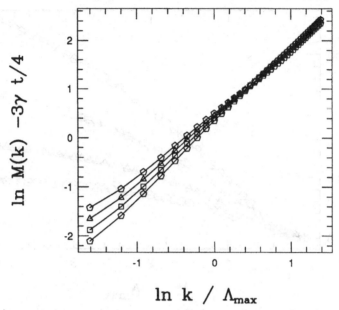

Figure 3. Data of Figure 2 on a log-log plot with growth factor $e^{3\gamma t/4}$ removed. We see that the spectrum approaches $M \sim k^{3/2}$ for small scales.

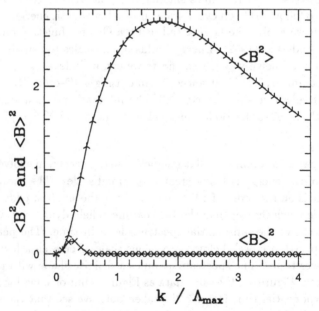

Figure 4. Comparison of $\langle B^2 \rangle$ and $\langle B \rangle^2$. The average total energy (top) is much greater than the energy in the average field (bottom) even after one turnover time.

down to the resistive scale.

Figure 4 shows both the energy in the mean field $\langle B \rangle^2$ and the total energy $\langle B^2 \rangle$ after one turnover time. The maximum alpha effect and a small Reynolds number $R_m = 25$ were used to make the energy in the mean field $\langle B \rangle^2$ as large as possible.

4 CONCLUSIONS

When a weak magnetic field is embedded in a turbulent fluid, what is the resulting spectrum of magnetic energy? We find that $M \sim k^{3/2} e^{3\gamma t/4}$ down to the resistive scale. That is, most of the magnetic energy is contained in features on the resistive scale, and the spectrum does not reach a steady state but grows exponentially until limited by nonlinear processes.

This investigation extends mean field dynamo theory, whose purpose is to find $\langle \mathbf{B}(t) \rangle$, to find the evolution of the mean magnetic energy $\langle B^2(t) \rangle$. The energy in the large-scale field $\langle B \rangle^2$ grows much more slowly than the total energy $\langle B^2 \rangle$, which consists mostly of small-scale fluctuations. (10^8 yr vs. 10^4 yr growth time). The growing magnetic noise completely obscures the weak average field and invalidates the kinematic assumption before the mean field has grown significantly. The origin of galactic magnetic fields needs to be reexamined since the simple versions of mean field dynamo theory do not apply for galactic parameters.

REFERENCES

Kazantsev, A.P. 1968 Enhancement of a magnetic field by a conducting fluid. *Sov. Phys. JETP* **26**, 1031–1034.

Kraichnan, R.H. & Nagarajan, S. 1967 Growth of turbulent magnetic fields. *Phys. Fluids* **10**, 859–870.

Kulsrud, R.M. & Anderson, S.W. 1992 The spectrum of random magnetic fields in the mean field dynamo theory of the galactic magnetic field. *Astrophys. J.* **396**, 606–630.

Pouquet, A., Frisch, U. & Léorat, J. 1976 Strong MHD helical turbulence and the nonlinear dynamo effect. *J. Fluid Mech.* **77**, 321–354.

Ruzmaikin, A.A., Shukurov, A.M. & Sokoloff, D.D. 1988 *Magnetic Fields of Galaxies.* Kluwer Academic Press.

Vainshtein, S.I. 1970 The generation of a large-scale magnetic field by a turbulent fluid. *Sov. Phys. JETP* **31**, 87–89.

On the Oscillation in Model Z

A.P. ANUFRIEV

Geophysical Institute
Bulgarian Academy of Science, Acad. Bonchev str., bl. 3
1113 Sofia, Bulgaria

I. CUPAL & P. HEJDA

Geophysical Institute
Czechoslovak Academy of Science, Boční II
141 31 Prague 4, Czech Republic

The paper deals with nonlinear decaying oscillations appearing in model Z. A method, based on the balance equations, is proposed which allows us to estimate whether or not the time behaviour of the solutions is correct. For this purpose the balance equation of energy and a new variable $J = B_\phi/s$ are used. The equation for J has conservative form. The oscillatory solution is characterized by two time scales. We speculate that the small time scale (the period of the oscillations) is connected to diffusion of azimuthal field through the boundary layer while the large time scale (the decay time of the oscillations) is linked to the diffusion of the meridional field (created in the boundary layer) into the volume of the core. The large meridional convection at the core–mantle boundary (CMB) plays a crucial role in this process.

---

1 INTRODUCTION

The solution of model Z has been found in many cases with account taken of both viscous and electromagnetic core–mantle coupling (Braginsky 1978; Braginsky & Roberts 1987; Braginsky 1988; Braginsky 1989; Cupal & Hejda 1989). Apart from Braginsky (1989), the time evolution of the solution was used simply as an aid to obtain the steady-state solution. Cupal & Hejda (1992) found numerically a transient solution of model Z having the form of a decaying oscillation. The accuracy of such solutions depends on the numerical method used, on the density of space and time discretization, and for that matter, on the character of the solution itself. An important question is which characteristics of the time behaviour of the solution reflect the real (physical) behaviour of the system and which follow from the limitations of the numerical method. Therefore, the balance equation of energy and a new variable, J, are

9

M.R.E. Proctor, P.C. Matthews & A.M. Rucklidge (eds.)
Theory of Solar and Planetary Dynamos, 9–17
©1993 Cambridge University Press.

used to estimate the accuracy of the solution at every time step. Apart from some multiplying factors, $J(= B/s)$ is a mean meridional electrical current across a circle with radius s in any plane parallel to the equatorial one. The calculation of individual terms in the balance equations is not connected directly with the numerical process by which the equations of the model are solved. Hence, we anticipate that the balance equations provide a good test of the solution. Our experience confirms this assumption. Whereas the energy balance equation plays the central role in this test, the balance equation for J helps us to understand the process by which the solution evolves with time.

The time-stepping method of the solution is usually started from a relatively arbitrary distribution of azimuthal, B, and meridional, $\mathbf{B}_p$, magnetic fields. During this initial phase of the calculation, the imbalance (the difference between the l.h.s. and r.h.s. of the energy balance equation) was large in our preliminary calculation. However, when the initial fields and other variables are 'adapted' the imbalance decreases (see Figure 1(b)). If the density of the space grid is increased the imbalance decreases essentially and also it decreases if a finer time step is used. Therefore, we hope that the transient oscillatory solution obtained by Cupal & Hejda (1992) has physical sense. The qualitative estimates of that solution and consequently some new features of model Z are found in section 4.

2 GOVERNING EQUATIONS

The equations governing the nearly symmetric hydromagnetic dynamo in dimensionless form were derived by Braginsky (1975); see also Braginsky (1993). Relative to spherical coordinates r, θ, ϕ or cylindrical coordinates s, ϕ, z, we introduce the stream function of poloidal field, ψ, and of poloidal velocity, χ, $(\mathbf{B}_p = s^{-1}\nabla\psi \times \mathbf{1}_\phi,\ \mathbf{v}_p = s^{-1}\nabla\chi \times \mathbf{1}_\phi)$ while v denotes the azimuthal velocity. In terms of the new variable $J = B/s$ the equations can be re-written in the form:

$$\frac{\partial \psi}{\partial t} = \nabla \cdot (-\psi \mathbf{v}_p + s^2 \nabla s^{-2}\psi) + s^2 \alpha J, \tag{1a}$$

$$\frac{\partial J}{\partial t} = \nabla \cdot (-J\mathbf{v}_p + s^{-2}\nabla s^2 J + \zeta \mathbf{B}_p), \tag{1b}$$

$$v_s = s^{-1}\nabla \cdot (s^2 J \mathbf{B}_p), \tag{1c}$$

$$\zeta = f + J^2 + \omega, \tag{1d}$$

where $\zeta(s,z) = v(s,z)/s$. $f(s,z)$ is the Archimedean or thermal wind, $J^2(s,z)$ is the magnetic wind and $\omega(s) = v(s,z_1)/s$ is the geostrophic shear determined by the equation

$$\omega = \frac{2\sqrt{z_1}}{\varepsilon s^3}\frac{d}{ds}\left(s^3 \int_0^{z_1} JB_s\, dz\right), \tag{1e}$$

where $z_1 = \sqrt{1 - s^2}$. Both the viscous coupling parameter ε and our scaling used coincide with Braginsky (1993). For given Archimedean wind f and α-effect with amplitudes f_0 and α_0 respectively, the system (1) can be solved subject to the boundary conditions of model Z (see e.g. Cupal & Hejda 1989). The α-effect can be prescribed in a slightly generalized form (Cupal & Hejda 1992) which makes it possible to investigate the effect of changing the thickness, δ, of the α-layer on the solution (Cupal & Hejda 1992).

Appropriate functions, which measure the amplitude of the magnetic field in the volume of the core, are the toroidal field energy, $E_B(t) = \int_{V+} B^2 dV$, and the squared poloidal magnetic flux, $E_\psi(t) = \int_{V+} \psi^2 dV$, integrated throughout the volume, V^+, of the northern hemisphere of the core. The time behaviour of the mean meridional current J in the northern hemisphere may be characterized by the volume integral $\Im(t) = \int_{V+} J dV$. Using (1b) we obtain the following balance equations for energy and $\Im$:

$$\frac{\partial E_B}{\partial t} = Q_A - Q_J - Q_\nu, \qquad (2a)$$

$$\frac{\partial \Im}{\partial t} = G_B + T_\omega - T_f. \qquad (2b)$$

In (2a) only the Archimedean forces, $Q_A = -2 \int_{V+} s f v_s dV$, do work. Energy in our $\alpha\omega$ dynamo is lost by Joule dissipation and by viscous dissipation in the Ekman layer:

$$Q_J = 2 \int_{V+} s^{-2} [\nabla(sB)]^2 dV, \qquad Q_\nu = 2\pi\varepsilon \int_0^1 (s\omega)^2 / \sqrt{z_1}\, s ds.$$

The terms on the r.h.s. of (2b) are

$$G_B = 2\pi [\int_0^{\frac{\pi}{2}} \frac{\partial(rB)}{\partial r} d\theta - \int_0^1 \frac{\partial B}{\partial z} ds],$$

$$T_\omega = 2\pi [\int_0^{\frac{\pi}{2}} \omega B_r \sin\theta\, d\theta - \int_0^1 \omega B_z\, s ds], \qquad T_f = 2\pi \int_0^1 f B_z\, s ds.$$

3 CALCULATED MODELS

Cupal & Hejda (1992) calculated several cases of model Z. Several solutions tended to oscillate before they reached their steady state. The oscillatory solution for $f_0 = 500$, $\varepsilon = 0.01$, $\alpha_0 = 25$ and $\delta = 0.3$ may be particularly useful in demonstrating the time behaviour of the solution. A general characteristic of all solutions is the quick adjustment of the azimuthal field to its steady state value about which in some cases it vacillates. The meridional field reaches its steady value relatively slowly and in the oscillatory case it vacillates about its changing mean value. Figure 1 represents the time evolution of the oscillatory solution. E_B and scaled values of E_ψ, Q_A are drawn in

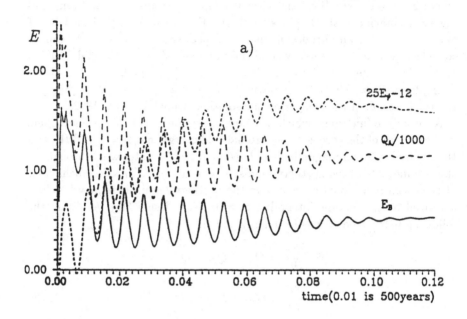

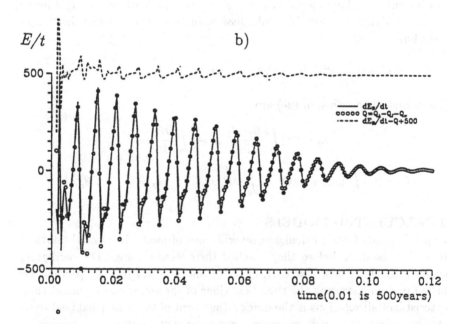

Figure 1. Time development of the oscillatory solution for $\alpha_0 = 25, \delta = 0.3$. (a) Energies E_B (solid line), E_ψ (short dash line) and the work of Archimedean forces Q_A (long dash line). (b) $\partial E_B/\partial t$ (solid line), $Q_A - Q_J - Q_\nu$ (circles) and the difference of both sides of (2a) (dashed line).

the upper graph (a). It should be noted that the equality of the long decay time for oscillations and the time to establish the poloidal field was confirmed by the numerical experiments under various conditions.

The time evolution of the solution demonstrated in Figure 1(a) is interesting. Therefore, we must ask whether it is not a numerical artifact, but rather reflects physical reality. Hence we tested the energy balance (2a) at each time step. The lower graph (b) in Figure 1 shows the behaviour of $\partial E_B / \partial t$ and the r.h.s. of (2a). The imbalance, to which a constant 500 is added to aid comparison, is in fact the difference between the l.h.s. and r.h.s. of (2a). The start of the graph is characterized by a large imbalance and it is not sensible to discuss the time behaviour of the solution. After time $t = 0.02$ the imbalance begins to decrease and the time behaviour illustrated by the solution makes physical sense. However even later e.g. after $t = 0.11$ the imbalance remains small, but non-zero as a consequence of the space discretization.

The space discretization and the length of time step can influence the solution, but artificial time behaviour of the numerical origin can easily be distinguished (Anufriev *et al.* 1993). The balance equation (2a) isolates this problem and so we know which numerical results are unreliable. The numerical experiments also confirmed that the imbalance generally decreases when a finer space grid or shorter time step are used. The detailed analysis of a period of the oscillatory solution confirmed again that the imbalance shows some substantial non-zero values at the times 0.0216 and 0.0276 of maximum E_B and that this imbalance generally decreases when a finer grid is used. This detailed numerical analysis (Anufriev *et al.* 1993) allows us to conclude that (2a) is a good test of the solution and that the oscillations are not a product of our numerical procedure.

The contour maps of the solution at time $t = 0.0216$ of maximum E_B and at time $t = 0.0248$ of its minimum are drawn in Figure 2. It can be seen that the mainly poloidal velocity $\mathbf{v}_p = s^{-1} \nabla \chi \times \mathbf{1}_\phi$ is large near the CMB at times of maximum of E_B. The complicated behaviour of ζ in the generating region at time of maximum E_B is a consequence of the geostrophic shear having peaks in this region. Numerical experiments with a finer grid showed that the peaks of ω disappear and as a result also ζ loses its complicated character in the generating region, while the pattern of the other isolines remain unchanged.

4 ANALYSIS OF THE SOLUTION

The balance of energy (2a) tells us at which time the solution represents the time behaviour of the physical system. The balance equation (2b) for $\Im$ was also checked numerically and the results were roughly the same as in the case of (2a). However, the checking role of the balance equation (2b) is not so significant as the energy balance (2a) because the equation is directly connected to (1b) and, thus, it cannot be simultaneously an independent

(a)

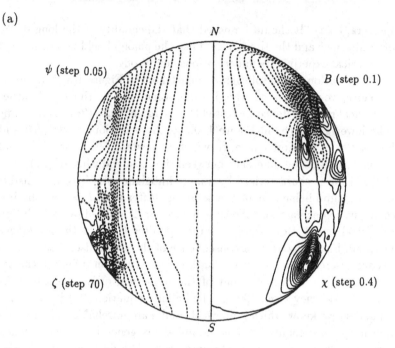

(b)

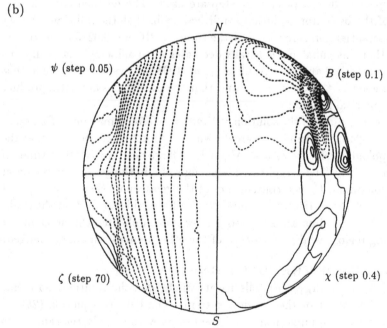

Figure 2. Contour maps of the solution referring to (a) maximum E_B and to (b) minimum E_B. The negative isolines are plotted by dashed lines and the positive and zero isolines are plotted by solid lines.

test of (1b). Nevertheless, the balance relation (2b) helps to explain the mechanism of field generation. The numerical analysis of (2b) shows that J, and therefore B, is created mainly at the CMB due to geostrophic shear, ω, but, at the same time it is 'destroyed' by geostrophic and Archimedean velocity, $\omega + f$, at the equatorial plane. An auxiliary role of the J-diffusion in this process is in spreading the azimuthal field through the CMB boundary layer. There meridional circulation moves it to the equatorial region where again it diffuses to the equatorial plane at which it is destroyed. Therefore, the azimuthal velocity not only creates azimuthal magnetic field, but also destroys it.

The main characteristic of the steady solution of model Z is the extremely large azimuthal angular velocity ζ. Such a large ζ should lead to a large term involving ζ in (1b), which is larger in order of magnitude than other terms on r.h.s. of (1b). Consequently an extremely large value of $\partial J / \partial t$ is to be expected. The only possibility to eliminate this (and it is also the case of model Z) is to keep the term $\nabla \cdot (\zeta \mathbf{B}_p)$ small and so $\zeta = \zeta(\psi)$. It does mean that the isolines of ζ and ψ must be parallel in the main volume of the core where the isorotation law is true (see e.g. Braginsky 1965). In fact, this phenomenon suggests model Z might be better named model ζ. Of course, it is not true at the CMB where large gradients in J equilibrate the term $\zeta \mathbf{B}_p$. Another important characteristic of model Z, which is usually ignored, is the large meridional velocity $\mathbf{v}_p$ near the CMB. Specifically, the magnetic field, whose isolines approach the region of closed streamlines, tends to disappear from this area (see e.g. Moffatt 1978). The numerical solution of model Z shows that such an expulsion of the poloidal field occurs near the CMB in the generation region also in this model (see Figure 2). Indeed it may explain why the solution with non-zero α-effect in the main volume of the core failed (Braginsky 1978). Firstly poloidal field is expelled from the generation region and consequently the last term in (1b) weakens in this region and so B decays. It would be interesting to test this idea numerically.

We can also observe the direct influence of poloidal velocity on the azimuthal magnetic field. We noted that the numerical analysis of (2b) shows that toroidal field is created by geostrophic shear at the CMB and is destroyed at the equatorial plane by the sum of the geostrophic and Archimedean velocity. However, the transfer of the toroidal field between these two regions is realized by the large poloidal velocity and not by the slow mechanism of the diffusion. This also explains the quick adaptation of the azimuthal field noted by Braginsky (1991). This process is accompanied by the development of the 'magnetic' layer of thickness δ_m at the CMB which continues at the inner boundary of the α-region where the α-effect falls to zero. The thickness of this layer can be found by comparing the convective and diffusion terms in (1b). If $J\mathbf{v}_p \sim s^{-2}\nabla(s^2 J)$ then $\mathbf{v}_p \sim \delta_m$ in the layer and so $\delta_m \sim 1/\mathbf{v}_p$.

Figure 1 represents a typical behaviour of the model Z solution. Starting from arbitrary initial magnetic field the solution oscillates (vacillates) or not, but it is quickly stabilized to its steady state level. The model Z solution is a very stable one. Two time scales can be observed in Figure 1. The long one is connected with the process of establishing the poloidal field which coincides with the decay time of the oscillations. The oscillation practically disappears during the time ≈ 0.1 and Figure 1 suggests the characteristic decay time $\lesssim 1/20$ which is greater than the characteristic decay time of the basic poloidal mode π^{-2} (see e.g. Moffatt 1978). The decay time is in agreement with Braginsky (1989) where he estimated it $\approx 1/23$ for the solution when the α-effect was periodically switched off and on. Braginsky (1989) used that value to estimate the conductivity of the Earth's core. This time scale is connected directly with the diffusion of the poloidal field generated near the CMB in the α-region into the main volume of the core, however, this process is accelerated by the meridional velocity and thus the decay time of the oscillation is shorter than the dipole decay time. The short time scale is the period of the oscillations. The meridional velocity is generated by poloidal and toroidal fields [see (1c)] and so this process is essentially nonlinear and, for some solutions, leads to nonlinear oscillations. The detailed oscillation mechanism is not completely clear, though clearly the large meridional velocity and the diffusion of the azimuthal field through the magnetic layer δ_m play the main roles. The steady-state equilibrium is disturbed in some way and leads to vacillations. It should be noted that the short time scale oscillations in Figure 1(a) are very similar to those obtained by Braginsky (1989). The decaying oscillation of the same character also appears when the amplitude of the α-effect is changed during the calculation (Cupal & Hejda 1992).

5 CONCLUSION

The balance equation (2a) helps us to decide for which time intervals the time-dependent behaviour of our solution makes physical sense. After several initial time steps, where the imbalance caused by the numerical process is large, the later time steps reflect the true time behaviour of the solution. The oscillating solutions are characterized by the short oscillation period time scale, and by the long decay time scale of the oscillations. This is also the time to establish the poloidal field at its steady level. The small time scale seems to be a diffusion time for toroidal field through the magnetic boundary layer while the long time scale is a diffusion time of poloidal field from the generating region into the main volume of the core. The second time scale can also be observed in the non-oscillating solutions. The balance equation (2b) helps us to understand the development of the solution and its time behaviour gives the possibility to imagine in more detail the complicated mechanism of generation in model Z. Generally, the investigation of the time behaviour of the solution in model Z makes it possible for us to understand much better the process of creating the layer structure in model Z and the role of the large geostrophic velocity.

REFERENCES

Anufriev, A.P., Cupal, I. & Hejda, P. 1993 Time evolution of the solution of model Z. In *IAU-Symposium Series* (in press). Kluwer Academic Publishers, Dordrecht.

Braginsky, S.I. 1965 Self-excitation of a magnetic field during the motion of a highly conducting fluid. *Soviet Phys. JETP* **20**, 726–737. (English translation).

Braginsky, S.I. 1975 The nearly symmetric hydromagnetic dynamo model Z I. *Geomag. Aeron.* **15**, 122–128. (English translation).

Braginsky, S.I. 1978 The nearly symmetric hydromagnetic dynamo model Z II. *Geomag. Aeron.* **18**, 225–231. (English translation).

Braginsky, S.I. 1988 The Z model of the geodynamo with magnetic friction. *Geomag. Aeron.* **28**, 407–412. (English translation).

Braginsky, S.I. 1989 The Z model of the geodynamo with an inner core and the oscillation of the geomagnetic dipole. *Geomag. Aeron.* **29**, 98–103. (English translation).

Braginsky, S.I. 1991 Towards a realistic theory of the geodynamo. *Geophys. Astrophys. Fluid Dynam.* **60**, 89–134.

Braginsky, S.I. 1993 The nonlinear dynamo and model Z. In *Lectures on Solar and Planetary Dynamos* (ed. M.R.E. Proctor & A.D. Gilbert), Cambridge University Press.

Braginsky, S.I. & Roberts, P.H. 1987 Model Z geodynamo. *Geophys. Astrophys. Fluid Dynam.* **38**, 327–349.

Cupal, I. & Hejda, P. 1989 On the computation of a model Z with electromagnetic core–mantle coupling. *Geophys. Astrophys. Fluid Dynam.* **49**, 161–172.

Cupal, I. & Hejda, P. 1992 Magnetic field and α-effect in model Z. *Geophys. Astrophys. Fluid Dynam.* **67**, 87–97.

Moffatt, H.K. 1978 *Magnetic Field Generation in Electrically Conducting Fluids.* Cambridge University Press.

Nonlinear Dynamos in a Spherical Shell

CARLO F. BARENGHI

Dept. of Mathematics and Statistics
University of Newcastle upon Tyne
Newcastle upon Tyne, NE1 7RU UK

Nonlinear models of the Geodynamo have been studied numerically using spectral methods. The axisymmetric magnetic induction equation has been solved in the geometry of a spherical shell
in rapid rotation under prescribed α and ω effects. The time dependence of the solutions is compared with the observed frequency
of reversals of the Earth's magnetic field.

1 INTRODUCTION

In the last few years there has been a renewed interest in dynamos in rapidly
rotating systems, of which the Geodynamo is the most important example.
In these systems the inertial and viscous terms in the fluid momentum equations can be considered asymptotically small and one is led to consider the
role played by Taylor's constraint (Jones 1991; Soward 1992). Numerical
calculations of such dynamos have been carried out by solving the axisymmetric magnetic induction equations for the toroidal and poloidal magnetic
field components under prescribed α and ω effects. A variety of models and
geometries have been explored. The studies which are more closely related
to the present work are the calculations of Abdel-Aziz & Jones (1988) and
Jones & Wallace (1992) in planar geometry, of Hollerbach & Ierley (1991) and
Hollerbach, Barenghi & Jones (1992) in a sphere, and of Barenghi & Jones
(1991) and Barenghi (1992a,b) in a spherical shell.

M.R.E. Proctor, P.C. Matthews & A.M. Rucklidge (eds.)
Theory of Solar and Planetary Dynamos, 19–26
©1993 Cambridge University Press.

The observed westward drift of some patches of the Earth's magnetic field suggests that in order to model the Geodynamo one should study the magnetic induction equation in the $\alpha\omega$ limit (Roberts 1988). In this limit the poloidal field is driven by the α effect and the large toroidal field is generated by the ω effect only. Because of the arbitrariness of the α and ω inputs the parameter space is very large and one cannot say that all possible models have been explored. Nevertheless it is becoming apparent that the $\alpha\omega$ dynamo equations tend to have solutions which oscillate very rapidly, a fact which is also well known from stellar models. Using a time scale δ^2/η based on the magnetic diffusivity η and the size δ of the Earth's core, the following values of the period T of the $\alpha\omega$ solutions have been found in different models and geometries: $T = 0.07$ to 0.12 (linear models, Roberts 1972); $T = 0.18$ and 0.08 (linear and nonlinear models, Barenghi & Jones 1991); $T = 0.1$ (nonlinear models, Jones & Wallace 1992); $T = 0.3$ to 0.1 and 0.1 to 0.2 (linear and nonlinear models, Abdel-Aziz & Jones 1988); $T = 0.045$ to 0.085 (nonlinear models, Hollerbach, Barenghi & Jones 1992). These values are in disagreement with the observation that the Earth's magnetic field has had long periods of fixed polarity or has undergone slow, irregular oscillations: in the last 45 My the periodicity of reversals has been approximately $T = 5$.

2 THE MODEL EQUATIONS

The dynamo model studied here consists of a spherical shell of inner radius R_1 and outer radius R_2 which contains incompressible fluid of constant kinematic viscosity ν and magnetic diffusivity η. In this geometry it is convenient to use spherical coordinates r, θ, ϕ together with the cylindrical variables $s = r \sin\theta$ and $z = r\cos\theta$. The shell rotates at constant angular velocity $\mathbf{\Omega} = \Omega\hat{\mathbf{z}}$ where $\hat{\mathbf{z}}$ is the unit vector in the z direction. Unless otherwise indicated, $\epsilon = R_1/R_2 = 0.3$ in order to represent the Earth's outer core. Using the length scale $\delta = R_2 - R_1$, the time scale δ^2/η and the magnetic scale $\sqrt{\mu\eta\rho\Omega}$ the magnetic induction equation becomes

$$\frac{\partial \mathbf{B}}{\partial t} = \nabla^2\mathbf{B} + \nabla \times (\mathbf{v} \times \mathbf{B} + \alpha\mathbf{B}), \tag{1}$$

where $\mathbf{B}$ is the magnetic field, $\mathbf{v}$ is the fluid velocity and α is the alpha effect. Here $\mathbf{B}$ represents the large scale, azimuthally averaged magnetic field, while α is determined by intermediate scale asymmetric effects like convective cells and wave motions, rather than small scale turbulence. The axisymmetric magnetic and velocity fields are decomposed into toroidal and poloidal components

$$\mathbf{B} = B\hat{\boldsymbol{\phi}} + \nabla \times (A\hat{\boldsymbol{\phi}}), \tag{2}$$

$$\mathbf{v} = v\hat{\boldsymbol{\phi}} + \nabla \times (\psi\hat{\boldsymbol{\phi}}), \tag{3}$$

where $\hat{\phi}$ is the unit vector in the ϕ direction, for which (1) splits into the two $\alpha^2\omega$ equations for $\partial A/\partial t$ and $\partial B/\partial t$ (Barenghi & Jones 1991).

The inertial and viscous terms in the fluid momentum equation are very small and can be neglected in the main body of the fluid; what is left in the equation represents a balance between pressure, buoyancy, Coriolis and Lorentz forces. One finds $v = v_M + v_T + v_G$ and $\psi = \psi_T + \psi_M$. Here v_T and ψ_T, the thermal wind and stream function, are determined by the buoyancy; the magnetic wind v_M and the magnetic stream function ψ_M,

$$v_M = \frac{1}{2} \int [M(B,B) + M(D^2A, A)]dz, \tag{4}$$

$$\psi_M = -\frac{1}{2} \int N(B, A)dz, \tag{5}$$

are determined by the Lorentz force; finally $v_G = v_G(s)$ must be added since only the z derivative of v is determined by the momentum equation. In order to determine v_G viscosity is reintroduced into the problem, but only as far as effects in the Ekman boundary layer between the core and the mantle are concerned. The effects of Ekman suction are discussed by Fearn, Roberts & Soward (1988). Following Barenghi & Jones (1991) a simplified model is adopted, which avoids numerical difficulties at the equator leading to the *ansatz*

$$v_G(s) = \frac{1}{s^3 E^{1/2}} \frac{d}{ds} \int B \left[\frac{\sin\theta}{r} \frac{\partial A}{\partial\theta} - \cos\theta \frac{\partial A}{\partial r} \right] dz, \tag{6}$$

where $E = \nu/\delta^2\Omega$ is the Ekman number. The path of integration in (4), (5) and (6) is given by the intersection of the spherical shell with the geostrophic cylinder of radius s. The explicit forms of the nonlinear differential operators M and N, which are quadratic in their arguments, are given in Barenghi & Jones (1991). The alpha and omega effects are arbitrary inputs $\alpha(r, \theta)$ and $\omega = v_T(r, \theta)/s$ which are parametrized by the Reynolds numbers R_α and R_ω. It has been known for a long time that a suitable imposed meridional circulation can steady otherwise oscillatory $\alpha\omega$ solutions (Roberts 1972); the drawback of this *ad hoc* approach, however, is that it introduces more arbitrary inputs into the equations. This is why, for the sake of simplicity, it is assumed $\psi_T = 0$.

The $\alpha^2\omega$ equations have two limits of interests: the α^2 limit, in which there is no differential rotation and both the poloidal and toroidal fields are maintained only by α effects, and the $\alpha\omega$ limit, in which the large toroidal field is generated only by the ω effect. In the latter case R_α and R_ω combine into a single driving parameter $D = R_\alpha R_\omega$ called the dynamo number. The boundary conditions are that $B = 0$ at $r = R_1$ and R_2 and that A matches a potential field outside the shell (Barenghi & Jones 1991). The equations for A and B are solved by time stepping using the spectral method devised by

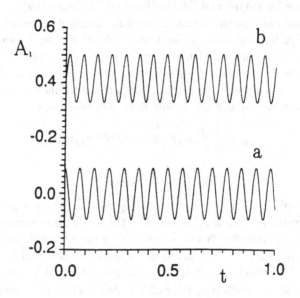

Figure 1. Plot of A_1 vs. t at (a) $D = 11600$ and (b) $D = 12000$.

Barenghi & Jones (1991). Further discussion of the numerical technique is also contained in Barenghi (1992a). All results presented here refer to dipolar symmetry for which $A(\pi - \theta) = A(\theta)$ and $B(\pi - \theta) = -B(\theta)$.

3 THE RESULTS

To resolve the disagreement between the typical rapid oscillations of $\alpha\omega$ models and the observed features of the Earth's magnetic field one has to find solutions of the magnetic induction equations which do not reverse or reverse very slowly, and which have a large toroidal field component. A factor which probably plays an important role in determining the time dependence of the the Geodynamo is the geometrical presence of the inner core: dynamo waves propagate better in a thin layer. Under the simple assumption $\alpha = \cos\theta$ and $\omega = r$ one finds that the $\alpha\omega$ dynamo equations have a nonreversing solution for positive dynamo number D in the geometry of a sphere (Hollerbach, Barenghi & Jones 1992). By slowly changing the radius ratio ϵ one can study the transition from a full sphere ($\epsilon = 0$) to the Geodynamo ($\epsilon = 0.3$). To illustrate the time dependence of the solutions it is convenient to plot the first coefficient A_1 of the spectral expansion of A as a function of time t. At $\epsilon = 0.01$, in the limit of a full sphere, the onset of dynamo action at $D = D_C$ is oscillatory. Above D_C, in the nonlinear Ekman regime, the solution is at first oscillatory (see Figure 1(a) at $D = 11600$); then, at higher dynamo number,

A_1 oscillates around nonzero mean (see Figure 1(b) at $D = 12000$). At higher value of D there is a transition to a Taylor state. The periodic, nonreversing solution of Figure 1(b), which can be referred to as a vacillation, is of course very interesting if one wishes to model the Geodynamo. Archaeomagnetic data suggest that the field can change considerably in few thousand years, while reversals take place on a longer time scale. Unfortunately the transition to vacillations disappears if the radius ratio is made as high as $\epsilon = 0.1$. The result however suggests that the inner core can have important effects. It must be stressed, however, that in order to take fully into account the dynamical role played by the inner core, with the possibility that it does not corotate with the mantle and the formation of Ekman and Stewartson layers, one should follow the approach of Hollerbach (1992) and Hollerbach & Jones (1992).

There are two possible ways to solve the conflict between the time dependence of $\alpha\omega$ models and the observed properties of the Geodynamo. The first way is based on Braginsky's discovery (Braginsky 1975) that the $\alpha\omega$ equations have a steady solution if the α distribution is sharply concentrated near the outer boundary and reverses the sign radially. Further calculations by Jones & Wallace (1992) in the simpler planar geometry seem to indicate that the key ingredient is not the fact that α has a sharp peak but rather that it changes sign. To investigate this issue the model $\omega = r$, $\alpha = \sin(\pi x)\cos\theta$, (where $x = 2r - (1 + \epsilon)/(1 - \epsilon)$ ranges between -1 and 1) is studied in the correct geometry for the Geodynamo ($\epsilon = 0.3$). The onset of the dynamo at D_C is time dependent, and by increasing the magnitude of D one finds a transition from oscillations to vacillations in the Ekman regime. In the oscillating case the magnetic field reverses completely throughout the shell; in the vacillating case (Figure 2) the dynamo waves are concentrated near the core–mantle boundary and the magnetic field maintains the same sign in the main body of the fluid. At higher dynamo number there is then a transition to a Taylor state.

The second route out of the difficulty is based on the recognition that the $\alpha\omega$ limit is perhaps an oversimplification. The estimated value of R_ω from the westward drift is only about 100 (Roberts 1988). Since one expects $1 < R_\alpha < 10$ to sustain the dynamo, R_ω cannot be more than one or two orders of magnitude bigger than R_α. This is probably not enough to be in the $\alpha\omega$ limit.

It seems then more appropriate to include some α effect in the toroidal field equation thus producing a full $\alpha^2\omega$ model; this ingredient should help in stabilizing the dynamo waves because α^2 dynamos tend to have steady solutions. Using the simple model $\alpha = \cos\theta$ and $\omega = r$ one finds that at $R_\omega = 30$, for example, a steady onset of dynamo action is followed at increasing values of R_α first by a steady and then by a slow, time dependent

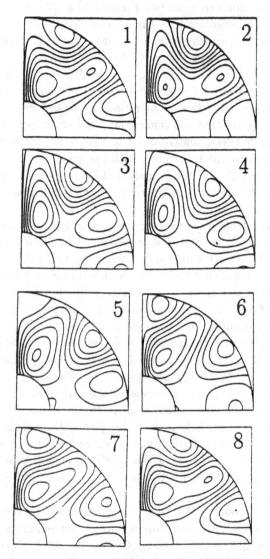

Figure 2. Vacillating solution at $D = -10000$: contour plots of A_1 at different times during a cycle (a negative value of D indicates dynamo waves propagating from the poles to the equator).

Ekman regime (typically $T = 5$ at $R_\alpha = 7.5$), with a subsequent transition to Taylor state. A more complete investigation of the parameter space has been carried out (Barenghi 1992b) using the model $\omega = 1 - x^2$, $\alpha = (1 - x^2)\cos\theta$. For $R_\omega < 90$ one finds at increasing values of R_α: steady onset, steady Ekman regime, transition to a steady Taylor state. For $R_\omega > 90$ the Ekman regime is at first steady, then vacillatory and oscillatory, and the equilibration past the Taylor transition is oscillatory with period ranging from $T = 16$ at $R_\omega = 93$, to 4 at $R_\omega = 100$ and 1 at $R_\omega = 160$. The fact that the Taylor states are steady or oscillate so slowly is consistent with the observed frequency of geomagnetic reversals. Moreover the peak value of the toroidal field during a cycle is one order of magnitude larger than that of the poloidal field and corresponds to an Elsasser number $\Lambda = B^2/\rho\mu\eta\Omega = 1$ which should favour the stable operation of a convective dynamo (Soward 1992).

4 CONCLUSIONS

The time dependence of a number of Geodynamo models has been investigated. Two routes have been suggested to cure the apparent conflict between the typical, rapid oscillations of $\alpha\omega$ solutions and the much slower frequency of reversals of the Earth's magnetic field. The first approach requires an α distribution which changes sign radially. At this stage too little is known about convection in the Earth's core to justify or reject this assumption. The second route consists in recognizing that the α effect in the toroidal field equation is important: the solutions of the resulting $\alpha^2\omega$ equations are consistent with the time dependence of the Geodynamo.

I wish to thank Professor C.A. Jones and Professor A.M. Soward for discussions and encouragement. This work was supported by the grant number GR/H 03278 of the Science and Engineering Research Council.

REFERENCES

Abdel-Aziz, M.M. & Jones, C.A. 1988 $\alpha\omega$ dynamos and Taylor's constraint. *Geophys. Astrophys. Fluid Dynam.* **44**, 117–139.

Barenghi, C.F. 1992a Nonlinear planetary dynamos in a rotating spherical shell. II. The post-Taylor equilibration for α^2-dynamos. *Geophys. Astrophys. Fluid Dynam.* **67**, 27–36.

Barenghi, C.F. 1992b Nonlinear planetary dynamos in a rotating spherical shell. III. $\alpha^2\omega$ models and the geodynamo. To be published.

Barenghi, C.F. & Jones, C.A. 1991 Nonlinear planetary dynamos in a rotating spherical shell. I. Numerical methods. *Geophys. Astrophys. Fluid Dynam.* **66**, 211–243.

Fearn, D.R., Roberts, P.H. & Soward, A.M. 1988 Convection, stability and the dynamo. In *Energy, Stability and Convection* (ed. G.P. Galdi & B. Straughan), **168**, pp. 60–324. Research Notes in Mathematics Series, Longmans.

Braginsky, S.I. 1975 Nearly axially symmetric model of the hydromagnetic dynamo of the Earth. *Geomag. Aeron.* **15**, 122–128.

Hollerbach, R. 1992 A direct spectral solution of the Ekman and Stewartson layers in a rotating spherical shell. To be published.

Hollerbach, R., Barenghi, C.F. & Jones, C.A. 1992 Taylor's constraint in a spherical $\alpha\omega$-dynamo. *Geophys. Astrophys. Fluid Dynam.* **67**, 3–25.

Hollerbach, R. & Ierley, G.R. 1991 A modal α^2 dynamo in the limit of asymptotically small viscosity. *Geophys. Astrophys. Fluid Dynam.* **56**, 133–158.

Hollerbach, R. & Jones, C.A. 1992 A geodynamo model incorporating a finitely conducting inner core. To be published.

Jones, C.A. 1991 Dynamo models and Taylor's constraint. In *Advances in Solar System Magnetohydrodynamics* (ed. E.R. Priest), pp. 25–50. Cambridge University Press.

Jones, C.A. & Wallace, S.G. 1992 Periodic, chaotic and steady solutions of $\alpha\omega$ dynamos. To be published.

Roberts, P.H. 1972 Kinematic dynamo models. *Phil. Trans. R. Soc. Lond. A* **272**, 663–703.

Roberts, P.H. 1988 Future of geodynamo theory. *Geophys. Astrophys. Fluid Dynam.* **44**, 3–32.

Soward, A.M. 1992 The Earth's dynamo. To be published.

The Onset of Dynamo Action in Alpha-lambda Dynamos

D.M. BARKER

Dept. of Mathematics
University of Manchester
Oxford Road, Manchester, M13 9PL UK

Solutions of the Navier–Stokes equation are computed in a deep, incompressible, spherical shell, including a parametrization of the Reynolds stresses arising from anisotropic turbulence. Thus the purely dynamical problem has solutions with marked differential rotation. The critical dynamo number for the onset of dynamo action is determined for different hydrodynamic models for both axisymmetric and nonaxisymmetric magnetic fields.

1 INTRODUCTION

Although kinematic dynamo models reveal some of the basic features of Solar and stellar magnetic fields, a fully satisfactory model must allow the dynamics to emerge as part of the solution of the governing system of equations. This has been attempted in a number of mean-field studies starting with Proctor (1977). It has been shown that solutions exist in which axisymmetric fields become saturated at a finite energy by the action of the macroscopic Lorentz force acting on the fluid.

The mean-field formalism is used in turbulent convection zones to parametrise the effects of the small scale dynamics on the magnetic field (Steenbeck *et al.* 1966). The influence of the small-scale turbulence on the macroscopic motions can be similarly modelled by the 'Λ-effect', representing the Reynolds stresses of anisotropic turbulence induced in a rotating, stratified medium (Rüdiger 1989). The resulting mean-field equations describe the evolution of quantities averaged over time or length scales greater than those of the

27

M.R.E. Proctor, P.C. Matthews & A.M. Rucklidge (eds.)
Theory of Solar and Planetary Dynamos, 27–34
©1993 Cambridge University Press.

turbulence.

It has not so far been possible in a dynamically consistent model (either by large scale simulations, e.g. Gilman & Miller (1981), Glatzmaier (1985), or in the mean-field formalism, Brandenburg *et al.* (1991; 1992)) to reproduce the disc-like angular velocity contours of the Sun, as revealed by helioseismology. Here we consider a rather deeper shell than the Solar convection zone, perhaps appropriate to giant stars. The simple model presented here indicates the complexity of the situation.

2 THE MEAN-FIELD EQUATIONS AND THEIR SOLUTION

We wish to compute solutions of the Navier–Stokes equation, which in the inertial frame takes the form

$$\rho \frac{\partial \mathbf{u}}{\partial t} = -\rho \mathbf{u} \cdot \nabla \mathbf{u} - \nabla \mathcal{P} + \mathbf{j} \times \mathbf{B} - \nabla \cdot (\rho \mathcal{Q} - \mathcal{B}), \tag{1}$$

and the dynamo equation

$$\frac{\partial \mathbf{B}}{\partial t} = \nabla \times (\mathbf{u} \times \mathbf{B} + \mathbf{E} - \eta_T \mu_0 \mathbf{j}), \tag{2}$$

where $\mathbf{u}$ and $\mathbf{B}$ are mean-field averages. Here, the reduced pressure $\mathcal{P}$ contains the gravitational term, ρ is the density, $\mathbf{j}$ the electric current, η_T the turbulent magnetic diffusivity and μ_0 the magnetic permeability.

The correlations $\mathcal{Q}_{ij} = \langle u_i' u_j' \rangle$ and $\mathcal{B}_{ij} = \langle j_i' B_j' \rangle$ are the turbulent Reynolds and Maxwell stresses respectively. We neglect for now the tensor $\mathcal{B}$, partly as little is known of its form in stellar convection zones, but also as magnetically induced flows appear to be less important than those due to turbulence, at least in the Solar case. We take the off-diagonal components of $\mathcal{Q}_{ij}$ as

$$\mathcal{Q}_{r\phi} = \nu_T (-\frac{r}{\Omega} \frac{\partial \Omega}{\partial r} + V^0) \Omega \sin \theta \tag{3}$$

and

$$\mathcal{Q}_{\theta\phi} = -\nu_T \frac{\partial \Omega}{\partial \theta} \sin \theta, \tag{4}$$

where Ω is the angular velocity, ν_T is the turbulent viscosity and V^0 is of order unity. For rapid rotation additional terms must be considered, but for now we consider only the fundamental mode of the Λ-effect described by Rüdiger (1989). The above expressions reduce to the Boussinesq relations for $V^0 = 0$, but for non-zero V^0 the component $\mathcal{Q}_{r\phi} \neq 0$, and thus solid body rotation is no longer a solution of the Navier–Stokes equation. The conditions $\nabla \cdot \mathbf{B} = 0$ and $\nabla \cdot \mathbf{u} = 0$ are automatically satisfied by writing the axisymmetric vectors as the curl of vector potentials and the nonaxisymmetric field in terms of superpotentials (Chandrasekhar 1961). The nonaxisymmetric

V^0	-3	-2	-1	0	1	2	3
Ta							
10^2	-146	-64	-16	0	4.9	6.7	8.1
10^4	-301	-201	-87	0	50	76	101
10^5	-934	-530	-219	0	149	244	317
10^6	-3021	-1605	-617	0	441	759	1070

Table 1. C_Ω values of hydrodynamic solutions of the Navier–Stokes equation for different values of the Λ-effect parameter V^0.

field is decomposed azimuthally into its Fourier components $e^{im\phi}$. In the study of growth rates that follows we only consider the $m = 0$ and $m = 1$ nonaxisymmetric field components.

The fluid velocities satisfy stress-free boundary conditions at both radial boundaries. The regime $r < 0.1R$ is assumed to be a perfect electrical conductor and the magnetic field is fitted smoothly onto a curl-free, external field at the surface $r = R$. We consider for simplicity an isotropic alpha tensor and so the electromotive force $\mathbf{E} = \alpha_0 \cos\theta \mathbf{B}$. We include a radial dependence for α_0 and V^0 so that they both go smoothly to zero at $r = 0.1R$ and are constant for $r > 0.3R$. Where numerical values of α_0 and V^0 are quoted it refers to the constant value in the outer part of the shell. For given V^0 the independent dynamo number $C_\alpha = \alpha_0 R/\eta_T$.

The equations are solved in an incompressible, spherical shell extending from $r = 0.1R$ to $r = R$ by a timestepping method. This is not very efficient for solving the linear dynamo problem considered here, but is well suited to fully nonlinear solutions, which are our ultimate aim. Further details of the method of solution of the axisymmetric hydrodynamical problem are given in Brandenburg *et al.* (1991).

3 HYDRODYNAMIC SOLUTIONS

The solution procedure can be envisaged as follows. The total angular momentum of the shell $r_0 \leq r \leq R$ is prescribed. The corresponding uniform angular velocity Ω_0 defines the Taylor number $Ta = (2\Omega_0 R^2/\nu_T)^2$. The magnetic Prandtl number $P_m = \nu_T/\eta_T = 1$. Given Ta and V^0, we can solve the Navier–Stokes equation, neglecting the magnetic field. The angular momentum retains its given value, but the solution is differentially rotating with meridional circulation. Table 1 gives the equatorial differential rotation parameter $C_\Omega = (\Omega(r = R, \theta = \pi/2) - \Omega(r = 0.1R, \theta = \pi/2))R^2/\eta_T$ for various models.

As Ta is increased it is expected that surfaces of constant Ω will become more cylindrical (the Taylor–Proudman theorem). This can be seen

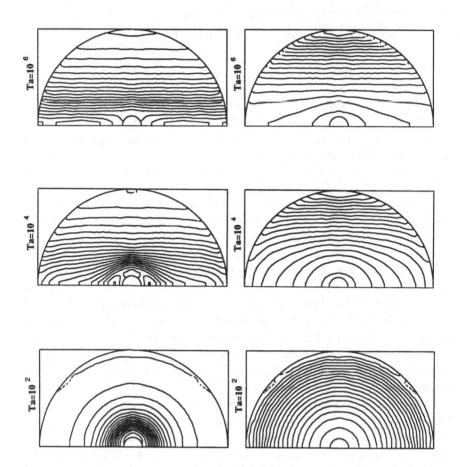

Figure 1. Contours of constant Ω for solutions of the Navier–Stokes equation for several values of Ta, with $V^0 = -2$ in the upper series and $V^0 = +2$ in the lower one.

in Figure 1. Streamlines of meridional circulation are shown in Figure 2. A positive V^0 corresponds to equatorward motion of fluid at the surface $r = R$. A similar result is found in thin-shell models of the Solar convection zone (Brandenburg *et al.* 1991), thus $V^0 > 0$ is therefore at least consistent with observations of sunspot migration. However, helioseismological evidence indicates $C_\Omega < 0$ which would require $V^0 < 0$ in thin-shell models, but the Sun is a relatively rapid rotator and further terms in the parametrization of the Λ-effect should be included. For giant stars with deep convection zones, there exist no observations against which to test the differential rotation model.

For large Ta and $|V^0|$, the energy in the meridional motions saturates as the circulation pattern becomes more complex. The ratio of meridional to toroidal kinetic energy (minus that of the initial solid body rotation) is

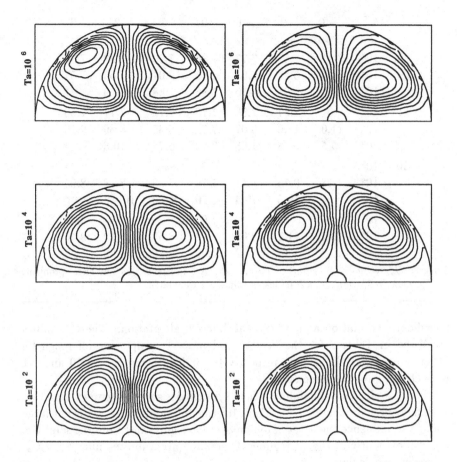

Figure 2. Streamlines of meridional circulation for hydrodynamic solutions of varying Ta. Again, $V^0 = -2$ in the upper series and $V^0 = +2$ in the lower one.

less than 1:100 in all cases considered. The next section indicates that the meridional motions still have a considerable influence on the dynamo.

4 THE ONSET OF DYNAMO ACTION

Having computed these detailed velocity fields, with significant radial and latitudinal differential rotation and meridional circulation, we evaluate the critical dynamo number $C_{\alpha c}$ of each of the above models for the excitation of the $m = 0$ and $m = 1$ components of the field. As we have ignored the Lorentz force in equation (1), the fluid velocities occurring in the dynamo equation (2) can now be regarded as given, and we are solving a linear dynamo problem. The $C_{\alpha c}$ values can, for example, be compared with those of kinematic models, which usually assume a simple form for the differential rotation and neglect

	V^0	-3	-2	-1	0	1	2	3
	Ta							
$S0$	10^2	7.11	7.57	7.77	7.86	7.90	7.93	7.86
	10^4	3.40	5.48	7.03	7.86	9.04	7.65	7.57
	10^5	*	*	8–10	7.86	10.8	10.9	*
$S1$	10^2	8.01	7.97	7.80	7.73	7.75	7.77	7.65
	10^4	11.0	11.0	9.07	7.73	8.35	8.86	9.16
	10^5	*	> 18	12.8	7.73	9.78	10.8	*
$A0$	10^2	–	–	–	7.70	–	–	–
	10^4	7.18	9.75	8.11	7.70	7.80	8.18	8.95
	10^5	*	12.5	10.4	7.70	8–10	10–12	*
$A1$	10^2	–	–	–	7.92	–	–	–
	10^4	10.4	11.1	9.08	7.92	8.55	9.12	9.40
	10^5	*	> 18	12.2	7.92	9.76	10.8	*

Table 2. $C_{\alpha c}$ values as functions of Ta and V^0. A * indicates that numerical problems were encountered, whilst a dash indicates values not calculated.

meridional circulation and latitudinal differential rotation. The $C_{\alpha c}$ values are listed in Table 2 for the $m = 0$ and $m = 1$ components of magnetic fields symmetric ($S0$ and $S1$ respectively) and antisymmetric ($A0$ and $A1$ respectively) about the equator.

The $C_{\alpha c}$ values for the different field symmetries show a complex variation with hydrodynamic model. For $Ta = 10^2$ the differential rotation is not large enough to influence the onset of dynamo action significantly and the solutions are still evidently in the α^2 regime. Therefore only a limited number of $C_{\alpha c}$ values were calculated.

More interesting behaviour can be seen for $Ta \geq 10^4$, when the $m = 0$ components show a particularly complex pattern of behaviour not seen in purely kinematic models with velocity fields of simple spatial structure. For negative V^0 ($C_\Omega < 0$) the $S0$ solution becomes easier to excite as $|C_\Omega|$ increases. However, for $V^0 > 0$ the $S0$ dynamo appears harder to excite for modest values of C_Ω. Consider in particular the case $Ta = 10^4$, $V^0 = +1$ ($C_\Omega \sim 50$), with $C_{\alpha c} = 9.04$ for the $S0$ solution. If the meridional circulation as calculated by our dynamical model is removed from the dynamo equation (2), the value of $C_{\alpha c}$ is reduced to 8.61. If the latitudinal differential rotation is also removed, then $C_{\alpha c} = 8.36$, compared with the value $C_{\alpha c} = 8.20$ for a kinematic dynamo model with uniform radial differential rotation $C_\Omega = d\Omega/dr = 50$. For the $A0$ solutions, a somewhat greater inhibition of dynamo action occurs. The inclusion of more complex dynamics than a simple radial differential rotation clearly has significant effects on the mechanism of magnetic field generation.

Oscillatory $m = 0$ fields are found for $|C_\Omega|$ values greater than about 150. The $C_{\alpha c}$ values bounded, but not fully determined, in Table 2 indicate oscillatory solutions whose period is of the order of a diffusion time. These long period solutions are found at the transition between steady and oscillatory modes and it is impractical to calculate the growth rates of such slowly varying modes using our code.

In general, the effect of the dynamics on the nonaxisymmetric $m = 1$ modes is to make them harder to excite. The effect of an azimuthal differential rotation is to wrap up field-lines. For nonaxisymmetric modes, this will lead to the juxtaposition of oppositely directed field-lines, enhancing reconnection. It therefore appears that the effect of the differential rotation on nonaxisymmetric fields is the dominant one. The meridional circulation does not seem as important here as for the $m = 0$ eigenmodes. A possible exception is the rather extreme case $V^0 = -3$, $Ta = 10^4$. Nowhere is it found that the nonaxisymmetric field is more easily excited than the axisymmetric one for any given hydrodynamic model.

5 CONCLUSIONS

We have computed a number of solutions of the Navier–Stokes equation in which the Reynolds stresses drive a significant differential rotation. At high Ta the results are consistent with the Taylor–Proudman theorem. The magnitude of the meridional circulation saturates as the complexity of the meridional flow pattern increases at high Ta. Köhler (1970) found a maximum of the meridional flow velocity for $Ta = 3 \times 10^7$ for a thin shell, a result verified by Brandenburg *et al.* (1991). The hydrodynamic results presented here for a thick shell are consistent with these previous results.

It has been shown that the computation of critical dynamo numbers can be misleading if meridional circulation and latitudinal differential rotation are ignored. There is accumulating observational evidence for the existence of large-scale nonaxisymmetric structures in giant stars with deep convective shells (see e.g. Moss *et al.* 1991). This has motivated the present preliminary study. The improvement of the hydrodynamic model must be a primary goal in future work on the study of stellar dynamos.

REFERENCES

Brandenburg, A., Moss, D., Rüdiger, G. & Tuominen, I. 1991 Hydromagnetic $\alpha\Omega$-type dynamos with feedback from large scale motions. *Geophys. Astrophys. Fluid Dynam.* **61**, 179–198.

Brandenburg, A., Moss, D., Rüdiger, G. & Tuominen, I. 1992 Stratification and thermodynamics in mean-field dynamos. *Astron. Astrophys.* **265**, 328–344.

Chandrasekhar, S. 1961 *Hydrodynamic and Hydromagnetic Stability.* Clarendon Press, Oxford.

Gilman, P.A. & Miller, J. 1981 Dynamically consistent nonlinear dynamos driven by convection in a rotating spherical shell. *Astrophys. J. Suppl.* **46**, 211–238.

Glatzmaier, G.A. 1985 Numerical simulations of stellar convective dynamos. II. Field propagation in the convection zone. *Astrophys. J.* **291**, 300–307.

Köhler, H. 1970 Differential rotation caused by anisotropic turbulent viscosity. *Solar Phys.* **13**, 2–18.

Moss, D., Tuominen, I. & Brandenburg, A. 1991 Nonlinear nonaxisymmetric dynamo models for cool stars. *Astron. Astrophys.* **245**, 129–135.

Proctor, M.R.E. 1977 Numerical solutions of the nonlinear α-effect dynamo equations. *J. Fluid Mech.* **80**, 769–784.

Rüdiger, G. 1989 *Differential Rotation and Stellar Convection: Sun and Solar-type Stars.* Gordon & Breach, New York.

Steenbeck, M., Krause, F. & Rädler, K.-H. 1966 Berechnung der mittleren Lorentz-Feldstärke $\mathbf{v} \times \mathbf{B}$ für ein elektrisch leitendes Medium in turbulenter, durch Coriolis-Kräfte beeinflusster Bewegung. *Z. Naturforsch.* **21a**, 369–376.

Multifractality, Near-singularities and the Role of Stretching in Turbulence

AXEL BRANDENBURG

Isaac Newton Institute for Mathematical Sciences
University of Cambridge, 20 Clarkson Rd., Cambridge, CB3 0EH UK

Current address:
HAO/NCAR, P.O. Box 3000
Boulder, CO 80307 USA

ITAMAR PROCACCIA & DANIEL SEGEL

Department of Chemical Physics
The Weizmann Institute of Science
Rehovot 76100, Israel

ALAIN VINCENT & MARCO MANZINI

CERFACS
42 Avenue Coriolis
F-31057 Toulouse, France

The squared vorticity field obtained from snapshots of different simulations is analysed and the generalised dimensions are estimated. The scaling behaviour is used to quantify the presence of 'near-singularities' in these fields. In three-dimensional simulations there is a crossover indicating the presence of near-singularities at small scales close to the Kolmogorov dissipation length. At larger scales these near-singularities appear weaker indicating more random behaviour of the large scale vorticity structure. Two-dimensional simulations show the opposite behaviour in that near-singularities seem to be absent at small scales. It is suggested that the presence of near-singularities is related to vortex stretching terms which occur only in the three-dimensional case. Finally the limitations of the approach due to contamination by noise is discussed.

M.R.E. Proctor, P.C. Matthews & A.M. Rucklidge (eds.)
Theory of Solar and Planetary Dynamos, 35–42
©1993 Cambridge University Press.

1 INTRODUCTION

The effect of magnetic field line stretching is thought to be essential for dynamo action (Batchelor 1950). Using the (imperfect) analogy between vorticity and magnetic field Batchelor argued that there will be dynamo action if the magnetic diffusivity η is smaller than the kinematic viscosity ν. His argument goes basically as follows: consider the evolution equations for enstrophy and magnetic energy density, assuming incompressibility and no forcing,

$$\frac{d}{dt}\langle \frac{1}{2}\omega^2 \rangle = \langle \omega_i \omega_j \partial_j u_i \rangle - \nu \langle |\nabla \times \boldsymbol{\omega}|^2 \rangle, \qquad (1)$$

$$\frac{d}{dt}\langle \frac{1}{2}\mathbf{B}^2 \rangle = \langle B_i B_j \partial_j u_i \rangle - \eta \langle |\nabla \times \mathbf{B}|^2 \rangle, \qquad (2)$$

where $\boldsymbol{\omega} = \nabla \times \mathbf{u}$ is the vorticity and $\mathbf{B}$ is the magnetic field density. It is known from laboratory turbulence measurements that the vortex stretching term $\omega_i \omega_j \partial_j u_i$, i.e. the rate of enstrophy production, is positive on average (Vincent & Meneguzzi 1993). For statistically steady turbulence the two terms on the right hand side of (1) are in balance. Even for large Reynolds numbers (small values of ν) the two terms are large in the sense that the dissipation time for vortex structures, $\tau = \langle \frac{1}{2}\omega^2 \rangle / \langle \nu |\nabla \times \boldsymbol{\omega}|^2 \rangle$ is short or comparable with the turnover time. By virtue of the analogy of (1) and (2) one may expect a balance between the two terms on the right hand side of (2) if $\eta = \nu$. If, however, $\eta < \nu$ then the average stretching term $\langle B_i B_j \partial_j u_i \rangle$ will dominate over $\eta \langle |\nabla \times \mathbf{B}|^2 \rangle$, and dynamo action is then possible.

The above considerations illustrate the importance of stretching terms in turbulence and dynamo theory. Locally the stretching term may be very large and this leads to exponential growth of ω^2 and $\mathbf{B}^2$ in such places. Indeed, it is known that the ω^2 and $\mathbf{B}^2$ fields are highly intermittent and the fields are only concentrated on a small fraction of the volume. Using data from direct simulations of turbulence we have found that the peaks of ω^2 and $\mathbf{B}^2$ are compatible with algebraic near-singular structures (Brandenburg *et al.* 1992, hereafter referred to as paper I). In the present paper we outline the basic idea and illustrate the method using artificial data. We also compare three and two dimensional simulations; in two dimensions stretching terms vanish and consequently there is no dynamo. The procedure adopted involves the study of the multifractal nature of the fields by the determination of the generalised dimensions D_q, which differ from each other for different orders q if the field possesses singularities (Halsey *et al.* 1986).

2 GENERALISED DIMENSIONS

An analysis of the generalised dimensions D_q is useful for several reasons. Their knowledge allows the study of the multifractal behaviour of the field under consideration (ω^2 in the present case). This is relevant because it is believed that corrections to the Kolmogorov scaling in the inertial range are related to the multifractal behaviour of the local rate of energy dissipation (Mandelbrot 1974; Frisch & Parisi 1984). Recent applications to laboratory and atmospheric turbulence have been presented by Meneveau & Sreenivasan (1991).

The existence and order of singularities are expressed by the dimensions D_q, which together contain information as to the range of exponents of the singularities in the field. For example the dimension belonging to the largest exponent, D_∞, gives information about the strength of the strongest singularity in these fields, which behaves locally like

$$|\mathbf{x} - \mathbf{x}_0|^{D_\infty - 3}, \tag{3}$$

where $\mathbf{x}_0$ is the position of the singularity and $|\mathbf{x} - \mathbf{x}_0|$ is the distance from the singularity.

Plausibly, viscosity always smoothes the fields at sufficiently small scales, and therefore we always talk about 'near-singularities'. The scale where these near-singularities are tamed is, however, short compared to the Kolmogorov dissipation scale. This came as a surprise to us and we therefore address in section 5 the possibility of 'numerical noise' being the reason for this behaviour.

Consider now the ω^2 field represented on a three-dimensional mesh with total volume V. We may divide the space into boxes of size r and subvolume $V_i(r)$. The weight or normalised field strength in each box is

$$p_i(r) = \frac{\int_{V_i} \omega^2 d^3 x}{\int_V \omega^2 d^3 x}. \tag{4}$$

These p_i are used to determine the generalised correlation integral

$$C_q(r) = \sum_i p_i(r)^q, \tag{5}$$

from which we estimate the 'local' generalised dimension

$$D_q(r) = \frac{1}{q-1} \frac{d \ln C_q(r)}{d \ln r}. \tag{6}$$

This particular definition of $D_q(r)$ allows us to determine whether there are ranges where power law behaviour occurs; see paper I for details.

In the following section we give numerical examples of one-dimensional fields, represented on a finite mesh, and containing only one singularity.

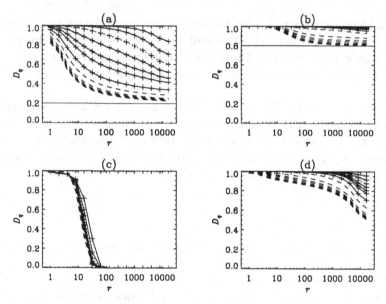

Figure 1. $D_q(r)$ versus r for different fields. The solid lines are for $q = 1/3$, 2/3, 4/3, 5/3, 2, and 3. The dotted line is for $q = 1$ and the dashed lines are for $q \geq 5$ (5, 10, 15, 20, 30, 50, 70, 100, ∞). (a) and (b) are for algebraic singularities with $\alpha = 0.2$ $\alpha = 0.8$, respectively. (c) and (d) are for a Gaussian and logarithmic peaks, respectively.

3 EXAMPLES

In order to illuminate the nature of the algorithm outlined above we consider a simple one-dimensional field

$$\omega^2(x) = |x - x_0|^{\alpha-1}, \qquad (7)$$

where α determines the strength of the singularity. In order to avoid numerical problems we place x_0 between mesh points.

The result for $D_q(r)$ is shown in the upper two panels of Figure 1 for two different values of α. For comparison we also consider non-algebraic peaks given by $\exp(-|x - x_0|^2/\delta x^2)$ and $-\ln(|x - x_0|/L)$, where δx is the mesh size and L the length of an array with $2^{15} = 32\,768$ mesh points. The results, shown in the lower panel of Figure 1, display quite different behaviour compared to the case of algebraic singularities. Therefore this technique allows us to characterise the nature of the peaks in the field. In the following section we consider data of actual simulations of turbulent flows.

4 TWO AND THREE DIMENSIONAL FLOWS

In this section we investigate the scaling behaviour of the squared vorticity using simulations of both three-dimensional and and two-dimensional homogeneous forced turbulence. The three-dimensional simulation is obtained by

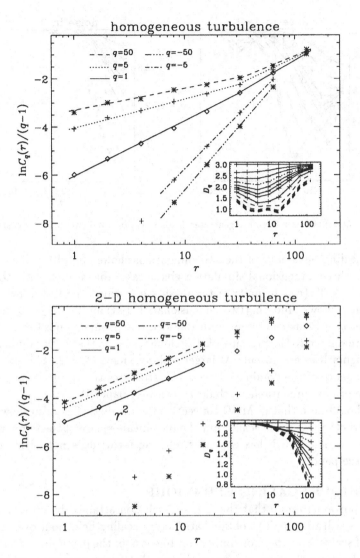

Figure 2. The scaling of $C_q(r)$ versus r for three-dimensional (upper panel) and two-dimensional homogeneous turbulence (lower panel). The scaling is only shown for selected values of q. The insets show $D_q(r)$ versus r for the same values of q as in Figure 1.

Vincent & Meneguzzi (1991) at a Reynolds number of about 1000 with a resolution of 240^3 collocation points, and a two-dimensional simulation is obtained by Manzini & Meneguzzi (private communication) at a Reynolds number of about 1500 with a resolution of 512^2 collocation points, using a modified version of the code described by Brachet *et al.* (1988). The results for the scaling behaviour of C_q and D_q are given in Figure 2.

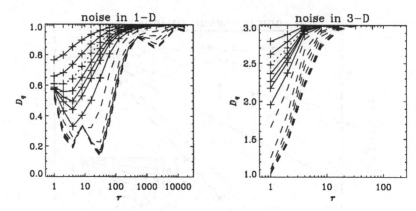

Figure 3. $D_q(r)$ versus r for noise generated fields in (a) one and (b) three dimensions.

The scaling behaviour of the two simulations shows opposite behaviour in that the three-dimensional simulation gives reasonable scaling with $D_q < 3$ for $1 \leq r \leq 32$ (r is in units of the mesh size), whilst the two-dimensional simulation shows smooth (two-dimensional) behaviour for $r \leq 16$ and all values of q. This is consistent with the idea that near-singularities at small scales are related to the presence of stretching terms, so that in two dimensions near-singularities are absent. At larger scales we have $D_q < 2$. This might be a consequence of the dominant large scale structures that typically appear in the case of two-dimensional turbulence (see also paper I).

We should note that in MHD convection the magnetic field also shows singular structures at small scales, but the root-mean-square of the temperature gradient (a field which has a weak stretching term) does not. For further details see paper I.

5 SPURIOUS SCALING BEHAVIOUR

It is important to note that the spatial resolution attained by current simulations is still too small to obtain satisfactory scaling behaviour over a wide range of scales. It is therefore important to consider the possibility of contamination by noise that may cause spurious scaling over some range. This has recently been addressed in great detail by Aurell *et al.* (1992), who pointed out that spurious scaling may occur at small scales with $r < |q(q-1)|$.

We computed $D_q(r)$ using instead of ω^2 positive random numbers with an exponential distribution. The results for one and three dimensional grids is shown in Figure 3. Note that spurious scaling appears for small r and large q. However, the one-dimensional case is more affected by this than the three-dimensional case, which may partly be explained by the larger number of data points available at each scale. This strengthens our hope that the interpretation of near-singular structures in fields whose evolution is governed by stretching terms is indeed justified.

In some cases we have checked this by direct inspection of the strongest peaks in such fields (see paper I). We have also shown that in simulations with smaller Reynolds number, near-singularities appear at larger scales, and that their strengths remain unchanged (cf. Figures 4 and 8 in paper I). Thus, even though the simulation at Reynolds number 1000 does not resolve scales smaller than the Kolmogorov cutoff scale, the quantitative consequences appear to be only of minor importance.

6 CONCLUSIONS

We have discussed evidence showing that fields whose evolution is governed by stretching terms show multifractal behaviour and singularities at small scales. There remains the question whether the crossover to nonfractal behaviour with $D_q \rightarrow 3$, indicated at large scales, is persistent in better resolved simulations. This would be in contrast to results of Meneveau & Sreenivasan (1991) who obtain good scaling in the inertial range. Of course, these data are obtained from a time series, and one should remember that the relation between spatially resolved data and those obtained from a time series is not entirely clear, although there is good evidence that Taylor's hypothesis is indeed applicable (see discussion in Meneveau & Sreenivasan). Spatially resolved turbulent flows can also be measured on the Solar surface (see Brandt 1991, and references therein) but, again, the theoretical interpretation may not be straightforward.

We thank Massimo Vergassola for clarifying discussions on this topic.

REFERENCES

Aurell, E., Frisch, U., Lutsko, J. & Vergassola, M. 1992 On the multifractal properties of the energy dissipation derived from turbulence data. *J. Fluid Mech.* **238**, 467–486.

Batchelor, G.K. 1950 On the spontaneous magnetic field in a conducting liquid in turbulent motion. *Proc R. Soc. Lond. A* **201**, 405–416.

Brachet, M.E., Meneguzzi, M., Politano, H. & Sulem, P.-L. 1988 The dynamics of freely decaying two-dimensional turbulence. *J. Fluid Mech.* **194**, 333–349.

Brandenburg, A., Procaccia, I., Segel, D. & Vincent, A. 1992 Fractal level sets and multifractal fields in direct simulations of turbulence. *Phys. Rev. A* **46**, 4819–4828 (paper I).

Brandt, P.N., Greimel, R., Guenther, E. & Mattig, W. 1991 Turbulence, fractals, and the Solar granulation. In *Applying Fractals in Astronomy* (ed. A. Heck & J.M. Perdang), pp. 77–96. Lecture Notes in Physics **3**, Springer.

Frisch, U. & Parisi, G. 1985 On the singularity structure of fully developed turbulence. In *Turbulence and Predictability in Geophysical Fluid Dynamics* (ed. M. Gil, R. Benzi & G. Parisi), pp. 84–88. North-Holland.

Halsey, T.C., Jensen, M.H., Kadanoff, L.P., Procaccia, I. & Shraiman, B.I. 1986 Fractal measures and their singularities: the characterisation of strange sets. *Phys. Rev.* **A 33**, 1141–1151.

Mandelbrot, B.B. 1974 Intermittent turbulence in self-similar cascades: divergence of high moments and dimension of the carrier. *J. Fluid Mech.* **62**, 331–358.

Meneveau, C. & Sreenivasan, K. R. 1991 The multifractal nature of turbulent energy dissipation. *J. Fluid Mech.* **224**, 429–484.

Vincent, A. & Meneguzzi, M. 1991 The spatial structure and statistical properties of homogeneous turbulence *J. Fluid Mech.* **225**, 1–20.

Vincent, A. & Meneguzzi, M. 1993 On the dynamics of vorticity tubes in homogeneous turbulence *J. Fluid Mech.* (in press).

Note on Perfect Fast Dynamo Action in a Large-amplitude SFS Map

S. CHILDRESS

Department of Mathematics
New York University, Courant Institute of Mathematical Sciences
New York, NY 10012 USA

We consider a simple extension of the SFS fast dynamo where the Liapunov exponent is $2N$ rather than 2, where N is a large integer. Fast dynamo action can be demonstrated for such a map for sufficiently large N, by making use of the properties of the adjoint eigenvalue problem.

1 INTRODUCTION

An interesting asymptotic limit in the theory of dynamical systems enforces a highly-developed chaotic structure by the assumption of large-amplitude particle excursion in flows and maps. An example of such a method applied to diffusion of a scalar is given by Rechester & White (1980). This important idea has been developed by Soward (1992) in the context of fast dynamo theory and in particular for the case of pulsed helical waves. Our purpose in this note is to apply the large-amplitude method of Soward (1992), to the simpler SFS map (Bayly & Childress 1987, 1988). In the SFS (stretch-fold-shear) map, a simple baker's map in the xy-plane is supplemented by a lateral shear in the z-direction. Numerical calculations indicate that, when the map operates in a perfectly conducting fluid on a magnetic field of the form $(B(y)e^{ikz}, 0, 0)$, the average of the field over planes z=constant can be made to grow exponentially for sufficiently large shear. This property of 'perfect' fast dynamo action has never been proved in the SFS problem, however,

43

M.R.E. Proctor, P.C. Matthews & A.M. Rucklidge (eds.)
Theory of Solar and Planetary Dynamos, 43–49
©1993 Cambridge University Press.

despite the existence of an especially simple adjoint eigenvalue problem, where the growing eigenfunctions, if they exist, are known to be smooth (Bayly & Childress 1988). Moreover, numerical studies show clearly the existence of these eigenfunctions for the perfect fast dynamo problem.

It is with the intention of filling in this gap of an analytic construction of the eigenfunctions that we study here a variant of the SFS map which stretches the cube to $2N$ times its original length, instead of twice the length as in the standard baker's map. By exploiting the adjoint formulation in conjunction with the assumption $N \gg 1$, we are able to establish perfect fast dynamo action. The same method can be used in related continuous maps, but is unfortunately not directly applicable to the pulsed flows studied asymptotically by Soward (1992).

2 A LARGE-AMPLITUDE SFS MAP

The idea is to consider an SFS-type situation where the cube is stretched by a factor of $2N$ where N is a large positive integer rather than 1. Such a map can be constructed in various ways, depending upon how the stretched elements are stacked to reform the original domain. Here N segments of length 1 are stacked with positive orientation to fill $0 < y < 1/2$, then N segments are stacked with negative orientation to fill $1/2 < y < 1$. This particular choice of stacking might be viewed as an unphysical one which omits folding in the two half-cubes. However, it is not unrealistic for approximating a large-amplitude pulsed wave by an SFS map, provided we extend the domain with period 1 in x, y. The reason is that a single pulsed wave, e.g. $(0, A \sin \pi x, 0)$, $A \gg 1$ will, if followed by a rotation $x, y \to y, -x$, tend to map a field in the x-direction into one which reverses sign at $x = 1/2$ but which consists in each half-cube of highly stretched field *from distant cubes* in the periodic array. Then, the elements stacked with like orientation in each half-cube should be regarded as coming from N different cubes.

Thus the field is transformed according to an operator T:

$$TB = \begin{cases} 2Ne^{i\alpha k(y-1/2)}B(2Ny - j), & j/2N < y < (j+1)/2N, \\ & j = 0, 1, ..., N-1, \\ -2Ne^{i\alpha k(y-1/2)}B(j+1-2Ny), & j/2N < y < (j+1)/2N, \\ & j = N, N+1, ..., 2N-1. \end{cases} \tag{1}$$

We now define the adjoint operator S. We have

$$\int_0^1 A^*(y)TB(y)\, dy = \int_0^1 (SA(y))^* B(y)\, dy \tag{2}$$

which gives

$$SB(y) = \sum_{j=0}^{N-1} B\left(\frac{j+y}{2N}\right)e^{-i\alpha k((j+y-N)/2N)} - \sum_{j=N}^{2N-1} B\left(\frac{j+1-y}{2N}\right)e^{-i\alpha k((j+1-y-N)/2N)}.$$

$$\tag{3}$$

We see that for large N we expect

$$SB \approx \bar{S}B \equiv 2N \int_0^{1/2} B(y)e^{-i\alpha k(y-1/2)} \, dy - 2N \int_{1/2}^1 B(y)e^{-i\alpha k(y-1/2)} \, dy. \quad (4)$$

We immediately have that 1 is an eigenfunction for $\bar{S}$ with eigenvalue

$$\lambda = 8Ni/\beta \sin^2 \frac{\beta}{4}, \qquad \beta = \alpha k. \quad (5)$$

The problem we have can thus be formulated generally in terms of a weight function $g(y)$ of bounded variation and an operator, now replacing $2N$ by N,

$$SB(y) \equiv \sum_1^N g(y_k)B(y_k),$$

where

$$y_k(y) = \begin{cases} (k-1+y)/N, & k = 1, 2, ..., N/2, \\ (k-y)/N, & k = N/2+1, ..., N. \end{cases} \quad (6)$$

We set $S' = S - \bar{S}$, where now

$$\bar{S}B = N \int_0^1 gB \, dy, \quad (7)$$

and introduce a norm

$$\|B\| = \left| \int_0^1 gB \, dy \right| + V_B(0,1). \quad (8)$$

Here $V_B(0,1)$ is the total variation of B on $[0,1]$, i.e.

$$V_B(0,1) = \sup_{\mathcal{P}} \sum_k |\Delta B_k|, \quad \mathcal{P} = \{y_0, y_1, ..., y_n\}, \quad \Delta B_k = B(y_k) - B(y_{k-1}), \quad (9)$$

where the supremum is over all partitions $\mathcal{P}$. In order to insure that (8) defines a norm, we shall assume that g has a non-zero integral from 0 to 1,

$$\left| \int_0^1 g \, dy \right| > 0. \quad (10)$$

Note also that, if P denotes the projection operator $PB = \int_0^1 B \, dy$, then $PS = \bar{S}$.

Let us now estimate the corresponding norms of $\bar{S}$ and S'. First $\bar{S}B$, being a constant, has zero variation and so we see that $\|\bar{S}\| \leq N|\int_0^1 g(y) \, dy| \equiv |\lambda|$. Second, $S'B$ may be bounded as follows: provided g is bounded and (10) holds, and for any $\epsilon > 0$, there is an N such that

$$\|S'\| \leq \epsilon N. \quad (11)$$

To prove this estimate we must examine, first, $\int_0^1 gS'B\,dy$. Now the theory of Riemann integration of functions of bounded variation (see e.g. Apostol 1957) insures that, for any $\delta > 0$, there is an N_1 such that for $N > N_1$

$$\left|\frac{1}{N}(\sum_1^N g(y_k)B(y_k)) - \int_0^1 gB\,dy)\right| \le \delta(M_g V_B(0,1) + M_B V_g(0,1)) \qquad (12)$$

where M_g, M_B are maxima of the absolute values of the indicated functions. Next consider $V_{S'B}(0,1)$. If we take differences we must deal with terms like

$$\left|\sum_{k=1}^N g(y_k(y_i))B(y_k(y_i)) - \sum_{k=1}^N g(y_k(y_j))B(y_k(y_j))\right| \qquad (13)$$

taken over many points y_i, y_j. But the result is just a refinement of the partition, so we get

$$V_{S'B}(0,1) \le M_g V_B(0,1) + M_B V_g(0,1). \qquad (14)$$

What we must now establish, to complete this first part of the proof, is that M_B can be bounded in terms of the norm of B, i.e. that there exists a positive constant C such that for all B with finite norm we have

$$M_B \le C\|B\|. \qquad (15)$$

Assuming for the moment that (15) holds, and combining (12), (14) and (15), we obtain the estimate

$$\|S'\| \le N(M_g\delta + 1/N)[M_g + (C+1)V_g] \qquad (16)$$

provided that $N > N_1$. If we choose δ so that $M_g[M_g + (C+1)V_g]\delta = \epsilon/2$ and then an $N \ge N_1$ so large that $[M_g + (C+1)V_g]/N \le \epsilon/2$ we obtain the required estimate $\|S'\| \le \epsilon N$.

We continue and now establish perfect fast dynamo action for this map. We first fix β, such that $\sin^2 \beta \ne 0$, and introduce an eigenvalue parameter μ of order N, such that $|\lambda - \mu|/N \ll 1$ for large N. Then $\mu^{-1}S'$ is small in the above norm, while $\mu^{-1}\bar{S}$ is of order unity. From slow dynamo theory it is known that if f is a constant and $b = (I - \mu^{-1}S')^{-1}f$, then $Sb = \mu b$ iff $\mu f = PS(I - \mu^{-1}S')^{-1}f$ (Childress 1970). This last equation is an equation for μ as a function of N (for given g). For large N this equation has a root close to λ. To demonstrate growth, we then take b as above as the flux weighting function, so we consider $(b, T^M B_0) = (\mu^*)^M(b, B_0)$. Choosing, e.g. $B_0 = b$ thus establishes perfect fast dynamo action.

The important property here is the estimate $\|S'\| \ll \|\bar{S}\|$, and we now turn to the proof of the bound (15) used in its derivation. The map has the property that field lines are stretched by a factor of $2N$ at each step and therefore, if the initial field is continuous, the maximum field M_B will be assumed to occur at some point x in $[0,1]$. Thus for a one-point partition we have

$$M_B - \left| \int_0^1 B(y)\,dy \right| \leq \left| B(x) - \int_0^1 B(y)\,dy \right| \leq V_B(0,1) \qquad (17)$$

and therefore

$$M_B \leq \left| \int_0^1 B(y)\,dy \right| + V_B(0,1) \qquad (18)$$

so it is sufficient to prove a bound of the form (15) for the mean of B.

To do this we first take the partition $[0, 1/K, 2/K, ..., 1]$, K a large even integer and try to bound $|g_1 B_1 + g_2 B_2 + ... + g_K B_K|$ from below. Set $B_j - B_{j+1} = A_j, j = 1, ..., K-1$, $A_K = B_1 + B_2 + ... + B_K$. With $\mathbf{B} = (B_1, B_2, ..., B_K)^T$ and similarly for $\mathbf{A}$ we have $\mathbf{A} = R\mathbf{B}$ where

$$R = \begin{pmatrix} 1 & -1 & 0 & \cdots & 0 \\ 0 & 1 & -1 & \cdots & 0 \\ \vdots & \vdots & \vdots & \ddots & \vdots \\ 1 & 1 & 1 & \cdots & 1 \end{pmatrix}. \qquad (19)$$

The inverse is

$$R^{-1} = \begin{pmatrix} \frac{K-1}{K} & \frac{K-2}{K} & \frac{K-3}{K} & \cdots & \frac{1}{K} & \frac{1}{K} \\ -\frac{1}{K} & \frac{K-2}{K} & \frac{K-3}{K} & \cdots & \frac{1}{K} & \frac{1}{K} \\ -\frac{1}{K} & \frac{-2}{K} & \frac{K-3}{K} & \cdots & \frac{1}{K} & \frac{1}{K} \\ \vdots & \vdots & \vdots & \vdots & \ddots & \vdots \\ -\frac{1}{K} & -\frac{2}{K} & -\frac{3}{K} & \cdots & -(\frac{K-1}{K}) & \frac{1}{K} \end{pmatrix}. \qquad (20)$$

From the latter we then see that

$$|g_1 B_1 + g_2 B_2 + ... + g_K B_K| \geq |m_g||A_K| - (K/2)M_g \sum_{k=1}^{K-1} |A_k| \qquad (21)$$

where

$$m_g = K^{-1} \sum_{k=1}^{K} g_k. \qquad (22)$$

Thus, letting $K \to \infty$ we obtain

$$\left| \int_0^1 B\,dy \right| \leq \left| \int_0^1 g\,dy \right|^{-1} \left[\left| \int_0^1 gB\,dy \right| + \frac{M_g}{2} V_B(0,1) \right]$$

$$\leq \left| \int_0^1 g\,dy \right|^{-1} (1 + M_g/2)\|B\|, \qquad (23)$$

which, in view of (11), gives the desired bound (15) with

$$C = \left| \int_0^1 g\,dy \right|^{-1} (1 + M_g/2) + 1. \qquad (24)$$

3 A CONTINUOUS 1-D MAP

Consider next a continuous 1-dimensional model, defined by

$$TB(y) = 2\pi\alpha \cos 2\pi y \, e^{i\beta \cos 2\pi y} B(\alpha \sin 2\pi y). \tag{25}$$

This map is an approximation to the the map of the pulsed flow studied in Soward (1992), by a one-dimensional model. (To retain the unit interval we have introduced 2π's.) Here the adjoint operator is more involved than for the SFS model, but a similar procedure is possible. Computing the adjoint, we see that we must define a new variable of integration $\alpha \sin 2\pi y = u$ where u varies between 0 and 1 modulus 1. We accordingly let $\alpha = N$ be a large positive integer and set

$$
\eta_k(u) =
\begin{cases}
\frac{1}{2\pi}\sin^{-1}((k-1)/N + u/N) & k = 1, ..., N \\
1/2 - \frac{1}{2\pi}\sin^{-1}((2N-k)/N - u/N) & k = N+1, ..., 2N \\
1/2 + \frac{1}{2\pi}\sin^{-1}((k-1)/N + u/N) & k = 2N+1, ..., 3N \\
1 - \frac{1}{2\pi}\sin^{-1}((4N-k+1)/N - u/N) & k = 3N+1, ..., 4N
\end{cases}
$$

Then (restoring y in place of u)

$$SB(y) = \sum_{k=1}^{4N} e^{-i\beta \cos 2\pi \eta_k(y)} B(\eta_k). \tag{26}$$

We then again define the projection P as an integral with respect to y over the unit interval, and set

$$\bar{S}B = 2\pi N \int_0^1 e^{-i\beta \cos 2\pi y} \cos 2\pi y B(y) \, dy. \tag{27}$$

The norm is now taken to be

$$\|B\| = \left| 2\pi \int_0^1 e^{-i\beta \cos 2\pi y} \cos 2\pi y B(y) \, dy \right| + V_B(0,1), \tag{28}$$

provided that β is properly chosen to make this a norm. With this formulation the results for the SFS model go through intact and we can again prove perfect fast dynamo action for large N.

4 CONCLUDING REMARKS

The adjoint formulation used in the present examples replaces the production of smaller scales associated with the direct baker's map by a sum which averages over the smaller scales. For the large-amplitude case this becomes a weighted Riemann sum over the field. The magnetic field, despite its complexity, remains Riemann integrable and this is the feature which we exploit.

The maps derived from continuous pulsed flows typically have adjoint formulations which offer no advantages of this kind (Bayly & Childress 1988), even at large amplitude. Thus there is an important difference between the continuous model of section 3 above and the pulsed flow, which must be associated with the 'folds' in the field. It is surprising that this must be a concern at large amplitude, because after a single pulse the field varies rapidly in one direction (in the direction of the wave-number vector for helical waves), and slowly in the other, where the field has been stretched out. The slow variation nevertheless causes, with subsequent pulses, the difficulties in estimating S' for pulsed flows.

The present method, together with recent calculations by Gilbert (private communication) on a related family of cat maps where the direct and adjoint maps also are essentially the same, offer some encouragement that large-amplitude methods can be successfully exploited for rigorous results in the analysis of fast dynamos.

This work was supported at New York University, Courant Institute of Mathematical Sciences, under Grant DMS-8922676 from the National Science Foundation.

REFERENCES

Apostol, T.M. 1957 *Mathematical Analysis.* Addison-Wesley Publishing Company, Inc.

Bayly, B.J. & Childress, S. 1988 Construction of fast dynamos using unsteady flows and maps in three dimensions. *Geophys. Astrophys. Fluid Dynam.* **44**, 211–240.

Bayly, B.J. & Childress, S. 1989 Unsteady dynamo effects at large magnetic Reynolds number. *Geophys. Astrophys. Fluid Dynam.* **49**, 23–43.

Childress, S. 1970 New solutions of the kinematic dynamo problem. *J. Math. Phys.* **11**, 3063–3076.

Rechester, A.B. & White, R.B. 1980 Calculation of turbulent diffusion for the Chirikov-Taylor model. *Phys. Rev. Lett.* **44**, 1586–1589.

Soward, A.M. 1992 An asymptotic solution of a fast dynamo in a two-dimensional pulsed flow. *Geophys. Astrophys. Fluid Dynam.* (submitted).

A Thermally Driven Disc Dynamo

A.Y.K. CHUI & H.K. MOFFATT

Department of Applied Mathematics and Theoretical Physics
University of Cambridge, Silver St., Cambridge, CB3 9EW UK

A new simple idealised model for the geodynamo is proposed. In this model, the energy required for the self-exciting dynamo action is provided by thermal convection in a Welander loop; this is coupled with two Bullard disc dynamos. It is found that two discs with opposite 'twists' are required in order to achieve reversals of the magnetic field, and the model can then be described by a dynamical system of 5th order. This system is found to be identical to the 5th order system of Kennett (1976) resulting from a truncation of the full magnetohydrodynamic equations. The bifurcation structure of the system is studied, and numerical solutions are presented.

1 INTRODUCTION

A simple mechanical model for self-exciting dynamo action was proposed by Bullard (1955). In this model, a conducting disc (driven by a constant external torque G, say) rotates in the presence of a magnetic field at angular velocity $2\pi\Omega(t)$; the current $I(t)$ induced by Faraday's law flows in a wire connecting the rim of the disc (through a sliding contact) and the axle. This system is traditionally described by a second order dynamical system

$$L\dot{I} = -RI + M\Omega I,$$
$$C\dot{\Omega} = G - k\Omega - MI^2,$$
(1)

where L and R are the self-inductance and resistance of the total current circuit respectively, M is the mutual inductance between the wire loop and the rim of the disc, C is the moment of inertia of the disc about the axle, and k is a frictional resistance parameter. When G is positive and large enough, steady dynamo states with non-zero current can be found. Note that

M.R.E. Proctor, P.C. Matthews & A.M. Rucklidge (eds.)
Theory of Solar and Planetary Dynamos, 51–58
©1993 Cambridge University Press.

this model has a preferred parity (or 'handedness') in the sense that dynamo action cannot be found for negative values of G.

Idealised systems with a few degrees of freedom such as the one above are often studied in the hope that they may illuminate the essential physics for the real problem. However, the Bullard disc model is so simplistic that the magnetic field does not reverse; whereas magnetic field reversals are commonly found in astronomical bodies such as the Sun and the Earth. The first idealised model exhibiting reversals was proposed by Rikitake (1958); this consisted of two coupled discs resulting in a third order system. Other third order models (Malkus 1972; Moffatt 1979) derived from a single Bullard disc can also exhibit reversals; they are either identical with, or closely related to, the Lorenz system. Again, all such models have preferred parities so that there is no dynamo action if the direction of applied torque is in the opposite sense relative to the twist in the wire.

There is however an argument suggesting that a good model for geodynamo should have no preferred parity. Helicity (and so the α-effect) results from an interaction between buoyancy and Coriolis effects and may therefore be expected to be proportional to the pseudo-scalar $\mathbf{g} \cdot \mathbf{\Omega}$, (where $\mathbf{g}$ is the gravitational acceleration, and $\mathbf{\Omega}$ is the angular velocity of the Earth); it is therefore antisymmetric about the equatorial plane (Steenbeck, Krause & Rädler 1966). We therefore build an analogue model (Figure 1) with two coupled Bullard discs mounted on the same horizontal axle but with wires twisted in *opposite* directions in order to provide antisymmetry about the vertical centre-plane; and the discs are thermally driven by a fluid loop heated from below. This model has no global preferred parity, as for the geodynamo.

It turns out that the 5th order system governing this model is identical with that found by Kennett (1976) in a truncation of the full magnetohydrodynamic equations for a convecting layer.

2 THE MODEL

2.1 The mechanical device
The model consists of two ingredients (Figure 1); the first is a coupled Bullard disc system in which two discs are mounted on the same axle and allowed to rotate at the same angular velocity. The wires are twisted in opposite senses so that when the discs rotate in one sense, the right-hand disc acts as a dynamo and the left-hand disc acts as an anti-dynamo; if the discs rotate in the opposite sense, then the right-hand disc becomes an anti-dynamo and the left-hand disc becomes a dynamo. The second model ingredient is a circular fluid loop in a vertical plane, heated from below and cooled from above (Welander 1967); thermal convection must occur when the thermal forcing exceeds some critical value. Suppose now we combine these two ingredients

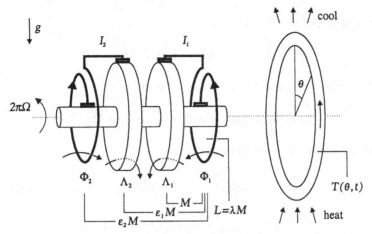

Figure 1. In this model, the coupled discs are driven by a fluid loop heated from below. The fluid loop must be imagined to be dynamically coupled to the rims of the discs, so that both fluid and discs have the same angular velocity $\Omega(t)$.

by identifying their angular velocities, which can be done by placing the fluid loop around the rim of the discs so that the thermal forcing is communicated to the discs; the composite system can now be described by a 5th order nonlinear dynamical system. A similar construction was first proposed by Moffatt (1992).

2.2 The model equations

Let $I_i\,(i = 1, 2)$ be the currents in the wires, positive when flowing outward from the axle; and Φ_i be the magnetic fluxes across the wire loops. The directions of fluxes are chosen (conventionally) so that the self inductances L_i of the wire loops are positive. Similarly we define Λ_i to be the fluxes across the discs, with the directions chosen so that the mutual inductance M_i between the wire loop and the disc in each circuit is positive (see Figure 1). For simplicity, we shall assume that the two circuits are identical except for the 'twists', hence $M_1 = M_2 \stackrel{\text{def}}{=} M$ and $L_1 = L_2 \stackrel{\text{def}}{=} \lambda M\ (\lambda > 1)$. If we further let $-\varepsilon_1 M$ be the mutual inductance between the wire loop and the disc on the different circuit, and $-\varepsilon_2 M$ be the mutual inductance between the two wire loops, then the fluxes and the currents are related by the following equations:

$$\begin{pmatrix} \Phi_1 \\ \Phi_2 \end{pmatrix} = M \begin{pmatrix} \lambda & -\varepsilon_2 \\ -\varepsilon_2 & \lambda \end{pmatrix} \begin{pmatrix} I_1 \\ I_2 \end{pmatrix}, \quad \begin{pmatrix} \Lambda_1 \\ \Lambda_2 \end{pmatrix} = M \begin{pmatrix} 1 & -\varepsilon_1 \\ -\varepsilon_1 & 1 \end{pmatrix} \begin{pmatrix} I_1 \\ I_2 \end{pmatrix}. \quad (2)$$

Note that the mutual inductances $-\varepsilon_1 M$ and $-\varepsilon_2 M$ are negative because the wires are twisted in opposite directions, hence ε_1 and ε_2 are positive. The induction equations for the two circuits are (cf. Moffatt 1979)

$$\dot{\Phi}_1 = -RI_1 + \Omega\Lambda_1, \quad \dot{\Phi}_2 = -RI_2 - \Omega\Lambda_2, \quad (3)$$

and the equation of motion is

$$C\dot{\Omega} = -k\Omega + G - I_1\Lambda_1 + I_2\Lambda_2 , \qquad (4)$$

where R, C, k and G are defined as in (1). Denoting

$$x = \Phi_1 + \Phi_2 , \quad y = \Phi_1 - \Phi_2 , \quad z = \Omega , \qquad (5)$$

we eliminate I_i and Λ_i using (2) and obtain

$$
\begin{aligned}
\dot{x} &= -\frac{R}{M(\lambda - \varepsilon_2)} x + \frac{1 + \varepsilon_1}{\lambda + \varepsilon_2} yz , \\
\dot{y} &= -\frac{R}{M(\lambda + \varepsilon_2)} y + \frac{1 - \varepsilon_1}{\lambda - \varepsilon_2} xz , \\
C\dot{z} &= G - kz - \frac{xy}{M(\lambda^2 - \varepsilon_2^2)} .
\end{aligned}
\qquad (6)
$$

Now let us derive equations for the fluid loop. Let $T(\theta, t)$ be the temperature perturbation along the fluid loop; then it satisfies the one dimensional heat diffusion equation

$$\frac{\partial T}{\partial t} + 2\pi\Omega\frac{\partial T}{\partial \theta} = D\frac{\partial^2 T}{\partial \theta^2} + S(\theta) , \qquad (7)$$

where D is an effective thermal diffusivity in the θ-direction and $S(\theta)$ is the differential heating, which for simplicity we may take to be

$$S(\theta) = -\sigma\sin\theta , \qquad \sigma > 0 . \qquad (8)$$

The solution of (7) then has the form

$$T(\theta, t) = u(t)\cos\theta + v(t)\sin\theta , \qquad (9)$$

where

$$\dot{u} = -Du - 2\pi vz , \qquad \dot{v} = -Dv - \sigma + 2\pi uz . \qquad (10)$$

The density perturbation is $-\alpha T$ where α is the coefficient of thermal expansion, and the gravitational torque acting on the fluid is then

$$G = \int_0^{2\pi} \alpha g T a \cos\theta (V/2\pi)\, d\theta = \frac{1}{2}\alpha g a V u(t) \overset{\text{def}}{=} \gamma u(t) , \qquad (11)$$

where V is the volume of the fluid.

For the composite system, we now identify G in (6c) and (11), and reinterpret C as the total moment of inertia and k as the total frictional resistance coefficient. By rescaling the variables suitably and shifting the origin, we obtain the model equations:

$$
\begin{aligned}
\dot{x} &= \alpha(-\eta x + \omega y z), \\
\dot{y} &= -\eta y + \omega x z, \\
\dot{z} &= \kappa(u - z - xy), \\
\dot{u} &= -u + \xi z - vz, \\
\dot{v} &= -v + uz,
\end{aligned}
\tag{12}
$$

where

$$
\xi = \frac{2\pi\sigma R}{MD^2}, \qquad \kappa = \frac{k}{CD}, \qquad \eta = \frac{R}{MD}\frac{1}{\lambda + \varepsilon_2}
\tag{13}
$$

are measures of thermal forcing, frictional resistance and magnetic diffusion respectively, and

$$
\alpha = \frac{\lambda + \varepsilon_2}{\lambda - \varepsilon_2}, \qquad \omega = \frac{1}{2\pi}\frac{\sqrt{1 - \varepsilon_1^2}}{\lambda + \varepsilon_2}
\tag{14}
$$

are geometrical parameters. Note that $\alpha > 1$ since $\lambda > \varepsilon_2 > 0$. We recall also the meaning of the dependent variables: x is the flux difference, y is the total flux, z is the angular velocity, and u, v give the temperature in the fluid loop, via (9).

2.3 Basic properties

The system has two symmetries: if (x, y, z, u, v) is a solution, then so are $(-x, -y, z, u, v)$ and $(-x, y, -z, -u, v)$. There are at most 7 steady states which can be divided into 3 classes: (i) the trivial state, pure heat conduction (one solution, no dynamo):

$$
x = y = z = u = v = 0;
\tag{14}
$$

(ii) pure thermal convection (2 solutions, no dynamo):

$$
x = y = 0, \quad u = z = \pm\sqrt{\xi - 1} \stackrel{\text{def}}{=} \pm z_0, \quad v = \xi - 1;
\tag{15}
$$

(iii) dynamo solutions (4 solutions, steady dynamo):

$$
z = \pm\frac{\eta}{\omega} \stackrel{\text{def}}{=} \pm z_1, \quad u = \frac{\pm\xi z_1}{1 + z_1^2}, \quad v = \frac{\xi z_1^2}{1 + z_1^2},
$$

$$
x = y = \pm\sqrt{\frac{\eta}{\omega}\left(\frac{\xi}{1 + z_1^2} - 1\right)} \stackrel{\text{def}}{=} \pm x_1.
\tag{16}
$$

Consider the case when the thermal forcing parameter ξ is increased slowly from zero. Figure 2 shows a typical scenario (with $\alpha = 3/2$, $\omega = 1$, $\eta = 4$ and $\kappa = 1$) for bifurcations from the trivial state. The first bifurcation occurs

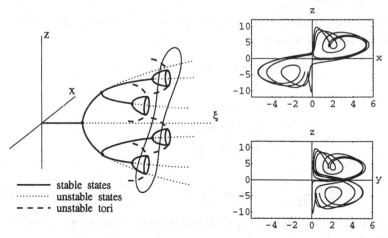

Figure 2. (left) Bifurcations through increasing thermal forcing ξ and (right) phase portraits of the strange attractor.

at $\xi = 1$, when the trivial state becomes unstable; stable thermal convection solutions are created in a pitchfork bifurcation and become preferred states. The second pitchfork bifurcation occurs at $\xi = 1+(\eta/\omega)^2$, at which point stable dynamo solutions are created. Note that Hopf bifurcations from thermal convection branches are also possible (when $\kappa > 2$), but we shall not consider them in this short account.

Typically, as ξ is increased further, the fixed points corresponding to dynamo solutions become unstable in a supercritical Hopf bifurcation; 4 stable periodic solutions (limit cycles) are created, one from each fixed point. They are non-reversing periodic dynamos because y does not change sign, where $y \propto \Phi_1 - \Phi_2$ is a measure of the total flux of the system. The amplitude of each periodic dynamo solution grows as ξ is increased, up to a point when a subcritical bifurcation to invariant torus takes place, the periodic solution becomes unstable and the system now has a strange attractor wandering about 2 unstable fixed points (steady dynamos). This strange attractor (Lorenz type) is believed to be associated with a homoclinic bifurcation at the origin, and always exhibits reversals in x, but not in y. A closer look at the roles taken by x and y in the model equations (12) reveals that it is necessary to make $\alpha < 1$ to allow reversals in y; this implies $\varepsilon_2 < 0$, which is not possible for the configuration of Figure 1.

2.4 A modified model for magnetic field reversals

The above dynamo model does not permit field reversals. Before we give a modified version that does exhibit reversals, let us first examine why the above model does not. Suppose the discs are rotating in an anti-clockwise direction, and the right-hand circuit, which acts as a dynamo, maintains an

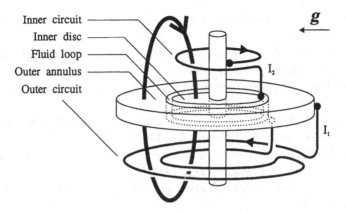

Inner circuit
Inner disc
Fluid loop
Outer annulus
Outer circuit

Figure 3. A modified model that exhibits reversals (gravity now acts horizontally).

positive flux through its disc. Some of the flux diffuses away, but some, however tiny it is, must pass through the left-hand disc. Now if the direction of rotation is reversed, the flux through the right-hand disc decays since the right-hand circuit now acts as an anti-dynamo; the flux through the left-hand disc increases but, because it 'remembers' its direction, it can never change sign. In order to achieve reversals, it is necessary therefore to modify the model so as to change the sign of the mutual inductance between two wire loops. A possible solution is illustrated in Figure 3: the fluid loop is now embedded in a conducting disc in such a way that it also acts as an insulator which separates the inner disc and the outer annulus. Wires with sliding contacts are added to complete two separated circuits. It is not difficult to show that the model equations remain unchanged, and the mutual inductance between the two wire loops $-\varepsilon_2 M$ is now positive (i.e. $\varepsilon_2 < 0$). Computations with $\varepsilon_2 < 0$ confirm that y does now show random reversals.

3 CONCLUSIONS

We have shown how a Welander thermal loop may be dynamically coupled with a pair of Bullard disc dynamos with opposite twists to yield a self-exciting dynamo system in which the power supplied is entirely of thermal origin. The bifurcations of the resulting 5th order nonlinear dynamical system have been studied, and the conditions for dynamo action are determined. Flux reversals do not occur if the mutual inductance between the twisted wire loops is negative; but if the geometrical arrangement is such that this mutual inductance is positive (as for example in the configuration of Figure 3), then flux reversals do occur. An analogous constraint on the distribution of helical eddies responsible for the α–effect may be necessary to explain the reversals of the Earth's magnetic field.

REFERENCES

Bullard, E.C. 1955 The stability of a homopolar dynamo. *Proc. Camb. Phil. Soc.* **51**, 744–760.

Kennett, R.G. 1976 A model for magnetohydrodynamic convection relevant to the Solar dynamo problem. *Stud. App. Math.* **55**, 65–81.

Malkus, W.V.R. 1972 Reversing Bullard's dynamo. *EOS, Trans. Am. Geophys. Union* **53**, 617.

Moffatt, H.K. 1979 A self-consistent treatment of simple dynamo systems. *Geophys. Astrophys. Fluid Dynam.* **14**, 147–166.

Moffatt, H.K. 1992 The Earth's magnetism: past achievements and future challenges. *I.U.G.G. Chronicle* **211**, 130–146.

Rikitake, T. 1958 Oscillations of a system of disk dynamos. *Proc. Camb. Phil. Soc.* **54**, 89–105.

Steenbeck, M., Krause, F. & Rädler, K.-H. 1966 Berechnung der mittleren Lorentz-Feldstärke **v** × **B** für ein elektrisch leitendes Medium in turbulenter, durch Coriolis-Kräfte beeinflusster Bewegung. *Z. Naturforsch.* **21a**, 369–376.

Welander, P. 1967 On the oscillatory instability of a differentially heated fluid loop. *J. Fluid Mech.* **29**, 17–30.

Magnetic Instabilities in Rapidly Rotating Systems

D.R. FEARN

Department of Mathematics
University of Glasgow
Glasgow, G12 8QW UK

Dynamo theory focuses on the generation of an axisymmetric (or mean) magnetic field by the action of a mean electromotive force (e.m.f.) and differential rotation. A topic that has received somewhat less attention is that of the stability of the field to non-axisymmetric perturbations. In mean-field dynamo theory, the field is *maintained* when the generation effect of the mean e.m.f. and differential rotation balance the decay due to ohmic diffusion. However, if the field is sufficiently strong and it satisfies certain other conditions then the field may be unstable. The instability extracts energy from the mean field so the generation mechanism may have a second sink of energy to counteract. Magnetic instabilities may therefore play an important role in determining what fields are observed and how strong they are. Theoretical and observational ideas are now converging. The idea that instability might be the mechanism for initiating a field reversal was suggested some time ago. Linear theory has established that the minimum field strength required for instability (though depending on many factors) is comparable with estimates of the Earth's toroidal field strength. More recently, a careful analysis of the reversal data has concluded that 'reversals are triggered by internal instabilities of the fluid motion of the core'. Here, we review the various classes of magnetic instability and the conditions required for instability.

M.R.E. Proctor, P.C. Matthews & A.M. Rucklidge (eds.)
Theory of Solar and Planetary Dynamos, 59–68
©1993 Cambridge University Press.

1 MOTIVATION

In the absence, so far, of fully hydrodynamic dynamo models representative of the Earth or the planets, the main focus of planetary dynamo theory remains with (axisymmetric) mean-field dynamo models in which the contribution from the nonlinear interaction of the non-axisymmetric components of the problem are parameterized through a prescribed α effect (see for example Roberts 1993). Such models have been very useful in exploring various aspects of the dynamo mechanism. Nonlinear effects have been included and the manner in which they act to equilibrate the field amplitude investigated (see Fearn 1993).

Mean-field dynamo theory gives information about whether the strength of given α- and ω-effects is sufficient to overcome ohmic losses and so generate a field. The inclusion of nonlinear effects gives information on the field strength that can be maintained. What mean-field theory is unable to address is the stability of the field to non-axisymmetric perturbations. A non-axisymmetric magnetic instability grows by extracting energy from the mean field. The rate of extraction of energy may be comparable with that due to ohmic diffusion of the mean field, in which case the field may still be maintained but at a lower amplitude than if the instability were absent. The instability typically takes the form of a travelling wave so magnetic instability is one possible mechanism for generating the geomagnetic secular variation (g.s.v.). [The identification of the g.s.v. with travelling hydromagnetic waves goes back to Hide (1966) and Braginsky (1964).] Alternatively, the effect of an instability may be rather more dramatic, with a rapid transfer of energy from the mean field to the instability. This would be a possible mechanism for initiating a magnetic field reversal. This idea was first put forward from theoretical considerations but more recently there has been support for it from a statistical analysis of the reversal data (McFadden & Merrill 1992).

There are of course other sources of energy that can drive non-axisymmetric instability, the principal ones being buoyancy and shear. Convective instability is a primary feature of the dynamics of planetary cores and is probably the source of the α-effect responsible for magnetic field generation. (The α-effect clearly cannot be due to magnetic instability since the mean field cannot be maintained by an instability of itself.) In this review we wish to concentrate on magnetic instabilities, but that does not mean that we believe them to be of greater importance than instabilities due to buoyancy or shear. Since space does not permit doing full justice to magnetic instabilities, it does not make sense to attempt to deal with the wider topic of non-axisymmetric hydromagnetic instabilities, especially when buoyancy and shear driven instabilities are dealt with elsewhere (see Proctor 1993).

Most of the work on magnetic field stability so far has been linear. This has identified the various classes of instability and the main result of this work has been to find how strong the field must be before the onset of instability. This critical field strength is a function of the field and the class of instability. What is clear is that there are two classes of instability (resistive and field-gradient) that have critical field strengths that are comparable with estimates of the Earth's toroidal field strength. They can therefore be expected to play a significant role in the evolution of the geomagnetic field.

In the following section, we give a general discussion of the type of models that have been used to investigate the problem of magnetic instability and we emphasise the importance of magnetic diffusion. In many instances we make specific mention of the Earth, but the basic theory relates to magnetic fields in rotating systems so has much wider applicability; to planetary cores and to stellar interiors. In section 3, we review the classes of instability that have been found and give details of what is known about the conditions required for instability in each case. Concluding remarks are made in section 4.

2 MODELS

Models used to investigate magnetic instability fall into two categories: those that neglect the effect of ohmic diffusion (*ideal* models) and those that include it. The former have the advantage that the problem is of lower differential order, permitting more progress to be made analytically. Their shortcoming is that important physical effects are absent, limiting the applicability of such models. Recent examples of ideal models include those of Friedlander (1987, 1989) and London (1992). London's work uses computer algebra to develop asymptotic solutions. The asymptotic scaling requires a weak magnetic field (such that the leading order force balance is geostrophic), whereas in the Earth's core it is believed that the Lorentz force is comparable in strength with the Coriolis force (magnetostrophic balance). The nature of the ideal problem has permitted some progress to be made on the nonlinear problem (Friedlander & Vishik 1990).

Ohmic diffusion plays two roles. When sufficiently strong, it will act to damp out any instability. Secondly, through permitting the reconnection of magnetic field lines and the motion of field lines relative to the fluid (not possible in an ideal fluid), it can promote instability. Such instability is usually referred to as *resistive* instability and is well known in non-rotating magnetohydrodynamics/plasma physics (see for example Bateman 1978).

The stabilising role of diffusion is important because it determines whether a given class of instability is relevant to the geo- and planetary dynamo problem. A non-dimensional measure of the strength of ohmic diffusion is Λ^{-1} where Λ is the Elsasser number defined by

$$\Lambda = \frac{\mathcal{B}^2}{2\Omega\mu\rho\eta} = \frac{T_\eta}{T_s},$$

(1)

where $\mathcal{B}$ is a typical magnetic field strength, Ω is the rotation frequency, μ the magnetic permeability, ρ the density of the core fluid, and η the magnetic diffusivity. The ohmic diffusion timescale is $T_\eta = L^2/\eta$ where L is a characteristic lengthscale (for example the radius of the core). The natural timescale of waves in an ideal system is the slow MHD timescale $T_s = 2\Omega/\Omega_A^2$ where $\Omega_A = \mathcal{B}/L\sqrt{\mu\rho}$ is the Alfvén frequency. The Elsasser number can thus be thought of in two equivalent ways: as a measure of the field strength or as an inverse measure of ohmic diffusion. The limit $\Lambda \to \infty$ is the perfectly conducting (ideal) limit. If a system is ideally unstable, the instability will grow exponentially according to linear theory, with a growth rate independent of Λ when $\Lambda \gg 1$. The effect of decreasing Λ (which can be achieved by decreasing $\mathcal{B}$ or by increasing η) will eventually be to bring the growth rate to zero (at $\Lambda = \Lambda_c$). At this point the energy supplied by the mean field to the instability is just sufficient to maintain the instability against the effects of ohmic diffusion. For $\Lambda < \Lambda_c$ the system is stable; any perturbation is damped out, there being insufficient energy available from the mean field to counteract ohmic damping. If Λ_c is very much larger than is characteristic of planetary cores then the corresponding instability is unlikely to be observed. Conversely if Λ_c is comparable with (or even less than) estimates of the value of Λ to be found in planetary cores, then the instability must play a role in the dynamics of the core. To establish the relevance of any instability to the dynamo problem, it is essential to know a typical value of Λ_c for that class of instability. Clearly this requires a diffusive (i.e. non-ideal) theory.

For resistive instability, ohmic diffusion plays a dual role. Its presence can act to destabilise a system but, as discussed above for ideal instability, if the diffusion is sufficiently strong, it has a stabilising effect; i.e. we can find a critical value Λ_c of Λ also for resistive instabilities. A characteristic property of resistive instability is that, since the instability is absent in the perfectly conducting limit, we expect the growth rate of the instability to approach zero in the limit $\Lambda \to \infty$. Indeed this is found to be the case in general (see Fearn 1984) but there is one exception (see Fearn & Weiglhofer 1992a, Fearn & Kuang 1993).

In recent years there has been considerable progress in our understanding of both resistive and ideal instability, through a series of studies. (Note: here, we shall use the term *ideal* to describe an instability to indicate that it exists in the absence of ohmic diffusion, and not to indicate that the effects of ohmic diffusion are absent.) All models consider an electrically conducting fluid permeated by a prescribed magnetic field $\mathbf{B}_0$ rotating rapidly with angular velocity $\Omega = \Omega 1_z$. In many cases the effects of buoyancy are neglected and the magnetostrophic approximation is made (i.e. inertial and viscous effects are neglected in the equation of motion). Acheson's (1983) work uses a local analysis for the diffusive problem that complements earlier non-diffusive work (Acheson 1972, 1973). He investigates (cylindrical) fields of the form

$$\mathbf{B}_0 = B_0(s)\mathbf{1}_\phi, \tag{2}$$

where (s, ϕ, z) are cylindrical polar coordinates and $\mathbf{1}_x$ is the unit vector in the x direction. His analysis can only deal with an ideal instability he calls the 'field-gradient instability'. Fearn (1983) used the same field in a cylindrical annulus and compared his numerical results with Acheson's. Good qualitative agreement was found. More recently, Lan *et al.* (1993) have improved on Acheson's analysis and found good quantitative agreement between their numerical and analytic results in the asymptotic limit $\Lambda, n \to \infty$ (where n is the axial wavenumber). Fearn's (1983) model has been investigated further in Fearn (1984, 1985, 1988) and Fearn & Weiglhofer (1992a) and extended to a spherical geometry and fields $\mathbf{B}_0 = B_0(s, z)\mathbf{1}_\phi$ by Fearn & Weiglhofer (1991a,b, 1992b). Kuang & Roberts (1990, 1991, 1992) have studied the force-free field $\mathbf{B}_0 = B_0(\cos qz/d, \sin qz/d, 0)$ in a plane-layer model, where q is a constant and d the layer depth. Their field lines are straight and change in direction only with height z. They exhibit resistive instability but not the field-gradient instability which depends on field-line curvature.

The above studies have established the importance of the field-gradient and resistive instabilities to the Earth. It is widely believed that, in the core, the toroidal component of the mean field is strong compared with the poloidal field that we observe at the surface. (The toroidal component is confined to the core.) This is the justification for neglecting the poloidal component and choosing to study fields of the form (2).

3 CONDITIONS FOR INSTABILITY
Here we simply summarise what has been learnt about the conditions required for instability. Typically, there are at least two necessary conditions. The first is that the function $\mathbf{B}_0$ must satisfy some condition, and the second is $\Lambda > \Lambda_c$. There may be additional requirements.

To gauge the relevance of an instability to the geodynamo problem it should be noted that $\Lambda = 1$ corresponds to a field strength of about 2×10^{-3}T (20 gauss) and $\Lambda = 100$ to 2×10^{-2}T. The poloidal field strength at the core-mantle boundary is $O(5 \times 10^{-4}$T$)$ and estimates of the toroidal field strength are a few hundred gauss or less. Elsasser numbers roughly in the range $O(1 - 100)$ are therefore of interest for the Earth.

3.1 Field-gradient instability
This is an ideal instability in the sense described in section 2. Acheson's (1983) local analysis describes the stability of the field (2) in terms of the

parameter

$$\Delta = \frac{s^3}{B_0^2}\frac{d}{ds}\left(\left(\frac{B_0}{s}\right)^2 - \frac{R}{\Lambda}\left(\frac{V_0}{s}\right)\right),\tag{3}$$

where the basic state includes the flow $\mathbf{V}_0 = V_0(s)\mathbf{1}_\phi$ and the magnetic Reynolds number $R = \mathcal{U}L/\eta$ is a non-dimensional measure of the flow speed $\mathcal{U}$. Acheson (1983) showed that instability is possible if

$$\Delta > m^2,\tag{4}$$

where m is the azimuthal wavenumber of the instability. Clearly $m = 1$ is the most unstable mode, and it is found that Λ_c increases rapidly with increasing m. If $B_0 \propto s^\alpha$ and $R = 0$ then field-gradient instability requires $\alpha > 3/2$. The critical Elsasser number depends strongly on the choice of B_0. In the cylindrical annulus choosing $B_0 \propto s^\alpha$ resulted in $\Lambda_c \geq O(200)$ for $m = 1$, with $m = 2$ having values of Λ_c typically an order of magnitude larger. Alternative choices of field gave values of Λ_c as low as 15 (see Fearn 1983, 1988).

Fearn & Weiglhofer (1991a,b) investigated the effect of a spherical geometry. They extended the local analysis to fields of the form $B_0 = B_0(s,z)$ (see also Fearn & Proctor 1983), and found that in general a z-dependence has a destabilising influence. However this is more than compensated for by the constriction of the spherical geometry (the cylindrical annulus was unbounded in z); a variety of choices of B_0 was investigated but the lowest value for Λ_c found was $O(200)$, an order of magnitude larger than for the cylindrical annulus.

3.2 Resistive instability

Resistive instability is usually associated with critical levels $\mathbf{k} \cdot \mathbf{B}_0 = 0$ where $\mathbf{k}$ is the wavevector of the instability. For fields of the form $\mathbf{B}_0 = B_0\mathbf{1}_\phi$, this condition reduces to $B_0 = 0$, i.e. resistive instability is associated with there being a zero of the field somewhere in the core. This condition is well known in the non-rotating literature and the main effect of rotation is to modify the timescale on which the instability operates (see Fearn 1984). Field-line curvature is unimportant; the instability has been found both for curved fields of the form (2) and for straight field lines in the model of Kuang & Roberts (1990). For the cylindrical model Fearn (1984, 1988) found $\Lambda_c \geq O(10)$, with a fairly weak dependence on m. Kuang & Roberts (1990), though concentrating mainly on an asymptotic analysis in the $\Lambda \to \infty$ limit, give an example with $\Lambda_c \approx 22$. Fearn & Weiglhofer (1991b) found the spherical geometry much less restrictive than for the field-gradient instability, finding examples with $\Lambda_c \approx 50$.

Both the classical resistive instability and the field-gradient instability are fairly insensitive to boundary conductivity. Recently, a new type of resistive instability has been found that is absent when the boundaries are perfectly conducting. This instability may be found when $\mathbf{k} \cdot \mathbf{B}_0 \neq 0$. It was first identified by Fearn & Weiglhofer (1992a) and is the subject of further investigation (see Fearn & Kuang 1993). For $B_0 = B_0(s)$ instability appears to be associated with a field satisfying $B_0''/B_0 < 0$ with the magnitude of the ratio being sufficiently large. A non-zero value of B_0' at the boundary also appears to be necessary. Field curvature is unimportant. Values of $\Lambda_c \gtrsim O(10)$ have been found.

3.3 Dynamic
Malkus (1967) investigated the field $B_0 \propto s$ and found instability for $\Omega_A > \Omega$. This requires a field strength of at least 500 T, very much larger than estimates of the Earth's toroidal field strength.

3.4 Exceptional
Roberts & Loper (1979) used the term 'exceptional' to describe an instability they found for the field $B_0 \propto s$. The instability required boundaries that are not perfect conductors and a core fluid that is not perfectly conducting. It also requires the presence of the inertial term in the momentum equation; i.e. the instability is filtered out if the magnetostrophic approximation is made. A very similar instability can be found if viscous effects are reinstated (even if inertia and ohmic diffusion are absent), see Fearn (1988). Roberts & Loper considered the asymptotic limit of high conductivity ($\Lambda \to \infty$), so were unable to determine Λ_c, nor were they able to say anything about how the instability depended on the choice of $\mathbf{B}_0$. Fearn (1988) used his cylindrical model with insulating boundary conditions to investigate fields of the form (2). This work resulted in two main conclusions. The first was the determination of typical values of Λ_c for the Roberts & Loper exceptional mode. The critical value depends on the strength of inertial (and/or viscous) effects (i.e. on the Rossby and Ekman numbers), but is typically $O(10^3)$ or larger, so this class of instability is unlikely to be relevant to the Earth.

The second result was finding one simple theory that explained the existence of several unusual instabilities, including the exceptional one. The condition (4) applies in rapidly rotating systems. In non-rotating systems it becomes $\Delta > m^2 - 4$, so the effect of rotation is stabilising; values of Δ in the range $m^2 - 4 < \Delta < m^2$ are stable in a rapidly rotating system and unstable in a non-rotating system. In a rotating system, if some effect can counteract the stabilising effect of rotation, then instability may be possible. This explanation is consistent with the Roberts & Loper exceptional instability, Fearn's (1988) viscous counterpart, the buoyancy-catalysed instability (Roberts & Loper 1979, Soward 1979, Acheson 1983) and the effect of adding an axial field to (2) (Fearn 1985). In none of these examples is instability found for $\Delta \leq m^2 - 4$.

4 CONCLUSION

In this brief review there has been no room for details of the various models investigated or the equations solved. For these the reader is referred to the original papers. The aim here has been to summarise the main results and emphasise the relevance of the field-gradient and resistive instabilities to the geodynamo problem. Many details have had to be omitted, for example information about the propagation direction of the instabilities. Nor has anything been said about the stabilising and destabilising roles of differential rotation [see (3), Fearn 1989, Fearn & Weiglhofer 1992b].

Acknowledgements

My work on magnetic field stability is supported by the Science and Engineering Research Council of Great Britain under grant GR/H 03506.

REFERENCES

Acheson, D.J. 1972 On the hydromagnetic stability of a rotating fluid annulus. *J. Fluid Mech.* **52**, 529–541.

Acheson, D.J. 1973 Hydromagnetic wavelike instabilities in a rapidly rotating stratified fluid. *J. Fluid Mech.* **61**, 609–624.

Acheson, D.J. 1983 Local analysis of thermal and magnetic instabilities in a rapidly rotating fluid. *Geophys. Astrophys. Fluid Dynam.* **27**, 123–136.

Bateman, G. 1978 *MHD Instabilities*, M.I.T. Press, Cambridge, Massachusetts.

Braginsky, S.I. 1964 Magnetohydrodynamics of the Earth's core. *Geomag. Aeron.* **4**, 698–712.

Fearn, D.R. 1983 Hydromagnetic waves in a differentially rotating annulus I. A test of local stability analysis. *Geophys. Astrophys. Fluid Dynam.* **27**, 137–162.

Fearn, D.R. 1984 Hydromagnetic waves in a differentially rotating annulus II. Resistive instabilities. *Geophys. Astrophys. Fluid Dynam.* **30**, 227–239.

Fearn, D.R. 1985 Hydromagnetic waves in a differentially rotating annulus III. The effect of an axial field. *Geophys. Astrophys. Fluid Dynam.* **33**, 185–197.

Fearn, D.R. 1988 Hydromagnetic waves in a differentially rotating annulus IV. Insulating boundaries. *Geophys. Astrophys. Fluid Dynam.* **44**, 55–75.

Fearn, D.R. 1989 Differential rotation and thermal convection in a rapidly rotating hydromagnetic system. *Geophys. Astrophys. Fluid Dynam.* **49**, 173–193.

Fearn, D.R. 1993 Nonlinear planetary dynamos. In *Lectures on Solar and Planetary Dynamos* (ed. M.R.E. Proctor & A.D. Gilbert), Cambridge University Press.

Fearn, D.R. & Kuang, W. 1993 Resistive instability in the absence of critical levels. *Geophys. Astrophys. Fluid Dynam.* (submitted).

Fearn, D.R. & Proctor, M.R.E. 1983 Hydromagnetic waves in a differentially rotating sphere. *J. Fluid Mech.* **128**, 1–20.

Fearn, D.R. & Weiglhofer, W.S. 1991a Magnetic instabilities in rapidly rotating spherical geometries I. From cylinders to spheres. *Geophys. Astrophys. Fluid Dynam.* **56**, 159–181.

Fearn, D.R. & Weiglhofer, W.S. 1991b Magnetic instabilities in rapidly rotating spherical geometries II. More realistic fields and resistive instabilities. *Geophys. Astrophys. Fluid Dynam.* **60**, 275–294.

Fearn, D.R. & Weiglhofer, W.S. 1992a Resistive instability and the magnetostrophic approximation. *Geophys. Astrophys. Fluid Dynam.* **63**, 111–138.

Fearn, D.R. & Weiglhofer, W.S. 1992b Magnetic instabilities in rapidly rotating spherical geometries. III. The effect of differential rotation. *Geophys. Astrophys. Fluid Dynam.* **67**, 163–184.

Friedlander, S. 1987 Hydromagnetic waves in the Earth's core. *Geophys. Astrophys. Fluid Dynam.* **39**, 315–337.

Friedlander, S. 1989 Conditions for instability for hydromagnetic waves in a contained rotating stratified fluid. *Geophys. Astrophys. Fluid Dynam.* **46**, 245–260.

Friedlander, S. & Vishik, M.M. 1990 Nonlinear stability for stratified magnetohydrodynamics. *Geophys. Astrophys. Fluid Dynam.* **55**, 19–46.

Hide, R. 1966 Free hydromagnetic oscillations of the Earth's core and the theory of the geomagnetic secular variation. *Phil. Trans. R. Soc. Lond. A* **259**, 615–647.

Kuang, W. & Roberts, P.H. 1990 Resistive instabilities in rapidly rotating fluids: Linear theory of the tearing mode. *Geophys. Astrophys. Fluid Dynam.* **55**, 199–239.

Kuang, W. & Roberts, P.H. 1991 Resistive instabilities in rapidly rotating fluids: Linear theory of the g-mode. *Geophys. Astrophys. Fluid Dynam.* **60**, 295–332.

Kuang, W. & Roberts, P.H. 1992 Resistive instabilities in rapidly rotating fluids: Linear theory of the convective modes. *Geophys. Astrophys. Fluid Dynam.* (in press).

Lan, S., Kuang, W. & Roberts, P.H. 1993 Ideal instabilities in rapidly rotating MHD systems that have critical layers. *Geophys. Astrophys. Fluid Dynam.* (in press).

London, S. 1992 Hydromagnetic waves in a rotating sphere. *Geophys. Astrophys. Fluid Dynam.* **63**, 91–110.

McFadden, P.L. & Merrill, R.T. 1992 Inhibition and geomagnetic reversals. *J. Geophys. Res.* (in press).

Malkus, W.V.R. 1967 Hydromagnetic planetary waves. *J. Fluid Mech.* **28**, 793–802.

Proctor, M.R.E. 1993 Convection and magnetic fields. In *Lectures on Solar and Planetary Dynamos* (ed. M.R.E. Proctor & A.D. Gilbert), Cambridge University Press.

Roberts, P.II. 1993 Fundamentals of dynamo theory. In *Lectures on Solar and Planetary Dynamos* (ed. M.R.E. Proctor & A.D. Gilbert), Cambridge University Press.

Roberts, P.H. & Loper, D.E. 1979 On the diffusive instability of some simple steady magnetohydrodynamic flows. *J. Fluid Mech.* **90**, 641–668.

Soward, A.M. 1979 Thermal and magnetically driven convection in a rapidly rotating fluid layer. *J. Fluid Mech.* **90**, 669–684.

Modes of a Flux Ring Lying in the Equator of a Star

A. FERRIZ-MAS & M. SCHÜSSLER

Kiepenheuer-Institut für Sonnenphysik
Schöneckstr. 6, D-7800 Freiburg, Germany

The oscillation modes of a toroidal flux tube in the equatorial plane of a differentially rotating star are investigated with the aid of the *thin flux tube approximation*. Both axially symmetric and non-axially symmetric perturbations are considered. The axisymmetric mode ($m = 0$) is essentially a radial mode oscillating with the *magnetic Brunt–Väisälä* frequency. The non-axisymmetric modes in the limit of rapid rotation are a pair of radial, inertial modes and a pair of longitudinal slow modes; in the general case, the four different modes do not have a definite character. The corresponding bifurcation diagrams are discussed.

1 INTRODUCTION

In this contribution we report on the results of a linear stability analysis of toroidal flux tubes lying in the equatorial plane of a differentially rotating star. The motivation for this analysis is twofold. On the one hand, magnetic buoyancy poses a serious problem for magnetic field storage in the convection zone (Parker 1975). Some authors have considered the possibility that a slightly subadiabatic region in the overshoot layer below the convection zone could be a favourable place for the storage of magnetic flux and the operation of the dynamo mechanism (cf. Stix 1991). On the other hand, we are interested in investigating the α-effect due to the instabilities of isolated flux tubes at the bottom of the convection zone or below (ongoing investigation). Although the net α-effect due to flux tubes in the equatorial plane would be zero, we focus our attention in this contribution on that simple case, for it permits an analytical investigation and can serve as a basis for understanding the more complicated case of a flux tube outside the equatorial plane.

M.R.E. Proctor, P.C. Matthews & A.M. Rucklidge (eds.)
Theory of Solar and Planetary Dynamos, 69–78

2 MODEL AND BASIC EQUATIONS

Consider an isolated flux tube in a rotating star which is in stationary equilibrium ($\partial/\partial t \equiv 0$) and is axisymmetric. We call the matter inside the flux tube the *internal medium* (and shall denote quantities inside the flux tube with the subscript 'i'), while the *external medium* is the matter outside the flux tube (subscript 'e'). The external equilibrium is determined by

$$\rho_e(\mathbf{v}_e \cdot \nabla)\mathbf{v}_e = -\nabla p_e + \rho_e\left[\mathbf{g} - \mathbf{\Omega} \times (\mathbf{\Omega} \times \mathbf{r})\right] + 2\rho_e\mathbf{v}_e \times \mathbf{\Omega}, \qquad (1)$$

where ρ_e is the density, p_e the pressure, and $\mathbf{g}$ is the acceleration of gravity at the position $\mathbf{r}$ measured from the center of the star. $\mathbf{\Omega}$ is the (constant) angular velocity of the reference frame. In this paper, the external velocity field $\mathbf{v}_e$ is assumed to be entirely due to stellar rotation. We shall employ cylindrical coordinates $\{r, \phi, z\}$ with the z-axis along the rotation axis, and let $\{\mathbf{e}_r, \mathbf{e}_\phi, \mathbf{e}_z\}$ be the corresponding unit vectors.

The dynamics of the flux tube is described with the aid of the *thin flux tube approximation* (Spruit 1981; Ferriz-Mas & Schüssler 1993). The momentum equation reads

$$\rho_i\frac{D\mathbf{v}_i}{Dt} = -\nabla\left(p_i + \frac{B^2}{8\pi}\right) + \frac{(\mathbf{B} \cdot \nabla)\mathbf{B}}{4\pi} + \rho_i[\mathbf{g} - \mathbf{\Omega} \times (\mathbf{\Omega} \times \mathbf{r})] + 2\rho_i\mathbf{v}_i \times \mathbf{\Omega}, \quad (2)$$

where $\mathbf{B}$ is the magnetic field in the flux tube, $\mathbf{v}_i$ the velocity, and ρ_i the density. The internal gas pressure p_i is related to p_e through the condition of instantaneous lateral pressure balance, $p_i + B^2/8\pi = p_e$. Equation (2) has to be complemented with the equations of continuity, induction and energy, together with an equation of state. We consider the limit of infinite electrical conductivity and as energy equation we use the condition of isentropic evolution, $DS/Dt = 0$, where S is the specific entropy (for a justification of this hypothesis in a deep stellar convection zone see, e.g., Moreno-Insertis 1986). As an equation of state we assume that of an ideal gas.

The equilibrium configuration is an axisymmetric toroidal flux tube (a flux ring) lying in the equator at a radial distance r_0 (all equilibrium quantities are denoted with the subscript '0'). We shall choose $\mathbf{\Omega}$ to be the angular velocity of the matter in the unperturbed flux tube. The external velocity field in the equatorial plane is $\mathbf{v}_e(r) = r\left[\Omega_e(r) - \Omega\right]\mathbf{e}_\phi$, where $\mathbf{\Omega}_e(r) = \Omega_e(r)\mathbf{e}_z$ is the angular velocity at $\mathbf{r}$. Further, assuming that the equator is a plane of symmetry, $\partial p_e(\mathbf{r})/\partial z \equiv 0$ at $z = 0$, the equilibrium of the external medium in the equatorial plane is determined by

$$\frac{\partial p_e}{\partial r} = -\rho_e(g - r\Omega_e^2). \qquad (3)$$

On application of the momentum equation (2) together with (3) to the equilibrium flux tube, the azimuthal and axial components vanish, while the radial component yields the equilibrium density ratio ρ_{e0}/ρ_{i0}:

$$\frac{v_A^2}{r_0 g_0} - \left(\frac{\rho_{e0}}{\rho_{i0}} - 1\right)\left(1 - \frac{r_0 \Omega_{e0}^2}{g_0}\right) + \frac{r_0}{g_0}(\Omega_{e0}^2 - \Omega^2) = 0. \tag{4}$$

Here $g_0 \overset{\text{def}}{=} g(r_0)$, $\Omega_{e0} \overset{\text{def}}{=} \Omega_e(r_0)$, and $v_A \overset{\text{def}}{=} B_0/\sqrt{4\pi\rho_{i0}}$ is the Alfvén speed. Equation (4) expresses a balance between curvature force, buoyancy force, and rotationally induced forces. In what follows we shall assume for simplicity that the matter in the unperturbed flux tube rotates with the same angular velocity as its surroundings, $\Omega = \Omega_{e0}$. In general, flux tube equilibria in the equatorial plane with $\Omega \neq \Omega_{e0}$ are possible; for a discussion see Ferriz-Mas & Schüssler (1993).

3 LINEARIZATION AND DISPERSION RELATION

The equations governing the dynamics of the flux tube are linearized about the equilibrium configuration. Let $\mathbf{r} = \mathbf{r}(s_0, t)$ be the vector function giving the path of the perturbed flux tube in terms of the Lagrangian coordinate (unperturbed arc-length $s_0 = r_0\phi_0$) and time. Then, $\mathbf{r}(s_0, t) = \mathbf{r}_0 + \boldsymbol{\xi}(s_0, t)$. We express all perturbations in terms of the components ξ_r, ξ_ϕ, ξ_z of the displacement vector $\boldsymbol{\xi}$. The resulting equations are a system of linear, homogeneous partial differential equations with constant coefficients. It turns out that the perturbations contained in the equatorial plane are decoupled from those perpendicular to the equator. It is useful to cast the linearized equations in dimensionless form by introducing the quantities $H \overset{\text{def}}{=} p_{i0}/(\rho_{i0}g_0)$ (pressure scale-height within the flux tube) and $\tau \overset{\text{def}}{=} \sqrt{2}\, H/v_A$ (the Alfvén travel-time across the height H). We define $f \overset{\text{def}}{=} H/r_0$ (a measure of the curvature of the equilibrium path of the flux tube), and $\tilde{\Omega} \overset{\text{def}}{=} \tau\Omega$.

The Fourier components are of the form $\boldsymbol{\xi} \sim \exp(i\omega t + im\phi_0)$, where ω is the (complex) frequency, and m the (integer) azimuthal wavenumber. The imaginary part of ω determines the stability of the equilibrium: if it is negative, the perturbation grows, if it is positive, the perturbation decays. The real part of ω indicates whether the instability is monotonic or oscillatory.

A discussion of the perturbations in latitude, which give rise to the *poleward slip instability* (Spruit & van Ballegooijen 1982) can be found in Moreno-Insertis *et al.* (1992). In this paper we shall concentrate on perturbations within the equatorial plane; the corresponding dispersion relation is a quartic equation in the dimensionless frequency $\tilde{\omega} \overset{\text{def}}{=} \tau\omega$,

$$\tilde{\omega}^4 + d_2\tilde{\omega}^2 + d_1\tilde{\omega} + d_0 = 0. \tag{5}$$

If the gas pressure in the flux tube is much larger than the magnetic pressure (i.e., $\beta \overset{\text{def}}{=} 8\pi p_{i0}/B_0^2 \gg 1$), which is a plausible assumption for the deep parts of the convection zones of cool stars like the Sun, the coefficients can be written in a particularly simple form:

$$d_2 = 2f^2(\sigma - 1 - 2m^2) + \frac{4}{\gamma}f - \frac{2}{\gamma}\left(\frac{1}{\gamma} - \frac{1}{2}\right) + \beta\delta - 4\tilde{\Omega}^2(1 + q),$$

$$d_1 = 16mf\left(f - \frac{1}{2\gamma}\right)\tilde{\Omega}, \tag{6}$$

$$d_0 = -2m^2f^2\left[2(\sigma + 3 - m^2)f^2 - \frac{4}{\gamma}f + \frac{1}{\gamma} + \beta\delta - 4q\tilde{\Omega}^2\right].$$

We have defined $q = r_0\Omega_e(r_0)/2\Omega'_e(r_0)$. The dependence of the acceleration of gravity with depth is expressed by the parameter $\sigma \overset{\text{def}}{=} [d\log g(r)/d\log r]_{\mathbf{r}_0}$. Further, $\delta = \nabla - \nabla_{\text{ad}}$, where $\nabla = [d\log T/d\log p]_{\mathbf{r}_0}$ and $\nabla_{\text{ad}} \simeq (\gamma-1)/\gamma$ is the corresponding value of ∇ for a homentropic stratification (loosely speaking, an *adiabatic* stratification).

4 THE AXISYMMETRIC MODE

For $m = 0$, the dispersion relation (5) reduces to $\tilde{\omega}^2(\tilde{\omega}^2 + d_2) = 0$. The axisymmetric mode is stable if $\tilde{\omega}^2 > 0 \iff d_2 < 0$, i.e., if

$$2(1 - \sigma)f^2 - \frac{4}{\gamma}f + \frac{2}{\gamma}\left(\frac{1}{\gamma} - \frac{1}{2}\right) - \beta\delta + 4\tilde{\Omega}^2(1 + q) > 0. \tag{7}$$

Following Moreno-Insertis (1986) (see also Moreno-Insertis *et al.* 1992), we can define a *magnetic Brunt–Väisälä frequency* for our problem as

$$\tilde{\omega}^2_{\text{MBV}} = 2(1 - \sigma)f^2 - \frac{4}{\gamma}f + \frac{2}{\gamma}\left(\frac{1}{\gamma} - \frac{1}{2}\right) - \beta\delta + 4\tilde{\Omega}^2(1 + q), \tag{8}$$

or, introducing dimensions,

$$\omega^2_{\text{MBV}} = -\frac{g_0\delta}{H} + 4\Omega^2(1 + q) + \frac{v_A^2}{2H^2}\left[\frac{2}{\gamma}\left(\frac{1}{\gamma} - \frac{1}{2}\right) - \frac{4}{\gamma}f + 2(1 - \sigma)f^2\right]. \tag{9}$$

The term $-g_0\delta/H$ is the square of the Brunt–Väisälä frequency in a plane-parallel atmosphere (in the absence of magnetic field) with constant gravitational acceleration g_0 and pressure scale-height H. With this definition, the necessary condition for stability (7) is $\omega^2_{\text{MBV}} > 0$. This condition replaces the condition $\delta < 0$ of the non-magnetic case (Schwarzschild criterion).

In the absence of rotation, for given values of the parameter set $\{\gamma, \sigma, f\}$, the eigenfrequencies $\tilde{\omega}_\pm$ depend only on the product $\beta\delta$. If rotation is included, for given values of the angular velocity and its gradient at r_0, the eigenfrequencies $\tilde{\omega}_\pm$ depend on δ and β separately. This follows from the definition of $\tilde{\Omega} = \tau\Omega = \Omega\sqrt{\beta H/g_0}$. In Figure 1, the eigenfrequencies $\pm\tilde{\omega}$ are

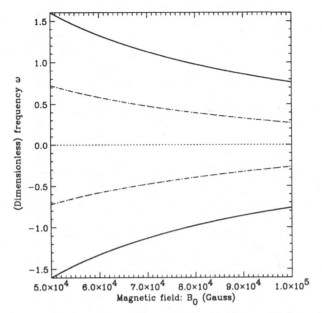

Figure 1. Frequency diagram for the axisymmetric modes: The solid lines are the eigenfrequencies $\tilde{\omega}_\pm = \pm\sqrt{\tilde{\omega}_{MBV}^2}$ in the stable regime. These are plotted versus the magnetic field strength B_0 for $\delta = -10^{-6}$. The dash-dotted lines are the corresponding frequencies $\tilde{\omega}_\pm$ when $\Omega = 0$. The values of the parameters entering the definition of $\tilde{\omega}_{MBV}^2$ are typical for the bottom of the Solar convection zone: $r_0 = 5 \cdot 10^5 \,\mathrm{km}$, $g_0 = 519 \,\mathrm{ms}^{-2}$, $\Omega = \Omega_{e0} = 2.7 \cdot 10^{-6} \,\mathrm{s}^{-1}$, $q = 0.06$, $f = 0.114$, $\sigma = -1.82$ and $\gamma = 5/3$.

plotted versus the field strength B_0 for given values of the superadiabaticity and of the angular velocity. The corresponding curves $\tilde{\omega}_\pm$ for the case of no rotation are also shown. From the amplitude relations it can be shown that, if rotation is present, the axisymmetric mode is not a purely radial (transversal) mode; i.e., it involves perturbations both in radial (transversal) and azimuthal (longitudinal) directions, in contrast to the case without rotation. This is a consequence of angular momentum conservation.

5 NON-AXISYMMETRIC MODES

5.1 Non-rotating case
We begin by discussing the non-axisymmetric modes in the absence of rotation, in which case the eigenfrequencies are either real numbers ($\tilde{\omega}^2 > 0$) or purely imaginary numbers ($\tilde{\omega}^2 < 0$). In Figure 2, the frequencies of the non-axisymmetric modes (here $m = 3$) are plotted versus $\beta\delta$. We have two pairs of modes which are symmetric with respect to $\tilde{\omega} = 0$. For $|\beta\delta| \gg 1$ and $\beta\delta < 0$, one mode pair takes on the character of the longitudinal tube mode and approaches $\tilde{\omega} = \pm\sqrt{2}mf$ (dotted lines), which corresponds to a

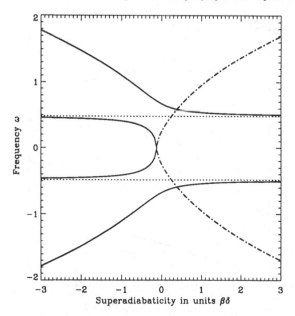

Figure 2. Frequency diagram for the non-axisymmetric modes $m = 3$ in the absence of rotation. The solid lines indicate the eigenfrequencies of the modes with $\tilde{\omega}^2 > 0$, while the dash-dotted lines represent the eigenfrequencies of the modes with $\tilde{\omega}^2 < 0$.

phase speed $\pm v_A$, the limit of the *tube speed* for the case $\beta \gg 1$ considered here. The other pair of modes approaches the transversal tube mode in this limit. The onset of instability at the critical value $\beta\delta = -0.12$ corresponds to a merging of the two longitudinal modes. As $\beta\delta$ increases further, the former transversal modes take on more and more the character of the longitudinal modes, while the unstable modes become mainly transversal (dashed-dotted lines in Figure 2).

5.2 Asymptotic behaviour in the limit of rapid rotation

We shall call *limit of rapid rotation* the limit in which $4\tilde{\Omega}^2$ dominates over all other terms in the definition of the Brunt–Väisälä frequency, i.e., $\tilde{\omega}_{\mathrm{MBV}}^2 \simeq 4\tilde{\Omega}^2$. With this assumption, $d_2 \simeq -4\tilde{\Omega}^2$, while d_1 and d_0 remain unchanged in (6). Following van Ballegooijen (1983), we can filter out the high-frequency inertial oscillations ($\tilde{\omega} \simeq \pm 2\tilde{\Omega}$) by neglecting the term $\tilde{\omega}^4$, so that the dispersion relation (5) reduces to a quadratic equation in $X = \tilde{\Omega}\tilde{\omega}$:

$$X^2 - 4mf\left(f - \frac{1}{2\gamma}\right)X - \frac{d_0}{4} = 0. \tag{10}$$

In this approximation, two modes have frequencies $\tilde{\omega} \simeq \pm 2\tilde{\Omega}$ (inertial waves); these are shown with dashed lines in Figure 3. The two other modes are obtained from (10) with $\tilde{\omega} = X_\pm/\tilde{\Omega}$ (solid curves in Figure 3); they correspond

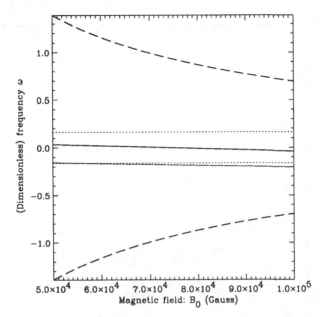

Figure 3. The frequencies of the four non-axisymmetric modes corresponding to $m = 1$ are plotted against the magnetic field strength B_0 in the *approximation of rapid rotation*: $4\tilde{\Omega}^2 \gg |\beta\delta|$. Here $\delta = -10^{-6}$.

to the *magnetostrophic waves* which appear for continuous field distributions (e.g., Schmitt 1987). If there were no rotation, the *magnetostrophic waves* would degenerate into longitudinal modes with frequency $\tilde{\omega} = \pm\sqrt{2}mf$; i.e., modes whose phase speed is given by $\pm v_A$ (dotted lines in Figure 3).

5.3 General case

The treatment of the general case for $m \neq 0$ is not as straightforward as the case $m = 0$, for the full fourth-order dispersion relation has to be considered. Since the coefficients are real, its roots (the frequencies $\tilde{\omega}$) are either real numbers or pairs of complex conjugates. It can be shown (Ferriz-Mas & Schüssler 1993) that a necessary and sufficient criterion for stability is:

$$-\frac{4}{27}d_0\left(d_2^2 - 4d_0\right)^2 + d_1^2\left(\frac{d_2^3}{27} - \frac{4}{3}d_0d_2 + \frac{d_1^2}{4}\right) < 0. \qquad (11)$$

In Figure 4 we show the stability diagram in the (B_0, δ)-plane for the non-axisymmetric modes $m = 1$ using values typical for the bottom of the Solar convection zone (see Figure 1). The dotted regions indicate stability. We observe that up to $B_0 \gtrsim 10^5$ G, the regions of stable (region I) and unstable configurations are separated by a single dividing line (*curve of marginal stability*). For larger field strengths, a new region of stability appears (II), whose

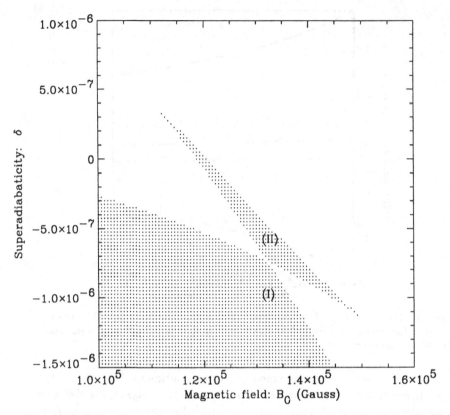

Figure 4. Stability diagram for the non-axisymmetric modes corresponding to $m = 1$. The regions of stability are marked with dots.

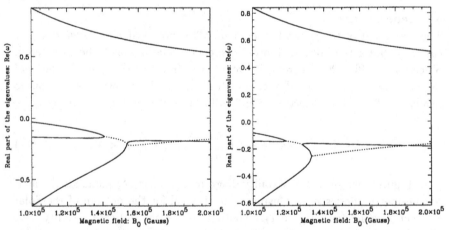

Figure 5. Bifurcation diagrams for the non-axisymmetric modes corresponding to $m = 1$. The real parts of the eigenfrequencies are plotted against the magnetic field strength for (a) $\delta = -1.25 \cdot 10^{-6}$ and (b) $\delta = -5 \cdot 10^{-7}$.

form and location depend on the parameter values (in any case, there is a common boundary point to both stability regions, I and II). As can be seen in Figure 4, this second stability region can extend into the superadiabatic lower part of the convection zone proper and provide a small range of field strengths within which rather strong fields can be stored.

Next we turn to the character of the four modes described by the dispersion relation (5). When rotation is considered, we still have two pairs of modes which, however, are no longer symmetric with respect to $\tilde{\omega} = 0$. In the limit of rapid rotation, one pair takes on the character of a (transversal) inertial mode ($\tilde{\omega} = \pm 2\tilde{\Omega}$), while the second pair evolves into predominantly longitudinal magnetostrophic modes. In general, when the limit of rapid rotation does not apply, the modes do no longer have a definite longitudinal or transverse character. In Figure 5 we have plotted the real part of the four eigenfrequencies $\tilde{\omega}$ versus the magnetic field for two given values of the superadiabaticity. The dotted parts of the curves indicate instability, i.e., non-vanishing imaginary part of $\tilde{\omega}$. This corresponds to two horizontal cuts through Figure 4, one crossing the second stability region (II) and one not. We see that, similarly to the non-rotating case, an unstable mode is created by the merging of the two longitudinal modes. However, with rotation the frequency of the unstable mode retains a finite real part, so that the instability attains the character of a wave with growing amplitude. From Figure 5(b) we see that region II in Figure 4 corresponds to the range of field strengths between two different regimes of instability which are due to the merging of different modes.

6 CONCLUSIONS

We have carried out a study of the stability and the nature of the axisymmetric and non-axisymmetric modes of toroidal flux tubes lying in the equator of a differentially rotating star. Under conditions which presumably prevail in the overshoot region below the Solar convection zone, flux tubes with field strength up to about 10^5 G are stable. For non-axisymmetric modes, which are easier to excite than the axisymmetric ones, a region of stability [on the (B_0, δ)-plane] has been found which may extend into the superadiabatically stratified lower part of the convection zone. The interpretation of this stability region is that it is a transition between regimes of instability corresponding to the merging of different modes. The range of variation of the parameters for which stability (or otherwise) is possible, as well as the oscillation frequencies and the growth rates, are very sensitive to differential rotation and also to the difference of rotation rates between the equilibrium flux tube and its surroundings (Ferriz-Mas & Schüssler 1993), although these parameters do not change the topology of the stability diagram on the (B_0, δ)-plane.

REFERENCES

Ferriz-Mas, A. & Schüssler, M. 1993 Instabilities of magnetic flux tubes in a stellar convection zone. I. Equatorial flux rings in differentially rotating stars. *Geophys. Astrophys. Fluid Dynam.* (in press).

Moreno-Insertis, F. 1986 Nonlinear time evolution of kink-unstable magnetic flux tubes in the convective zone of the Sun. *Astron. Astrophys.* **166**, 291–305.

Moreno-Insertis, F., Schüssler, M. & Ferriz-Mas, A. 1992 Storage of magnetic flux tubes in a convective overshoot region. *Astron. Astrophys.* **264**, 686–700.

Parker, E.N. 1975 The generation of magnetic fields in astrophysical bodies. X. Magnetic buoyancy and the Solar dynamo. *Astrophys. J.* **198**, 205–209.

Schmitt, D. 1987 An $\alpha\omega$-dynamo with an α-effect due to magnetostrophic waves. *Astron. Astrophys.* **174**, 281–287.

Spruit, H.C. 1981 Motion of magnetic flux tubes in the Solar convection zone and chromosphere. *Astron. Astrophys.* **98**, 155–160.

Spruit, H.C. & van Ballegooijen, A.A. 1982 Stability of toroidal flux tubes in stars. *Astron. Astrophys.* **106**, 58–66.

Stix, M. 1991 The Solar dynamo. *Geophys. Astrophys. Fluid Dynam.* **62**, 211–228.

van Ballegooijen, A.A. 1983 On the stability of toroidal flux tubes in differentially rotating stars. *Astron. Astrophys.* **118**, 275–284.

A Nonaxisymmetric Dynamo in Toroidal Geometry

M. FOTH

Universitäts–Sternwarte Göttingen
Geismarlandstraße 11
D-3400 Göttingen, Germany

The gaseous components in spiral galaxies often have the form of
a torus. Magnetic fields are observed which are axisymmetric or
bisymmetric. To study nonaxisymmetric magnetic fields in such
rings an $\alpha^2\omega$ dynamo will be investigated. The main problem is
the representation of the magnetic induction **B** and the treatment
of the functions describing **B**.

---

1 SPIRAL GALAXIES AND THEIR MAGNETIC FIELDS

The observations of spiral galaxies show magnetic fields. Most are of axisymmetric structure, for example M31 (Beck 1982) or IC342 (Krause *et al.*
1989a), but also nonaxisymmetric magnetic fields were observed in M51 (Sofue *et al.* 1980) and in M81 (Krause *et al.* 1989b).

The observed ring-shaped distribution of the interstellar medium (Unwin
1980; Lesch *et al.* 1989) suggests the choice of toroidal coordinates $(\eta, \vartheta, \varphi)$ for
the description of the magnetic field (see Figure 1). The coordinate η varies
from zero to infinity. The surface of a torus is determined by a constant η_0, the
larger the value of η_0 the thinner is the torus. The coordinate ϑ varies from
$-\pi$ to π; surfaces of constant ϑ are spherical bowls through the equatorial
plane. The azimuthal coordinate φ varies from 0 to 2π.

Magnetic fields can be generated by plasma motions in the electrically conducting interstellar gas (see, for example, Krause & Rädler 1980). We consider the magnetic field generated by the action of a dynamo in a torus. The

79

M.R.E. Proctor, P.C. Matthews & A.M. Rucklidge (eds.)
Theory of Solar and Planetary Dynamos, 79–82
©1993 Cambridge University Press.

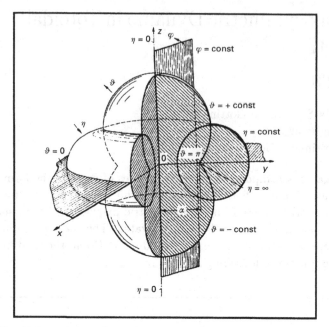

Figure 1. Toroidal coordinates $(\eta, \vartheta, \varphi)$.

axisymmetric dynamo in toroidal coordinates has been investigated (Grosser 1988; Deinzer *et al.* 1990, 1992; Schmitt 1990; Kehrer 1991; Richter 1992). Here, we represent the description of the nonaxisymmetric dynamo in toroidal coordinates.

2 THE DYNAMO EQUATION AND THE NUMERICAL METHOD

The dynamo equation

$$\frac{\partial \mathbf{B}}{\partial t} = \nabla \times (\mathbf{v} \times \mathbf{B} + \alpha \mathbf{B}) - D \nabla \times (\nabla \times \mathbf{B}) \qquad (1)$$

is to be solved; D is the diffusivity, which is assumed constant. The magnetic induction $\mathbf{B}$ is represented by two scalar functions S and T:

$$\mathbf{B} = \nabla \times (\mathbf{e}_\eta \times \nabla S + \mathbf{e}_\eta T), \qquad (2)$$

where $\mathbf{e}_\eta$ is the unit vector of the η coordinate. The functions S and T are to be determined, from which the magnetic field is derived.

First, the external magnetic field is determined. It is assumed that the torus is embedded into vacuum. The vacuum field is described by a scalar potential Φ which obeys Laplace's equation ($\nabla \Phi = \mathbf{B}$, $\Delta \Phi = 0$). The solution of Laplace's equation is known:

$$\Phi = (\cosh \eta - \cos \vartheta)^{\frac{1}{2}} \sum_{m,n} P^m_{n-\frac{1}{2}}(\cosh \eta) \left\{ A^m_n \cos n\vartheta + B^m_n \sin n\vartheta \right\} e^{im\varphi}. \qquad (3)$$

$P^m_{n-\frac{1}{2}}(\cosh\eta)$ are the toroidal functions. The solutions with $Q^m_{n-\frac{1}{2}}(\cosh\eta)$ which are singular for $\eta \to 0$ are omitted. The function S is written as

$$S = (\cosh\eta - \cos\vartheta)^{\frac{1}{2}} \sum_{m,n} \left\{ A^m_n \sum_\nu \alpha^m_{n\nu}(\eta)\cos\nu\vartheta + B^m_n \sum_\nu \beta^m_{n\nu}(\eta)\sin\nu\vartheta \right\} e^{im\varphi},$$

$$(4)$$

and similarly for T. The coefficients $\alpha^m_{n\nu}(\eta)$ and $\beta^m_{n\nu}(\eta)$ of S and the coefficients of T are needed at a fixed η_0, the surface of the torus, and are calculated by an inhomogenous system of equations with the scalar potential Φ as the inhomogenous part. The equations for the coefficients of S and of T are coupled, in contrast to the axisymmetric case dealing with the φ component of the vector potential A_φ and B_φ.

Inside the torus, where the dynamo acts, the functions S and T have to be represented by a complete system of functions. The resulting magnetic field **B** must fit continuously at the surface of the torus at η_0.

$$S = (\cosh\eta - \cos\vartheta)^{\frac{1}{2}} e^{\frac{1}{2}(\eta_0-\eta)} \sum_{m,n,l} \left\{ a^m_{nl} \sum_\nu \frac{\alpha^m_{n\nu}(\eta_0)}{J_\nu(r^m_{n\nu l})} J_\nu(r^m_{n\nu l}e^{\eta_0-\eta})\cos\nu\vartheta \right.$$

$$\left. + b^m_{nl} \sum_\nu \frac{\beta^m_{n\nu}(\eta_0)}{J_\nu(s^m_{n\nu l})} J_\nu(s^m_{n\nu l}e^{\eta_0-\eta})\sin\nu\vartheta \right\} e^{im\varphi} \quad (5)$$

satisfies the above conditions, if $A^m_n = \sum_l a^m_{nl}$, $B^m_n = \sum_l b^m_{nl}$, and if $r^m_{n\nu l}$, $s^m_{n\nu l}$ ($l = 1, 2 \ldots$) are the roots of equations which follow from the conditions of S at η_0. A similar form is used for T. Every single expansion function (5) with fixed indices m, n, l already satisfies the boundary conditions.

The time dependence of the functions is written as $e^{\Omega t}$. Dealing with these functions, the dynamo equations for S and T represent an eigenvalue problem. Ω is the eigenvalue, a^m_{nl} and b^m_{nl} form the eigenvector along with the coefficients of T.

The question is whether nonaxisymmetric magnetic fields can be more easily excited than axisymmetric fields if the induction effects are axisymmetric or must the excitation conditions be nonaxisymmetric.

Before the full dynamo equation (1) is solved the diffusion equation will be solved. Then, starting from the decay modes, the induction effects will be added with increasing amplitude to attain the critical point where the dynamo works (cf. Schmitt 1990). In this way, in comparison with the axisymmetric solutions, we shall answer the above question.

REFERENCES

Beck, R. 1982 The magnetic field in M31. *Astron. Astrophys.* **106**, 121–132.

Deinzer, W., Grosser, H. & Schmitt, D. 1990 Torus–dynamo. In *Proc. IAU–Symp.* **140** (ed. R. Beck, P. Kronberg & R. Wielebinski), pp. 95–96. Kluwer.

Deinzer, W., Grosser, H. & Schmitt, D. 1992 *Astron. Astrophys.* (to be submitted).

Grosser, H. 1989 Magnetic decay modes for a slender torus. *Astron. Astrophys.* **199**, 235–241.

Kehrer, K. 1991 Der axialsymmetrische α^2- und $\alpha^2\omega$-Torus–Dynamo. *Diploma–Thesis*, Göttingen.

Krause, F. & Rädler, K.H. 1980 *Mean-Field Magnetohydrodynamics and Dynamo Theory*. Pergamon.

Krause, M., Hummel, R. & Beck, R. 1989a The magnetic field structures in two nearby spiral galaxies. *Astron. Astrophys.* **217**, 4–16.

Krause, M., Beck, R. & Hummel, R. 1989b The magnetic field structures in two nearby spiral galaxies. *Astron. Astrophys.* **217**, 17–30.

Lesch, H., Crusius, A., Schlickeiser, R. & Wielebinski, R. 1989 Ring currents and poloidal magnetic fields in nuclear regions of galaxies. *Astron. Astrophys.* **217**, 99–107.

Richter, S. 1992 Ein axialsymmetrischer $\alpha^2\omega$-Torus–Dynamo eingebettet in ein unendlich leitfähiges Medium. *Diploma–Thesis*, Göttingen.

Schmitt, D. 1990 A torus–dynamo for magnetic fields in galaxies and accretion disks. *Rev. Mod. Astron.* **3**, 86–97.

Sofue, Y., Takano, T. & Fujimoto, M. 1980 Bisymmetric open–spiral configuration of magnetic fields in the galaxies M51 and M81. *Astron. Astrophys.* **91**, 335–340.

Unwin, S.C. 1980 Neutral hydrogen in the Andromeda nebula. *Mon. Not. R. Astron. Soc.* **192**, 243–262.

Simulating the Interaction of Convection with Magnetic Fields in the Sun

P.A. FOX

High Altitude Observatory, National Center for Atmospheric Research
Boulder, CO 80307-3000 USA

M.L. THEOBALD & S. SOFIA

Center for Solar and Space Research, Yale University
New Haven, CT 06511-6666 USA

The detailed dynamics of the Solar dynamo presents a significant challenge to our understanding of the interaction of convection and magnetic fields in the Solar interior. In this paper we discuss certain aspects of this interaction, such as modification of convective energy transport, and turbulent dissipation of magnetic fields. The latter controls the spatial distribution of the magnetic field and its time dependence. We also discuss how these results may influence current Solar dynamo calculations.

---

1 MOTIVATION

Solar activity manifests itself in many forms but perhaps most importantly through the presence of a magnetic field. The topic of this meeting is that of dynamos, in Solar and planetary contexts. In the case of the Sun the dynamo, which seems likely to be responsible for at least part of the Solar activity we observe, acts on a global scale. That is, the period of the dynamo is 22 years (a timescale distinct from those usually encountered on the Sun), sunspots appear within latitude bands and their numbers (in terms of monthly or yearly running means) increase and decrease over one cycle. There is however, a strong asymmetry of the Solar cycle in time, i.e. the growth phase is shorter (and dependent of the amount of activity) than the decay, or descending phase. In addition, the polar field of the Sun is observed to reverse around Solar maximum, again with a distinct asymmetry between hemispheres. Despite these global-scale features, the Solar magnetic field has

83

M.R.E. Proctor, P.C. Matthews & A.M. Rucklidge (eds.)
Theory of Solar and Planetary Dynamos, 83–90
©1993 Cambridge University Press.

many spatial components (Stenflo 1991) and the majority of the magnetic flux appears in small elements.

Because of the nature of the global Solar cycle, in particular sunspots and their active longitudes, the most likely form for the magnetic field which gives rise to sunspots is toroidal. The site of this field during the Solar cycle is likely to depend on its strength, spatial extent, how the field propagates in position, etc. In any case, the magnetic flux that eventually appears at the surface must pass through the Solar convection zone, whose influence upon bundles of magnetic flux is still poorly understood.

Some important questions that arise are: Can small-scale energetic convection provide large-scale ordering of the Solar magnetic field? Can a dynamic dynamo provide the detailed comparison to observation that is imposed by the Solar cycle? What are the details of magnetic fields in regions of stable stratification? The aim of this paper will be to provide some new information on these three questions.

2 UNDERSTANDING THE CONVECTION–MAGNETIC FIELD INTERACTION

2.1 Posing the problem
Since numerical simulations are an invaluable experimental tool in this type of study, we will pose a series of experiments and discuss them in the context of the Solar dynamo. These experiments are not the 'ideal' ones that could be conducted, but serve to illustrate the main points. We will study the evolution of initially concentrated magnetic flux under the influence of turbulent convection. We wish to examine the dynamical interaction, the degree of concentration of magnetic fields, the spatial structure of both field and flow, etc.

2.2 Equations and constraints
In order to investigate the dynamical evolution of the magnetic field, we solve the equations describing stratified compressible magnetohydrodynamical flow with as few simplifying assumptions as possible. In the case of the present simulations, these equations together with all auxiliary quantities are presented in Fox, Theobald and Sofia (1991) and are solved numerically using the ADISM method. Although we have computed a number of three dimensional simulations, to illustrate the main points we will restrict our attention to two dimensional simulations. Thus, it is important to note some assumptions that we make and the boundary conditions we impose.

Since the effects of turbulence (as estimated by non-dimensional quantities, such as the Reynolds number) are important for the Solar convective region, we adopt the Large Eddy Simulation approach, where the transport coefficients are expressed as subgrid scale (SGS) terms, i.e. for the dynamic viscosity μ, thermal diffusivity κ, and the magnetic resistivity η:

$$\mu = \rho \left(c_\mu \Delta\right)^2 \left(2\mathbf{s} : \mathbf{s}\right)^{\frac{1}{2}}, \quad \kappa = \frac{\mu}{\rho Pr}, \quad \eta = \left(c_\eta \Delta\right)^2 |\mathbf{J}| + \eta_0, \tag{1}$$

where ρ is the density, $c_\mu = 0.4$ is the Deardorff coefficient, Δ is the local numerical grid size, $\mathbf{s}$ is the usual rate of strain tensor, $Pr = 1/3$ is the Prandtl number, $c_\eta = 1$ is the resistivity coefficient, $\mathbf{J}$ is the electric current, and η_0 is the bulk resistivity (very small compared to the first term). The derivation of the SGS resistivity, which is a new concept, is discussed in detail in Theobald, Fox & Sofia (1992). One consequence of these variable diffusion coefficients is that the ratio of magnetic to thermal diffusion, $\zeta = \eta/\kappa$, the magnetic Prandtl number, varies in space and time.

The primary goal of Solar convection is to transport an incident heat flux from the lower layers through the domain, to the upper surface. At any particular depth in the convecting region the local rate of change of total energy can be described in terms of a total flux $\mathbf{F}$ made up of enthalpy, kinetic energy, viscous, Poynting (magnetic), and diffusive fluxes:

$$\mathbf{F} = \mathbf{F}_e + \mathbf{F}_k + \mathbf{F}_v + \mathbf{F}_B + \mathbf{f}.$$

The degree to which each of these components dominates at a particular location influences the nature of the convective flow, how magnetic fields (particularly intense fields such as those that participate in the Solar cycle) are organized, etc. In the case where neither the flow, nor the field completely dominates, the exchange between kinetic energy flux and Poynting flux

$$\mathbf{F}_k = \left(\frac{1}{2}\rho V^2\right)\mathbf{V}, \qquad \mathbf{F}_B = \mathbf{E} \times \mathbf{B}, \tag{2}$$

becomes prominent and will be discussed in the next section.

2.3 Cases to be studied

For this paper we present two dimensional models using (x, z), and ignoring variation and flow in the toroidal (y) direction. This simplifies the approach so that only one component (A_y) of the vector potential is needed to determine the magnetic field. Since we are emphasizing the nature of the convection–magnetic field interaction, we can examine many of these features with two dimensional calculations. We have performed three dimensional calculations for many of the experiments presented here but for the points we stress here, the overall qualitative results are the same (this does not mean that all features are similar for two and three dimensional models).

In the present models the boundary conditions on the flow quantities are stress free and fixed, i.e. all normal velocities are zero. The thermodynamic variables have zero flux in the horizontal direction, fixed values at the top and the heat flux is imposed at the lower boundary. By specifying the normal

(a)
(b)

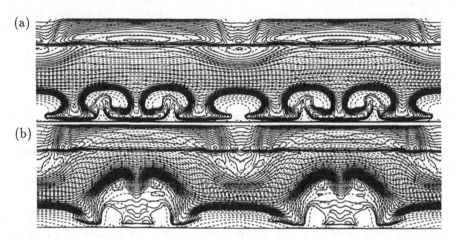

Figure 1. Lines of magnetic force (dashed) and temperature (solid) for experiment A.
(a) $t=4.16$, (b) $t=6.06$.

derivative of A_y at the boundary, we fix the tangential component of the
magnetic field at the boundary and let the normal component vary. Different
tangential fields can be separately specified on each boundary; for example,
$B_z = B_0$ at the left and right boundaries, and $B_x = 0$ at top and bottom.
The equation of state is that of an ideal gas, so that $p = \rho T = (\gamma - 1)e$,
where $\gamma = 5/3$ is the ratio of specific heats.

The three cases considered in this paper are: (A) an embedded horizontal
flux concentration in the middle of an unstably stratified layer with a stably
stratified layer on top, (B) an embedded horizontal flux concentration in the
lower part of a stably stratified layer, and (C) an embedded vertical flux
concentration in the center of an unstably stratified layer. Each of these
concentrations was specified with a Gaussian profile for the vector potential,
ensuring continuity of the magnetic field, electric current and thus electric
field. The time unit in each of the figures is approximately 4 to 8 sound
crossing times depending on vertical or horizontal propagation.

3 RESULTS

3.1 Dynamical interactions
The initial stages of each calculation exhibit a series of transient features, the
most notable being the modal structure of the finite amplitude flow instabilities compared to those in the magnetic field. Although this effect is best
displayed in an animation sequence, Figure 1 shows the evolution of lines of
magnetic force (dashed contours) and those of the temperature field (solid
contours) in experiment A. The key feature is that the wavenumber of the
flow disturbance is higher than the magnetic field will allow; in essence, two or

Figure 2. Lines of magnetic force (dashed) and temperature (solid) for experiment B. (a) $t=1.95$, (b) $t=2.92$.

more of the hot 'plumes' combine to push the field up. This feature certainly depends on the strength and position of the magnetic flux concentration above the lower boundary, from which the plumes emanate, although the behavior is the same if a stable layer is included in the lower part of the domain.

Figure 2 shows an equivalent representation for experiment B. The primary difference in this case is that the finite amplitude thermal perturbations form *above* the magnetic field concentration and, for the given strength, drag the field lines up out of the stable layer with the *same* modal structure as the flow (in contrast to A). Clearly further tests must be performed to understand how strong the field must be to change this result and also determine how deep the field needs to be placed so that it will remain unperturbed by the thermal flows.

After the initial transients, the convection tries to establish its regular up-flow/downflow pattern in the presence of magnetic flux concentration. What eventuates is that the magnetic field is gathered into strong concentrations which may rise buoyantly depending on whether they are anchored (usually by means of a boundary condition – which is an artificial constraint) or not. Figures 3 and 4 show instants from later in the calculations of experiments A and B respectively.

Perhaps the most distinctive feature of these simulations, and many others like them, is that convection allows for some large scale structure in the magnetic field, in addition to significant time dependence of smaller scale, weaker strength flux concentrations. This is most striking in the results of experiment C (initially vertical flux concentration). Its evolution is quite distinct from the horizontal flux concentration. Figure 5 shows two instants from the calculation. Again, there is considerable evolution of large and small scale structures.

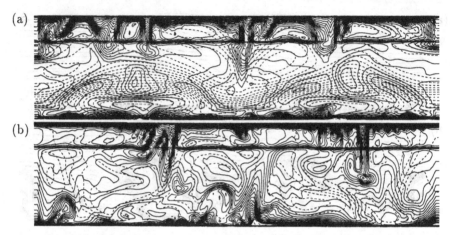

Figure 3. Lines of magnetic force (dashed) and temperature (solid) for experiment A. (a) t=10.74, (b) t=19.57.

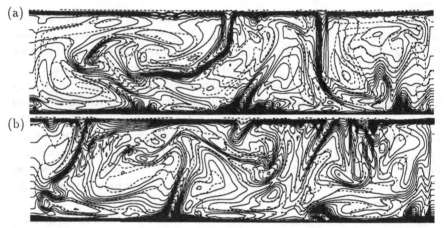

Figure 4. Lines of magnetic force (dashed) and temperature (solid) for experiment B. (a) t=9.00, (b) t=10.82.

3.2 Poynting flux

In the previous section we briefly mentioned buoyant rise of magnetic flux elements. One noticeable feature of the solutions is the amount of heat (flux) drawn into strong magnetic field regions. To quantify this effect we can examine the Poynting flux (2) at particular instants in the calculation. Unfortunately this vector quantity is hard to visualize in a time sequence but up to 30% of the total heat flux can be found in $\mathbf{F}_B$, which is always perpendicular to the field lines and in the present case, directed into regions of strong concentration, i.e. heating the gas inside (this occurs for cases A, B and C). In turn, the extra heating can contribute to the buoyant rise of the

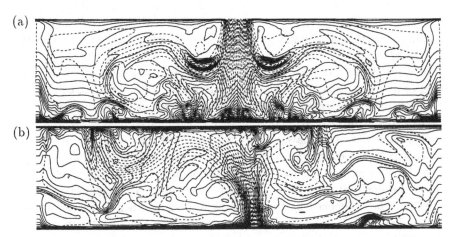

Figure 5. Lines of magnetic force (dashed) and temperature (solid) for experiment C. (a) $t=4.44$, (b) $t=23.38$.

flux concentration. This effect, which is not accounted for in most studies of flux tube dynamics because the energy equation is overly simplified, may place an important constraint on the development and dynamics of magnetic flux concentrations in the Solar interior, and in particular those thought to be part of the Solar dynamo field. This in-situ modulation of the flux concentrations can disorder a global-scale field that is not strongly stabilized (by some other means) and may explain the greater spatial distribution of magnetic elements reaching the Solar surface. Of course the magnetic fields that give rise to sunspots then must be protected from this convective influence.

In the case of magnetic fields within stable layers, the Poynting flux can still play a role in heating field concentrations, particularly if thermal perturbations (hot and cold regions along the interface) provide preferred points where the field can obtain energy and become locally buoyant. There are many other factors which could influence the dynamical evolution in such a situation and this is a clear area for further study.

4 DISCUSSION

Recently there has been much discussion of the role of magnetic diffusivity in dynamo calculations. Our approach to dissipation of magnetic fields is to allow as much dynamical interaction between the velocity and magnetic fields as possible and only provide diffusion (in the forms prescribed by (1)) in regions of high electric current. Essentially this allows magnetic field structures to form, almost down the size of the smallest resolvable scale (i.e. numerical grid). If we compare this result with the mean field concept, i.e. a large and uniform magnetic diffusion, which produces a very limited and low wavenumber range of magnetic field scales, then our formulation would allow significantly greater spatial structure to occur and perhaps provide a natural explanation of the global and fine scale elements of magnetic flux that appear at the surface.

Although there has been much discussion about the base of convection zone, or below the convection zone as a site for the Solar dynamo, the issues of how the field is created, how and why it reverses, how the sunspot fields make it to the surface, etc. are still very much undealt with. There are a few considerations that have arisen from our work that must impact upon the choice of this region in the Sun as a site for the dynamo. One of these is the movement of magnetic flux tubes through the highly turbulent Solar convection zone (at least the uppermost layer) which we have already indicated may be influenced by the Poynting flux. The other is the validity of a mean-field approach to the problem, especially in a stably stratified layer where the conventional α-effect could not operate.

As a final remark, we note that as a consequence of variable resistivity and diffusivity, the magnetic Prandtl number varies within the domain, even in stable regions. It has been known for some time that ζ is one control on the onset of overstable modes of convection (in both the magnetic and velocity fields). The timescale of the oscillation is determined by the relative strength of the field to its surroundings. Furthermore, the onset of this overstable mode in the presence of rotation (which is usually stabilizing), can be achieved even when the layer is locally stable to convective motions (Van der Borght, Murphy & Spiegel 1972). Such conditions are available in the lower part, or beneath the Solar convection zone. Clearly further calculations are warranted to determine if such an oscillatory mode of convection can occur, and if its period is comparable to that of the Solar activity period.

PAF acknowledges partial support from NATO to attend the ASI. MLT and SS thank NASA and AFOSR for support. We also thank Tom Bogdan for reading the manuscript. The National Center for Atmospheric Research is sponsored by the National Science Foundation.

REFERENCES

Fox, P.A., Theobald, M.L. & Sofia, S. 1991 Compressible magnetic convection I: formulation and two-dimensional models. *Astrophys. J.* **383**, 860–881.

Stenflo, J.O. 1991 Diagnostics of the Solar dynamo using the observed pattern of surface magnetic fields. In *The Sun and Cool Stars: Activity, Magnetism and Dynamos* (ed. I. Tuominen, D. Moss & G. Rüdiger), pp. 193–212. Springer.

Theobald, M.L., Fox, P.A. & Sofia, S. 1992 A sub-grid-scale resistivity for magnetohydrodynamics. *Physics Fluids B* (submitted).

Van der Borght, R.F.E., Murphy, J.O. & Spiegel, E.A. 1972 On magnetic inhibition of thermal convection. *Aust. J. Phys.* **25**, 703–718.

Experimental Aspects of a Laboratory Scale Liquid Sodium Dynamo Model

A. GAILITIS

Institute of Physics
Latvian Academy of Sciences
LV-2169 Salaspils 1, Riga, Latvia

This paper describes a laboratory dynamo experiment with liquid sodium moved through a specially designed dynamo model by a huge pump. As the theory and the main experimental results are published (Gailitis 1989; Gailitis *et al.* 1987) we describe here some unpublished aspects such as laboratory resources available, optimum requirements for model and pump, technology for stable steel–sodium electrical contact and post-experimental air test.

1 LABORATORY RESOURCES COMPARED WITH THE GEODYNAMO

Our aim is to reproduce the dynamo phenomenon – the magnetic field generation process known for natural objects, say the Earth. Evidently the reproduction cannot be perfect as the liquid conductor must be moved by an external pump to reach a similar value of the generation parameter $R_m = \mu_0 \sigma v L$:

		Earth	Laboratory	Ratio
	μ_0		1.257×10^{-6}	
	σ	3×10^5	7.5×10^6	0.04
length (m)	L	10^6	0.07	1.4×10^7
velocity (m s^{-1})	v	10^{-3}	30	3×10^{-5}
$R_m = \mu_0 \sigma v L$		400	20	20

M.R.E. Proctor, P.C. Matthews & A.M. Rucklidge (eds.)
Theory of Solar and Planetary Dynamos, 91–98
©1993 Cambridge University Press.

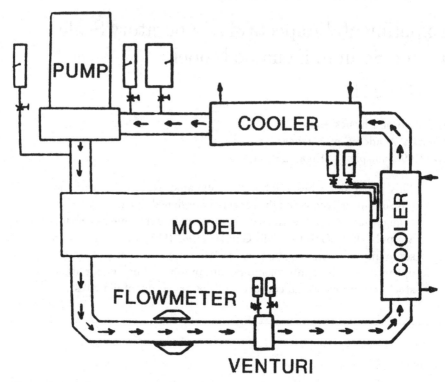

Figure 1. Test loop with pump CEMP 3/1200: device at the Efremov Institute of Electrophysical Apparatus, St. Petersburg, Russia. The reported test was carried out there in 1986.

We can conclude from the above estimates:

1. The only appropriate fluid can be liquid sodium.
2. The experiment is at the margin of laboratory capability and so all aspects must be optimized.

2 OPTIMUM RELATION BETWEEN A MODEL AND A PUMP

The experimental situation (see Figure 1) is governed by three parameters: the field generation parameter $R_m = \mu_0 \sigma v L$, the flow rate $Q = v L^2$ and the hydraulic pressure losses $p = \lambda \rho v^2 / 2$. The three formulae lead to a condition with model properties on the left and pump ones on the right:

$$\lambda R_m^4 = 2(\mu_0 \sigma)^4 / \rho \times p Q^2.$$

The following conclusions can be drawn from the above formula:

1. The model geometry must be optimized for generation at minimum λR_m^4.
2. The model size must be adjusted for the pump to work at its maximum pQ^2.

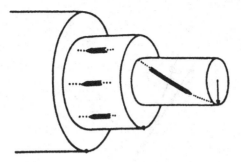

Figure 2. Theoretical model: two coaxial cylinders moving as solid bodies inside a motionless shell.

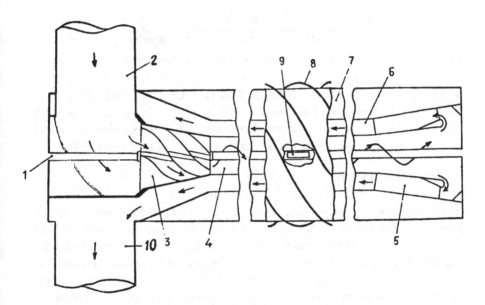

Figure 3. Experimental model. 1–Measuring channel. 2–Sodium entrance.
3–Helical labyrinth. 4–Main channel. 5–Reverse system. 6–Counterflow channel.
7–Immobile sodium. 8–Helical coil for outside excitation.
9–Coil, measuring transverse field. 10–Sodium exit.

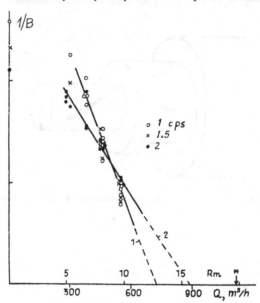

Figure 4. Measured $1/B$ signal versus flow rate Q for three frequencies.

3 HELICAL DYNAMO MODEL WITH A COUNTERFLOW

Comparing known theoretical models with the above criterion we decided upon a helical stream with a counterflow (Figure 2). The experimental model was an all-welded cylinder, 0.5 m in diameter and 3 m long (Figure 3).

4 DYNAMO TEST

To predict the critical R_m that must be achieved for generation to occur the whole model was inserted in a helical 3-phase coil fed by a low frequency generator. The frequency was tuned to obtain the maximum magnetic field signal B measured at the centre of the model. The $1/B$ curve must reach the R_m axis at generation point (Figure 4). Mechanical vibration damaged the model and the generation was not reached. It looks likely that the critical R_m can be even lower than the calculated one (indicated by a * in Figure 4).

We conclude from the dynamo test that the model can reach generation with this pump if vibration is prevented.

5 STAINLESS STEEL–SODIUM ELECTRICAL CONTACT

The model has 5 m^2 of stainless steel–sodium surfaces where a good electrical contact is important for dynamo success. Before the experiment a contact test was made. Three sodium-filled cells were prepared: (i) with 12 chemically cleaned 1 mm thick radial walls; (ii) with 12 uncleaned walls; and (iii) without any walls. Each had a small coil inside and they were all inserted into a common solenoid fed from 400 cps net. Voltages induced into inside coils

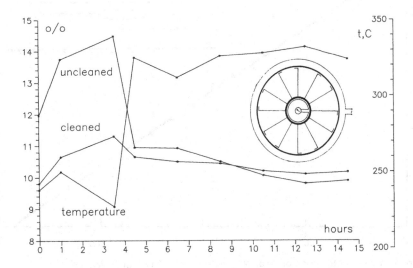

Figure 5. Stainless steel–sodium contact under heating. Test cell and wall part (per cent) in its electrical resistance.

were compared with 400 cps net. The resistance differences deduced from phase shifts, together with a temperature diagram are presented in Figure 5.

We conclude from the contact test:

1. No chemical cleaning is needed.
2. Before an experiment the stainless steel must be in contact with the sodium at $T > 300°C$ for at least ten hours.
3. The experiment itself can succeed at 150–200°C.
4. A successful restart after the steel is separated from sodium and kept in argon can be achieved without reheating to 300°C.

6 POST-EXPERIMENTAL AIR TEST

After the first stage of the experiment the model was disassembled and the sodium channel tested by an air flow. The flow rotation was checked by a vorticity meter (Figure 6). The helical nature of the flow is adequately maintained by inertia, but some slowdown can be observed.

A Pitot type instrument was used to measure velocity details (Figure 7 and 8). At the beginning the helical flow has some concentration near the axis. With flow expansion some rotation is lost.

Air pressure on the sodium channel wall is presented in Figure 9. Pressure losses appear to be higher than those calculated.

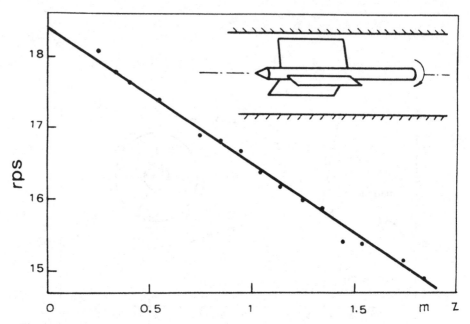

Figure 6. Air test – Rotation of a vorticity meter inserted in a main channel.

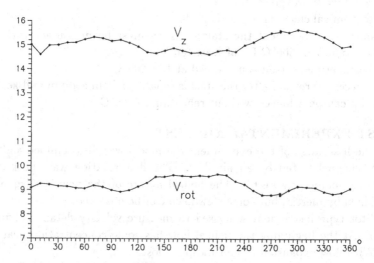

Figure 7. Air test – Azimuthal profiles of axial (above) and rotational (below) velocities ($z = 50$ cm and $2r = 10$ cm).

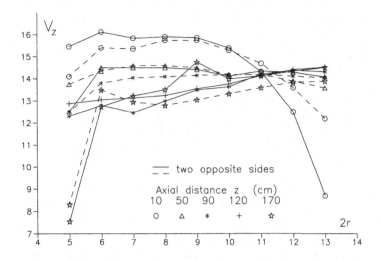

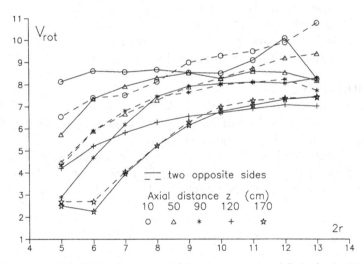

Figure 8. Air test – Radial profiles for axial (above) and rotational (below) velocities (solid and dashed lines – two opposite sides of one diameter).

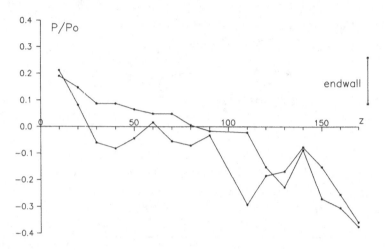

Figure 9. Air test – Pressure on main wall related to entrance pressure (two opposite sides).

We conclude from the air test:
1. To improve velocity field and to minimize pressure losses the entrance labyrinth needs to be redesigned.
2. Some mechanical parts need to be strengthened.

We are looking forward to continuing the experiment.

REFERENCES

Gailitis, A.K., Karasev, B.G., Kirillov, I.A., Lielausis, O.A, Luzhanskii, S.M., Ogorodnikov, A.P. & Preslitskii, G.V. 1987 Liquid metal MHD dynamo model experiment. *Magn. Gidrodin.* **4**, 3–7.

Gailitis, A. 1989 The Helical MHD Dynamo. In *Topological Fluid Mechanics, Proceedings of the IUTAM Symposium, Cambridge, August 13–18, 1989* (ed. H.K. Moffatt & A. Tsinober), pp. 145–156. Cambridge University Press.

Influence of the Period of an ABC Flow on its Dynamo Action

B. GALANTI

Institute of Computational Fluid Dynamics
1-22-3 Haramachi, Meguro-ku
Tokyo 152, Japan

A. POUQUET & P.L. SULEM

CNRS URA 1362
OCA, Observatoire de Nice
BP 229, 06304 Nice Cedex 4, France

Numerical simulations of the kinematic dynamo problem produced by an ABC flow are presented. It is shown that, at fixed magnetic Reynolds number, the dynamo action is strengthened when the spatial period of the ABC flow is chosen moderately smaller than (although commensurable with) the largest available period of the magnetic field.

---

1 INTRODUCTION

The kinematic dynamo refers to the growth of a seed magnetic field $\mathbf{b}$ due to the motion with a prescribed velocity field $\mathbf{v}$ of an electrically conducting fluid of magnetic diffusivity η. The phenomenon is governed by the induction equation

$$\partial_t \mathbf{b} = \nabla \times (\mathbf{v} \times \mathbf{b}) + \eta \nabla^2 \mathbf{b} . \tag{1}$$

Much effort has been devoted in recent years to the case of smooth deterministic flows with chaotic streamlines, in particular when velocity and magnetic field vary on comparable scales. Simple examples of such flows are provided by the so-called 'ABC-flows', whose velocity takes the form

$$\mathbf{v}_{\mathrm{ABC}} = (A \sin k_0 z + C \cos k_0 y,\, B \sin k_0 x + A \cos k_0 z,\, C \sin k_0 y + B \cos k_0 x), \tag{2}$$

where A, B and C are constant coefficients. The value of the wavenumber k_0 measures the ratio of the largest magnetic scale to that of the flow. These

99

M.R.E. Proctor, P.C. Matthews & A.M. Rucklidge (eds.)
Theory of Solar and Planetary Dynamos, 99–103
©1993 Cambridge University Press.

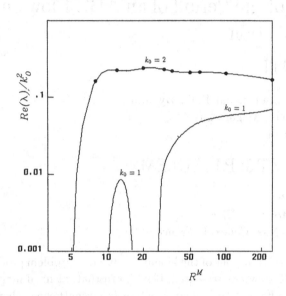

Figure 1. Growth rate of the magnetic field, scaled by the rate of strain of the flow, as a function of Reynolds number for $k_0 = 1$ and $k_0 = 2$. Note the substantially larger rate in the latter case.

flows are Beltrami, consisting of the superposition of three helical waves. Numerical simulations presented in Arnold & Korkina (1983) and Galloway & Frisch (1986) were performed in a 2π-periodic box with $k_0 = 1$, a condition which prevents the development of magnetic structures at scales larger than those of the velocity field. Values of the A, B and C amplitude coefficients for which the dynamics of the fluid particles is apparently chaotic (Dombre *et al.* 1986 and references therein) were considered. For $A = B = C = 1$, a dynamo was in particular shown to exist in two windows in the magnetic Reynolds number $R^M = 1/k_0\eta$. The first window extends from $R^M \approx 8.9$ to $R^M \approx 17.5$, while the second window begins at $R^M \approx 27$ and extends beyond 550. Whether the growth rate remains finite ('fast dynamo') or tends to zero ('slow dynamo') as $R^M \to \infty$, is not yet settled. Visualizations in physical space reveal that the fastest growing eigenmode corresponds to cigar-like structures centered on heteroclinic orbits connecting stagnation points of the flow.

2 INFLUENCE OF THE SPATIAL PERIOD OF THE FLOW

When the spatial scale of the flow is much smaller than that of the magnetic field, the problem is amenable to a multiple-scale analysis and the magnetic field is amplified by the alpha-effect. Here, in contrast, we concentrate on the influence of a moderate decrease in the spatial period of the ABC flows on the growth rate of the magnetic field. We assume $A = B = C$. In order to keep

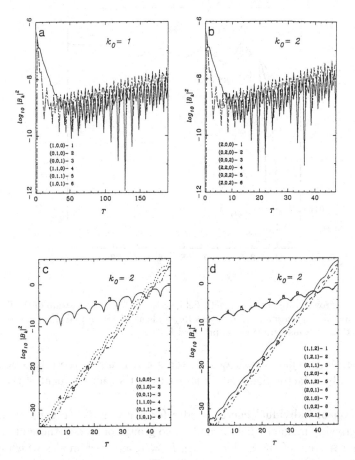

Figure 2. Temporal evolution of a few magnetic modes at $R^M = 12$, for $k_0 = 1$ (a) and $k_0 = 2$ (b)-(d); see the insert for a nomenclature of the various modes.

the magnetic Reynolds number based on the flow scale $(2\pi/k_0)$ at the fixed value η^{-1}, while changing k_0, we take $A = B = C = k_0$. As a consequence, the rate of strain $\nabla \mathbf{v}$ scales like k_0^2.

Figure 1 shows in lin-log coordinates, the growth rate of the r.m.s. magnetic field $\mathbf{B} = \sqrt{\int \mathbf{b}^2 dx}$ in the cases $k_0 = 1$ and $k_0 = 2$ when the magnetic Reynolds number R^M is increased up to 250. It improves on a similar figure presented in Galanti *et al.* (1992) where the maximal magnetic Reynolds number was limited to 40. Two features are striking for $k_0 = 2$, when compared to $k_0 = 1$: (i) the minimum magnetic Reynolds number for existence of a dynamo is substantially lower; (ii) the dynamo growth rate reaches a much larger value and this, at a smaller magnetic Reynolds number. At Reynolds numbers for

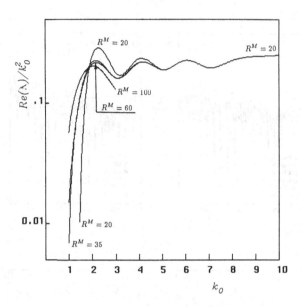

Figure 3. The growth rate of the magnetic field, scaled by the rate of strain of the flow, as a function of scale separation k_0, for various Reynolds numbers that label the curves. The scaled growth rate is insensitive to k_0, provided $k_0 \geq 2$.

which there is a dynamo with $k_0 = 1$, the growth rate of the magnetic field is increased by a factor significantly larger than the enhancement of the rate of strain.

Inspection of individual energy modes for a given R^M (Figure 2) shows that the growing magnetic field includes two groups of modes, those obtained by simple rescaling of the problem with $k_0 = 1$, and new modes which grow faster and dominate the rate of instability. The growth of the magnetic field for all k_0 is oscillatory, corresponding to a Hopf bifurcation, as already noted by Galloway & Frisch (1986) for $k_0 = 1$.

The variation with k_0 of the global growth rate of the dynamo is sketched in Figure 3, for $1 \leq k_0 \leq 10$ and different magnetic Reynolds numbers. Calculations of growth rates at the highest magnetic Reynolds number were limited to a few values of k_0 due to limitation of computational resources. Specifically, for $R^M \leq 20$, we explored $1 \leq k_0 \leq 10$, whereas for $R^M = 35$, we concentrate on $1 \leq k_0 \leq 5$, for $R^M = 60$, on $1 \leq k_0 \leq 4$ and finally for $R^M = 100$, on $1 \leq k_0 \leq 3$. We nevertheless observe similar variations for all cases, the results being seemingly insensitive to the Reynolds number. In particular, the rate of growth of the magnetic field scales approximately like the rate of strain as soon as $k_0 > 1$. The slight oscillation visible in Figure 3 is due to slightly different fastest growing modes for even/odd k_0.

For sufficiently large scale-separation ($k_0 \geq 6$), it is noteworthy that the lowest magnetic mode of the system ($k = 1$) grows at a substantially smaller rate than modes of wavenumbers comparable to k_0, and furthermore in a non-oscillatory fashion. This point requires further investigation.

In physical space, the magnetic patterns keep the same global structure as in the case $k_0 = 1$, up to an approximate rescaling.

3 CONCLUSION

We have thus shown that a moderate decrease of the spatial period of the flow velocity generally leads to an enhancement of the growth rate of the magnetic field, when compared, for the same magnetic Reynolds number, to situations where magnetic field and velocity have the same period. The increase is especially significant when the flow wavenumber increases from $k_0 = 1$ to $k_0 = 2$. For further increase of k_0, the growth rate of the magnetic field scales like the rate of strain of the flow. Nevertheless, the magnetic field is found to be dominant at scales comparable to those of the velocity.

The nonlinear saturation of the growing magnetic field and its reaction on the velocity field, both for $k_0 = 1$ and $k_0 \neq 1$, has been studied in Galanti *et al.* (1992) when the full MHD equations are taken into account, with in particular proper forcing included in the momentum equation. At the moderate Reynolds numbers that have been simulated, the saturation level of the magnetic energy is low (of the order of a few percent) for $k_0 = 1$, while there is a tendency to achieve equipartition between kinetic and magnetic energies when k_0 is increased. Another striking result is that in the late saturation regime, for all k_0, it is the $k = 1$ energy mode that dominates, possibly through some kind of inverse cascade mechanism.

Acknowledgements
Computations were performed on the CM-2 and Convex of the Institute of Computational Fluid Dynamics (Tokyo) and on the Cray-2 of the Centre de Calcul Vectoriel pour la Recherche (Palaiseau).

REFERENCES
Arnold, V.I. & Korkina, E.I. 1983 The growth of a magnetic field in a steady incompressible flow. *Vestn. Mosk. Univ. Mat. Mekh.* **3**, 43–46.

Dombre, T., Frisch, U., Green, J.M., Hénon, M., Mehr, A. & Soward, A.M. 1986 Chaotic streamlines in the ABC flows. *J. Fluid Mech.* **167**, 353–391.

Galanti, B., Sulem, P.L. & Pouquet, A. 1992 Linear and non-linear dynamos associated with ABC flows. *Geophys. Astrophys. Fluid Dynam.* **66**, 183–208.

Galloway, D.J. & Frisch, U. 1986 Dynamo action in a family of flows with chaotic stream lines. *Geophys. Astrophys. Fluid Dynam.* **36**, 53–83.

Numerical Calculations of Dynamos for ABC and Related Flows

D.J. GALLOWAY & N.R. O'BRIAN

School of Mathematics and Statistics
University of Sydney
Sydney, NSW 2006 Australia

We summarise a series of numerical experiments examining kinematic dynamo action associated with the spatially periodic ABC flows

$$\mathbf{u} = (A\sin z + C\cos y, B\sin x + A\cos z, C\sin y + B\cos x),$$

and with some related flows. First, new results are given for the standard case $A = B = C$, pointing out some interesting difficulties. Second, high magnetic Reynolds number runs for the case $A : B : C = 5 : 2 : 2$ are described. Third, results are presented for the case $A = B = C$ with cosine terms omitted; this has no mean helicity. Finally we refer to 'two-and-a-half dimensional' time dependent flows which so far provide the most convincing numerical demonstrations of fast dynamo action.

---

1 INTRODUCTION

Since necessary background information on fast dynamos is being given elsewhere in these proceedings (see also Childress 1992), this section will be very short. We concern ourselves with numerical solutions of the scaled kinematic induction equation

$$\frac{\partial \mathbf{B}}{\partial t} = \nabla \times (\mathbf{u} \times \mathbf{B}) + \frac{1}{R_m}\nabla^2 \mathbf{B},$$

where the prescribed velocity fields $\mathbf{u}$ are as specified in the abstract. Details of the numerical methods together with earlier results are found in Galloway & Frisch (1986), and Galloway & Proctor (1992). Here we describe some additional calculations with particular emphasis on the display of eigenfunctions. Complementary work has recently been carried out by Galanti *et al.* (1993) and Ponty *et al.* (1993), and is reported in these proceedings; see also Galanti *et al.* (1992). Additional computations for the case $A = B = C$ are

M.R.E. Proctor, P.C. Matthews & A.M. Rucklidge (eds.)
Theory of Solar and Planetary Dynamos, 105–113
©1993 Cambridge University Press.

reported in Lau & Finn (1992). All flows used here have regions with chaotic particle paths; we subdivide them into four categories which are described in the following sections.

2 ABC FLOWS WITH $A = B = C$: SOME PERPLEXING ASPECTS

This is much the most studied case, starting with Arnold & Korkina (1983), who identified a window of dynamo action from $R_m = 8.9$ to 17.5, just above which the magnetic field decayed. In their paper, and in Arnold (1984), much was made of the symmetries to be expected in the magnetic field as a consequence of the symmetries of the flow (Dombre *et al.* 1986). Galloway & Frisch (1986) extended Arnold & Korkina's calculations to higher R_m, and in doing so discovered another window of dynamo action extending upwards from around $R_m = 27$. Its eigenfunctions bore some resemblance to those of the first window, and in particular had features ('cigars') localised around those stagnation points with two-dimensional attracting and one-dimensional repelling manifolds. However, these solutions broke certain of the symmetries just referred to, whereas those in the first window subscribed to them. Moreover, solutions started with eigenfunctions from the first window decayed for all R_m values above 17.5. The two types of solution are thus quite different animals. Growth persisted in the second window, with a gradual lengthening of the oscillation period until by $R_m = 300$ the solutions appeared to be steady. Lau & Finn (1992) report that there is apparently another mode changeover associated with this. These features made it impossible to decide whether the $A = B = C$ case was a fast dynamo or not, though by using higher resolution and extrapolating, Lau & Finn now claim that it is.

In fact the combination of symmetry breaking and oscillatory behaviour makes it extremely difficult to identify the true structure of the eigenfunctions. This is because of degeneracy; by definition there must be more than one way of breaking symmetry, implying that there are different eigenfunctions with the same growth rate. When starting the calculation, one generically excites all these modes with an unknown ratio of amplitudes, and unknown phases. At any one time, an interference pattern is seen, but identifying the basic components is next to impossible. Figure 1 shows an isosurface visualisation in the second window of growth at $R_m = 100$. In this and similar plots, the reader looks through the top surface $z = 2\pi$ to the base $z = 0$, shown gridded. The origin is at the bottom left of the base plane. The box is slightly tilted to show features aligned along principal diagonals. Note that the cigars are double structures, with a narrow gap between them. Given the appreciable diffusivity, this is only possible if the features in fact have oppositely aligned fields, which cancel via strong currents in the gap. Figure 2 gives contours of B_z crossing the plane $z = 7\pi/4$, showing that this indeed happens. Also

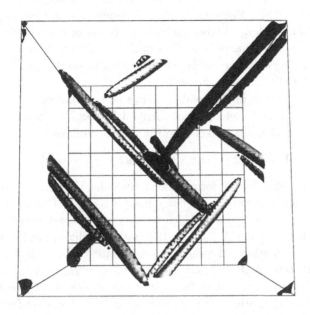

Figure 1. Isosurface plot of the surface $|\mathbf{B}|^2 = 0.17B_{max}^2$, for the mode in the second dynamo window with $R_m = 100$ and $A : B : C = 1 : 1 : 1$. A moment has been chosen where double cigars are clearly visible. The periodicity is apparent, and the two cigars associated with the stagnation point at $(7\pi/4, 7\pi/4, 7\pi/4)$ appear at top right.

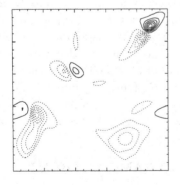 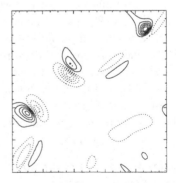

Figure 2. Contours of B_z for the run in Figure 1 in the plane $z = 7\pi/4$, showing polarity of the structures crossing the neighbourhood of $(7\pi/4, 7\pi/4, 7\pi/4)$ (top right). In the left frame, the double cigar is comprised of opposite polarities. In the right, approximately one quarter cycle later, one cigar is passing through zero.

shown in Figure 2 is the same thing a quarter of an oscillation period later; now only one feature is seen, the other presumably being in the process of passing through zero. These difficulties of degeneracy and phasing have no real bearing on deciding whether the dynamo is fast or not, but they obscure the physics of what is happening. It is in principle possible to rig the initial conditions so that only one mode is excited, but to do this one has to know certain aspects of the answer beforehand.

One case where the initial conditions *can* be so doctored is for the mode that grows in the first window. Both Arnold & Korkina (1983) and Gilbert (1991) point out that this arises from the initial condition

$$\mathbf{B} = (\sin z - \cos y, \sin x - \cos z, \sin y - \cos x).$$

Gilbert (1991, 1992) has performed calculations for the perfectly conducting case with this initial condition, by numerically solving the Cauchy initial value problem. He finds field growth which he ascribes to 'constructive folding' of magnetic structures. For $R_m > 17.5$ this initial condition always decayed in our diffusive calculations, showing that something odd is going on. We obtained growth rates with real part around -0.03 for $R_m = 100, 200$ and 400, suggesting that this mode is in fact a 'fast non-dynamo'. Figure 3 gives an isosurface of its magnetic energy. Each cigar in fact consists of four features, three offset from the heteroclinic orbit, with one polarity, and one sitting right on it, opposite in sign. Contours of B_z in the plane $z = 7\pi/4$ show this polarity structure clearly in the neighbourhood of the stagnation point at $x = y = z = 7\pi/4$ (Figure 4). The solution is oscillating with a period similar to its value in the dynamo window, around 10 time units based on the turnover timescale. As R_m increases, the central feature shrinks, but nonetheless its dissipative influence on the whole dynamo remains crucial. The perfectly conducting solution has no such feature; it has shrunk to nothing (see Gilbert's 1991 plot). Thus it seems likely that the $R_m \to \infty$ limit is singular in a wonderfully subtle way. However, there is another possibility: $A = B = C$ has very small chaotic regions, and it may be that the typically $R_m^{-1/2}$-sized structures that form are inefficient at exploiting the chaos when $R_m = 400$. In that case runs at much higher R_m (currently impossible) might show a restoration of growth. In retrospect the smallness of its chaotic regions as well as its additional symmetries make the standard case untypical when searching for fast dynamos.

3 THE CASE $A : B : C = 5 : 2 : 2$: A DYNAMO WITHOUT STAGNATION POINTS

In view of the key rôle played by the stagnation points in the standard case, it is natural to ask whether they are a necessary ingredient for fast dynamos; the results of Vishik (1989) are silent on this point. To this end we note

Figure 3. Isosurface $|\mathbf{B}|^2 = 0.32 B_{max}^2$ for standard ABC with symmetric initial conditions, at $R_m = 100$. The fragmentation of some features is an artifact of the plotting routine, which acts with reference to the coordinate axes.

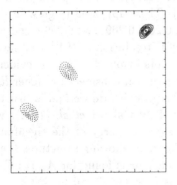 

Figure 4. Contours as for Figure 2, but with symmetric initial conditions.

Figure 5. Isosurface $|\mathbf{B}|^2 = 0.19B^2_{max}$ for $A : B : C = 5 : 2 : 2$; $R_m = 300$. Irregular edges are again a plotting artifact.

that if A^2, B^2 and C^2 fail to form a triangle, there are no stagnation points (Dombre *et al.* 1986). We therefore pushed one such case to the highest R_m possible, selecting an instance with a high growth rate at low R_m (Galanti *et al.* 1992, Figure 2; see also Galanti *et al.* 1993), and computing additional solutions for $R_m = 200, 300, 400$ and 800. These appeared to grow in a non-oscillatory way, with growth rates 0.206, 0.209, 0.209, and 0.209 respectively. The $R_m = 800$ run was performed at $(96)^3$ resolution; a $(64)^3$ run gave the value 0.206, giving some measure of the errors involved. Thus the evidence is that this is a fast dynamo, though of course runs at considerably higher R_m are necessary to support this conclusion. The growth rate we find is just below the largest Lyapunov exponent calculated by Galanti *et al.* (1992), which is around 0.23. An isosurface of the magnetic energy of the eigenfunction (presumed nondegenerate) is given in Figure 5, showing structures that are more sheetlike than cigarlike. Similar behaviour is found for $A : B : C = 2 : 1 : 1$ and $4 : 1 : 1$, suggesting that this is a general trend for flows without stagnation points.

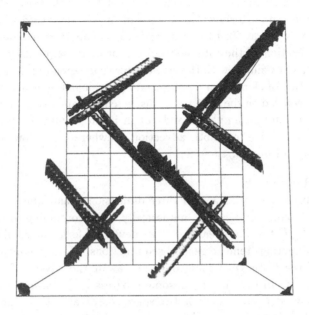

Figure 6. Isosurface $|\mathbf{B}|^2 = 0.1 B_{max}^2$ for the flow with sines only; $R_m = 450$.

4 DYNAMOS FOR THE FLOW $\mathbf{u} = (\sin z, \sin x, \sin y)$

The growth rates for this case were reported briefly in Galloway & Proctor (1992); we refer to Figure 2 there for the behaviour as R_m is varied. Here the chaotic flow regions permeate almost the whole space, except that planes such as $y = 0$ or π, where there is no flow z component, appear as unvisited lines in $z = $ constant Poincaré sections. The balance of evidence is that this dynamo is fast, although the growth rate is still increasing slightly up to $R_m = 800$. Originally this flow was chosen because having no cosines its magnetic field can be integrated almost twice as quickly as in the standard case. But it is different because its helicity, equal to $\sin z \cos y + \sin x \cos z + \sin y \cos x$, has zero mean. Conventional wisdom says that mean helicity promotes dynamo action, and this is certainly the case for mean field dynamos. But in these non-scale-separated crystalline dynamos, it apparently makes little difference.

Figure 6 shows the eigenfunction from this calculation, for $R_m = 450$. It has similar features to the standard case shown in Figure 1. Again the question arises as to whether the double structures represent one eigenfunction or an interference pattern resulting from two or more. In this case the growth is apparently non-oscillatory, removing one source of confusion. The equality (modulo opposite polarity) of each cigar of a pair suggests the former possibility, though it is not excluded that some symmetry lies buried in the initial conditions, which were chosen non-randomly. Obviously the stagnation points play an analogous rôle to those in the standard case—those which accumulate flux are at $(0,0,0),(0,\pi,\pi),(\pi,0,\pi)$ and $(\pi,\pi,0)$. Linearisation and diagonalisation of the rate-of-strain tensor at these points yields eigenvalues $-1/2, -1/2$ and 1, with the eigenvectors corresponding to 1 giving the correct direction of elongation for the cigars (e.g. along the principal diagonal for the cigar centred on the origin). As discussed in Galloway & Frisch (1986), this process is only an amplifier, and not in itself a dynamo. However, any dynamo operating due to chaotic advection elsewhere in the flow domain will act as a source of flux for these features.

5 CONCLUSION

The flows discussed so far are all three-dimensional, and illustrate well the difficulties of achieving high R_m numerically while continuing to resolve the small-scale $R_m^{-1/2}$-thickness features that inevitably form: in the limit $R_m \to \infty$ the eigenfunctions tend to generalised functions. More convincing demonstrations of the fastness of these dynamos await the advent of more powerful computers. In the meantime, some progress has been made with *time-dependent* flows depending only on two space coordinates; these can be highly chaotic, and their three-dimensional magnetic fields may be computed with a 2-D code. It thus becomes relatively easy to reach R_m values of 10000 or more, and the growth rate levels off persuasively. The best example is the flow family

$$\mathbf{u} = (A\sin(z+\sin\omega t)+C\cos(y+\cos\omega t), A\cos(z+\sin\omega t), C\sin(y+\cos\omega t)),$$

representing an $A(B=0)C$ flow which is stirred round in a circle. Since the results are published (Galloway & Proctor 1992, and in preparation), we omit discussion here. These proceedings contain extensions described by Ponty *et al.* (1993) and a physical mechanism involving heteroclinic tangling is discussed by Childress (1993).

When sought numerically, it seems that fast dynamos are generic for many classes of chaotic flows. This should comfort the astrophysicist, who depends on their existence to magnetise most stars. The crucial feature of flow non-integrability makes analytic investigations extremely hard, and an eventual rigorous proof for the existence of fast dynamos remains a major mathematical challenge.

Acknowledgements

We are grateful to the Australian Research Council for support under the grant 'Nonlinear Analysis and Computer Graphics', and to NCSA, Illinois, for the free availability of their ISOVIS plotting routines.

REFERENCES

Arnold, V.I. & Korkina, E.I. 1983 The growth of magnetic field in a steady incompressible flow. *Vest. Mosk. Un. Ta. Ser.1 Math. Mec.* **3**, 43–46.

Arnold, V.I. 1984 The evolution of magnetic field under the influence of transport and diffusion. In *Some Questions of Present-day Analysis* (ed. V.M. Tikhomirov), pp. 8–21. Moscow University.

Childress, S. 1992 Fast dynamo theory. In *Topological Aspects of the Dynamics of Fluids and Plasmas* (ed. H.K. Moffatt, G.M. Zaslavsky, P. Comte & M. Tabor), pp. 111–147. Reidel.

Childress, S. 1993 Note on perfect fast dynamo action in a large-amplitude SFS map. In this volume.

Dombre, T., Frisch, U., Greene, J.M., Hénon, M., Mehr, A. & Soward, A.M. 1986 Chaotic streamlines in the *ABC* flows. *J. Fluid Mech.* **167**, 353–391.

Galanti, B., Sulem, P.-L. & Pouquet, A. 1992 Linear and non-linear dynamos associated with ABC flows. *Geophys. Astrophys. Fluid Dynam.* **66**, 183–208.

Galanti, B., Pouquet, A. & Sulem, P.-L. 1993 Influence of the period of an ABC flow on its dynamo action. In this volume.

Galloway, D.J. & Frisch, U. 1986 Dynamo action in a family of flows with chaotic streamlines. *Geophys. Astrophys. Fluid Dynam.* **36**, 53–83.

Galloway, D.J. & Proctor, M.R.E. 1992 Numerical calculations of fast dynamos in smooth velocity fields with realistic diffusion. *Nature* **356**, 691–693.

Gilbert, A.D. 1991 Fast dynamo action in a steady chaotic flow. *Nature* **350**, 483–485.

Gilbert, A.D. 1992 Magnetic field evolution in steady chaotic flows. *Phil. Trans. R. Soc. Lond. A* **339**, 627–656.

Lau, Y.-T. & Finn, J.M. 1992 Fast dynamos with finite resistivity in steady flows with stagnation points. Preprint, University of Maryland.

Ponty, Y., Pouquet, A., Rom-Kedar, V. & Sulem, P.-L. 1993 Dynamo action in a nearly integrable chaotic flow. In this volume.

Vishik, M.M. 1989 Magnetic field generation by the motion of a highly conducting fluid. *Geophys. Astrophys. Fluid Dynam.* **48**, 151–167.

Local Helicity, a Material Invariant for the Odd-dimensional Incompressible Euler Equations

S. GAMA[1,2] & U. FRISCH[1,3]

[1]CNRS, Observatoire de Nice
BP 229, 06304 Nice Cedex 4, France

[2]FEUP, Universidade de Porto
R. Bragas, 4099 Porto Codex, Portugal

[3]Isaac Newton Institute for Mathematical Sciences
University of Cambridge, 20 Clarkson Rd., Cambridge, CB3 0EH UK

We show that in any odd number of dimensions, the Euler equations of ideal incompressible fluid flow have scalar invariants along any fluid particle trajectory. In three dimensions, the helicity is the space-integral of this invariant. Our proof is elementary. We give an alternative, much more compact, Lie group demonstration (suggested by B.A. Khesin) which is readily extended to the MHD case and provides D-dimensional generalizations of Elsasser's material invariant.

It is known that the D-dimensional Euler equations

$$\partial_t \mathbf{v} + \mathbf{v} \cdot \nabla \mathbf{v} = -\nabla p, \qquad (1)$$
$$\nabla \cdot \mathbf{v} = 0,$$

where $\mathbf{v} = (v_1, v_2, \ldots, v_D)$ is an incompressible velocity field, are equivalent to the equations (Kuz'min 1983; Oseledets 1989), here written with indices

$$D_t \gamma_i = -\gamma_j \partial_i v_j, \qquad (2)$$
$$\gamma_i = v_i + \partial_i \varphi, \qquad (3)$$
$$\partial_j v_j = 0.$$

Here, all indices vary from 1 to D and $D_t = \partial_t + v_j \partial_j$ is the Lagrangian derivative. Actually, the Kuz'min–Oseledets formulation is equivalent to choosing a special gauge in Arnold's Lie group formulation of the Euler equations (Arnold 1966). We shall return to this question at the end of the paper.

M.R.E. Proctor, P.C. Matthews & A.M. Rucklidge (eds.)
Theory of Solar and Planetary Dynamos, 115–119
©1993 Cambridge University Press.

We show in this Note that in an odd number of dimensions, the 'local helicity'

$$J = \gamma \wedge [\wedge^{(D-1)/2} d\gamma], \tag{4}$$

is a D-dimensional 'material invariant'. By this we understand that J is conserved along *any* fluid-particle trajectory, i.e., $D_t J = 0$.

The notation $\wedge^{(D-1)/2} d\gamma$ stands for $d\gamma \wedge d\gamma \wedge \ldots \wedge d\gamma$, with $n = (D-1)/2$ factors. J is actually a D-form, proportional to $dx_1 \wedge dx_2 \wedge \ldots \wedge dx_D$ and may thus be treated as a scalar. Note that by (3), we have $d\gamma = dv$. The 2-form dv is the anti-symmetrical tensor with components $\omega_{ij} = \partial_i v_j - \partial_j v_i$ (generalized vorticity). In three dimensions, the material invariance of the local helicity $\gamma \wedge d\gamma = \gamma_i \omega_i d^3 x$ (where ω is the usual vorticity) is proved in Kuz'min (1983) and Oseledets (1989). Their proof is essentially the same as that given by Elsasser (1956) for the material invariance of $\mathbf{a} \cdot \mathbf{b}$ in MHD ($\mathbf{a}$ and $\mathbf{b}$ are the magnetic vector potential and field, respectively). Note that in three dimensions our material invariant is quadratic, in five dimensions, it is cubic, etc.

Proof. We shall give an elementary derivation involving manipulations of indices in a Cartesian frame of reference. From (1) or (2), we easily obtain

$$D_t \omega_{ij} = \omega_{\ell i} \partial_j v_\ell - \omega_{\ell j} \partial_i v_\ell, \tag{5}$$

which is the D-dimensional generalization of the vorticity equation.

We now show that $I = D_t J = 0$. Indeed, I may be written as

$$I = D_t(\varepsilon_{ij_1 k_1 \ldots j_n k_n} \gamma_i \omega_{j_1 k_1} \ldots \omega_{j_n k_n}). \tag{6}$$

Here, the tensor $\varepsilon_{ij_1 k_1 \ldots j_n k_n}$ is the fundamental anti-symmetrical tensor. All repeated indices are summed over unless otherwise stated. Expanding (6) and using the anti-symmetrical property, we get

$$\begin{aligned} I = \ & \varepsilon_{ij_1 k_1 \ldots j_n k_n} D_t(\gamma_i) \omega_{j_1 k_1} \ldots \omega_{j_n k_n} \\ & + n\, \varepsilon_{ij_1 k_1 \ldots j_n k_n} \gamma_i D_t(\omega_{j_1 k_1}) \omega_{j_2 k_2} \ldots \omega_{j_n k_n}. \end{aligned}$$

Using (2) and (5), we now obtain

$$\begin{aligned} I = \ & -\varepsilon_{ij_1 k_1 \ldots j_n k_n} \gamma_\ell \omega_{j_1 k_1} \ldots \omega_{j_n k_n} (\partial_i v_\ell) \\ & + n\, \varepsilon_{ij_1 k_1 \ldots j_n k_n} \gamma_i \omega_{j_2 k_2} \ldots \omega_{j_n k_n} \omega_{\ell j_1} (\partial_{k_1} v_\ell) \\ & - n\, \varepsilon_{ij_1 k_1 \ldots j_n k_n} \gamma_i \omega_{j_2 k_2} \ldots \omega_{j_n k_n} \omega_{\ell k_1} (\partial_{j_1} v_\ell) \\ = \ & -\varepsilon_{ij_1 k_1 \ldots j_n k_n} \gamma_\ell \omega_{j_1 k_1} \ldots \omega_{j_n k_n} (\partial_i v_\ell) \tag{7} \\ & + 2n\, \varepsilon_{ij_1 k_1 \ldots j_n k_n} \gamma_i \omega_{j_2 k_2} \ldots \omega_{j_n k_n} \omega_{\ell j_1} (\partial_{k_1} v_\ell). \tag{8} \end{aligned}$$

(Exchange j_1 and k_1 to obtain the last equality.)

In the line (7) (resp., (8)) it is convenient to separate in the summations the terms where ℓ is equal to i (resp., k_1) from where it differs. In this way, we obtain

$$I = - \varepsilon_{ij_1k_1\ldots j_nk_n} \gamma_i \omega_{j_1k_1} \cdots \omega_{j_nk_n} (\partial_i v_i) \tag{9}$$

$$- \varepsilon_{ij_1k_1\ldots j_nk_n} \gamma_{\substack{\ell \\ i}} \omega_{j_1k_1} \cdots \omega_{j_nk_n} (\partial_i v_{\substack{\ell \\ i}}) \tag{10}$$

$$+\, 2\, n \times$$

$$[\ \varepsilon_{ij_1k_1\ldots j_nk_n} \gamma_i \omega_{j_2k_2} \cdots \omega_{j_nk_n} \omega_{k_1j_1} (\partial_{k_1} v_{k_1})$$

$$+ \varepsilon_{ij_1k_1\ldots j_nk_n} \gamma_i \omega_{j_2k_2} \cdots \omega_{j_nk_n} \omega_{\substack{\ell \\ k_1} j_1} (\partial_{k_1} v_{\substack{\ell \\ k_1}})\,], \tag{11}$$

where the three-times repeated indices i is summed over and where indices such as $\substack{\ell \\ i}$ are understood to omit the value i.

For the remainder, we first state some technical Lemmas, the proofs of which are straightforward.

Lemma 1. *Let* $\ell \neq i$ *be fixed. Then,*

$$\varepsilon_{ij_1k_1\ldots j_nk_n} \gamma_i \omega_{j_1k_1} \cdots \omega_{j_nk_n} [(\partial_\ell v_\ell) - (\partial_{k_1} v_{k_1})] = 0.$$

Lemma 2. *Let* $\ell \neq i$ *be fixed. Then,*

$$\varepsilon_{ij_1k_1\ldots j_nk_n} \gamma_i \omega_{j_1k_1} \cdots \omega_{j_nk_n} [(\partial_i v_\ell) - (\partial_i v_{j_1})] = 0.$$

Lemma 3. *Let* $\ell \neq i, j_1, k_1$ *be fixed. Then,*

$$\varepsilon_{ij_1k_1\ldots j_nk_n} \gamma_i \omega_{j_2k_2} \cdots \omega_{j_nk_n} [\omega_{\ell j_1}(\partial_{k_1} v_\ell) - \omega_{j_2j_1}(\partial_{k_1} v_{j_2})] = 0.$$

By Lemma 1, the term (11) is equal to (note that the dummy index i is repeated three times)

$$-\frac{1}{2n}\, \varepsilon_{ij_1k_1\ldots j_nk_n} \gamma_i \omega_{j_1k_1} \cdots \omega_{j_nk_n} (\partial_j v_j - \partial_i v_i),$$

which vanishes by (9).

Expanding (10) over $\substack{\ell \\ i}$ and using Lemma 2, I becomes

$$I = 2\, n \times$$

$$[\ \varepsilon_{ij_1k_1\ldots j_nk_n} \gamma_i \omega_{j_2k_2} \cdots \omega_{j_nk_n} \omega_{ij_1} (\partial_{k_1} v_i) \tag{12}$$

$$+ \varepsilon_{ij_1k_1\ldots j_nk_n} \gamma_i \omega_{j_2k_2} \cdots \omega_{j_nk_n} \omega_{\substack{\ell \\ ik_1} j_1} (\partial_{k_1} v_{\substack{\ell \\ ik_1}})$$

$$- \varepsilon_{ij_1k_1\ldots j_nk_n} \gamma_{j_1} \omega_{j_1k_1} \cdots \omega_{j_nk_n} (\partial_i v_{j_1})\,], \tag{13}$$

where $\substack{\ell \\ ik_1}$ omits both indices i and k_1. Now, we show that (12) and (13) cancel. Indeed, starting from (13), we have

$$\varepsilon_{ij_1k_1\ldots j_nk_n} \gamma_{j_1} \omega_{j_1k_1} \cdots \omega_{j_nk_n} (\partial_i v_{j_1}) \tag{14}$$

$$= \varepsilon_{j_1ik_1\ldots j_nk_n} \gamma_i \omega_{ik_1} \omega_{j_2k_2} \cdots \omega_{j_nk_n} (\partial_{j_1} v_i)$$

$$= - \varepsilon_{ij_1k_1\ldots j_nk_n} \gamma_i \omega_{ik_1} \omega_{j_2k_2} \cdots \omega_{j_nk_n} (\partial_{j_1} v_i) \tag{15}$$

$$= - \varepsilon_{ik_1j_1j_2k_2\ldots j_nk_n} \gamma_i \omega_{ij_1} \omega_{j_2k_2} \cdots \omega_{j_nk_n} (\partial_{k_1} v_i)$$

$$= \varepsilon_{ij_1k_1\ldots j_nk_n} \gamma_i \omega_{j_2k_2} \cdots \omega_{j_nk_n} \omega_{ij_1} (\partial_{k_1} v_i).$$

This sequence of equalities is obtained by exchanging i and j_1 in (14), and k_1 and j_1 in (15). So, we obtain

$$I = 2n\, \varepsilon_{ij_1k_1\ldots j_nk_n}\gamma_i\omega_{j_2k_2}\cdots\omega_{j_nk_n}\omega_{\overset{i}{ik_1}j_1}(\partial_{k_1}v_{\overset{i}{ik_1}}).$$

Writing all the terms with the constraint that $\overset{i}{ik_1}$ should be different from i and k_1, and that the contribution where $\overset{i}{ik_1} = j_1$ vanishes (since $\omega_{j_1j_1} = 0$), we obtain a sum with $D - 3 = 2n - 2$ terms. These terms, in virtue of Lemma 3, are all equal. Hence,

$$I = 2n\,(2n-2)\,\varepsilon_{ij_1k_1\ldots j_nk_n}\gamma_i\omega_{j_2k_2}\cdots\omega_{j_nk_n}\omega_{j_2j_1}(\partial_{k_1}v_{j_2}).$$

Exchanging in the last expression j_1 and k_2, we obtain

$$\begin{aligned} I &= \quad 2n\,(2n-2)\,\varepsilon_{ik_2k_1j_2j_1j_3k_3\ldots j_nk_n}\gamma_i\omega_{j_2j_1}\omega_{j_3k_3}\cdots\omega_{j_nk_n}\omega_{j_2k_2}(\partial_{k_1}v_{j_2}) \\ &= -\,2n\,(2n-2)\,\varepsilon_{ij_1k_1\ldots j_nk_n}\gamma_i\omega_{j_2k_2}\cdots\omega_{j_nk_n}\omega_{j_2j_1}(\partial_{k_1}v_{j_2}) \\ &= -\,I. \end{aligned}$$

Thus, $I = 0$. This establishes the material invariance of J. QED.

Note that in *even* dimensions, the material invariance of $\wedge^{D/2}dv$ has been proved by L. Tartar (1982), as quoted by Serre (1984). In *odd* dimensions, Serre (see also Khesin & Chekanov 1989) shows that the *integral* over the whole space of $J' = v \wedge [\wedge^{(D-1)/2}dv]$ is conserved. Since γ and v differ by a gradient, Serre's result is a consequence of our result. In three dimensions, we obtain the conservation of the space-integral of $\mathbf{v}\cdot\boldsymbol{\omega}$, i.e. the usual helicity.

A more compact derivation of our result may be given as follows. (The following paragraph is intended for the reader familiar with Arnold's 1966 Lie group approach and with Khesin & Chekanov 1989.) If γ is now considered to be a one-form, the Kuz'min–Oseledets form (2) of the Euler–Arnold equations may be simply written as

$$\partial_t\gamma = -L_v\gamma, \tag{16}$$

which expresses that γ is being transported by the flow. Here, $L_v = i_vd + di_v$ is the Lie derivative along the flow. Equation (16) is just Arnold's equation $\partial_t u = -L_v u + d\psi$ written in a special gauge. Since L_v commutes with the exterior derivative and with the wedge product, it follows the quantity J, defined in (4), is itself transported by the flow. J being a scalar (except for the factor $dx_1 \wedge dx_2 \wedge \ldots \wedge dx_D$), the quantity $(\partial_t + L_v)J$ is just the Lagrangian derivative of J, which thus vanishes.

This Lie group proof highlights the fact that the conservation of 'local helicity' does not make use of the particular relation (3) between the fields γ_i and u_i. It therefore also applies if we take γ to be the magnetic potential a of the ideal MHD equations (in the gauge chosen by Elsasser, 1956). For example, in five dimensions, $a \wedge da \wedge da$ is a material invariant.

We finally stress that the Lie group proof applies just in the same way to demonstrate the material invariance of $\wedge^{D/2}dv$ in even dimensions.

Acknowledgements

We are most grateful to B.A. Khesin for pointing out to us the shorter (Lie group) derivation of the result. We also thank V.I. Arnold, H.K. Moffatt and V. Zeitlin for useful discussions.

REFERENCES

Arnold, V.I. 1966 Sur la géométrie différentielle des groupes de Lie de dimension infinie et ses applications a l'hydrodynamique des fluides parfaits. *Ann. Inst. Fourier* **16**, 319–361. See also 1978 *Mathematical Methods of Classical Mechanics.* Springer.

Elsasser, W.M. 1956. Hydromagnetic dynamo theory. *Rev. Mod. Phys.* **28**, 135–163.

Khesin, B.A. & Chekanov, Yu.V. 1989 Invariants of the Euler equations for ideal and barotropic hydrodynamics and superconductivity in D dimensions. *Physica* **40D**, 119–131.

Kuz'min, G.A. 1983 Ideal incompressible hydrodynamics in terms of the vortex momentum density. *Phys. Lett.* **96A**, 88–90.

Oseledets, V.I. 1989 New form of the Navier–Stokes equation. Hamiltonian formalism (in Russian). *Moskov. Matemat. Obshch.* **44** (no. 3, 267), 169–170.

Serre, D. 1984 Invariants et dégénérescence symplectique de l'équation d'Euler des fluides parfaits incompressibles. *C. R. Acad. Sc. Paris* **298**, 349–352.

On the Quasimagnetostrophic Asymptotic Approximation Related to Solar Activity

MIHAI GHIZARU

Astronomical Institute of Romanian Academy
str. Cutitul de Argint 5, 75212, Bucharest, Romania

The quasimagnetostrophic equations are derived as a fourth-order asymptotic approximation of the ideal MHD equations written in spherical coordinates. A regular perturbation method is applied by expanding the nine dimensionless variables as asymptotic series in the Rossby number ($Ro = O(\epsilon)$). The order of magnitude of the ten nondimensional parameters describing the flow is estimated for suitably characterising the interaction of large-scale dynamic and magnetic features at the interface between the radiative interior and the convective zone in the Sun. The importance of interactions between different low frequency modes (magnetostrophic, gravity and Rossby waves) in determining the topology of Solar activity structures is discussed.

1 INTRODUCTION

Starting with the paper of Parker (1955), one of the main purposes of scientists working in Solar physics was directed towards the understanding of the mechanisms governing the generation and the maintenance of the magnetic field that plays a central role in the Solar activity process. Many characteristic features, for example the well-known 'butterfly' diagram, are well described by kinematic dynamo theories. Great efforts are now directed towards the derivation of dynamo models that could agree with recent helioseismological data, and there is accumulating evidence that the seat of the dynamo is in the overshoot layer at the base of the convection zone (see for example, De Luca & Gilman 1991).

However, even the most sophisticated theories based mainly on the magnetic field are not able to explain topological aspects of Solar activity structures like active longitudes.

M.R.E. Proctor, P.C. Matthews & A.M. Rucklidge (eds.)
Theory of Solar and Planetary Dynamos, 121–127
©1993 Cambridge University Press.

Other attempts to explain most of the Solar activity features (Wolff 1974; Wolff 1984; Wolff & Blizard 1986; Wolff & Hickey 1987; Wolff & Hoegy 1989) replace the magnetic field's central role by global low frequency modes of g (gravity) and r (Rossby) type. This approach became more consistent after the analysis of the Solar EUV flux monitored on Pioneer Venus Orbiter, presented by Wolff & Hoegy (1989).

But, as well as neglecting dynamics, kinematic dynamo models seem to be a too severe simplification; conversely, trying to explain the main Solar activity patterns only dynamically, without involving magnetic fields, could be an inconsistent procedure, due to the nonlinear coupling between magnetic and velocity fields in MHD equations. That is why, it is reasonable to find a suitable framework for explaining active longitudes as well as other large-scale features by different interactions between low frequency dynamic and magnetic modes.

A simplification of the ideal MHD equations using asymptotic methods will be performed, after presenting some qualitative statements as hints in explaining the presence of active longitudes in the Solar activity cycle.

2 TOWARDS A FRAMEWORK FOR EXPLAINING ACTIVE LONGITUDES

Resonant interactions between magnetostrophic and r or g modes having comparable velocities in the overshoot region at the base of the convection zone, could give rise to some magneto-cyclogenetic zones, explaining the magnetic flux tube formation at preferred longitudes.

Magnetostrophic waves excited by magnetic buoyancy were mentioned by Moffatt (1978) as a possible source for dynamo action and Schmitt (1984) investigated this idea further by developing a dynamo model with the α-effect sustained by magnetostrophic low frequency modes (10^{-7}s^{-1}).

Rossby modes were placed in Solar context by Ward (1965) who suggested the presence of a Rossby regime rather than one of Hadley type for the large-scale Solar photospheric flow. Along these lines, Gilman (1967; 1969a,b) investigated in more detail the specific behaviour of Rossby waves in a background magnetic field and proposed a dynamo mechanism driven by Rossby waves extended deep in the convection zone.

The theory of r modes (as a generalisation of Rossby waves from geophysical fluid dynamics) was extensively studied in papers by Papaloizou & Pringle (1978), Provost *et al.* (1981), Smeyers *et al.* (1981) and Saio (1982), and was applied in Solar context by Wolff & Blizard (1986). The g mode model (developed in the radiative core) was completed by Wolff & Hickey (1987) with r modes trapped in the convective envelope for explaining periodicities in Solar irradiance and consequently associated large-scale patterns like active longitudes.

But in explaining active longitudes, a more consistent approach would be to consider interactions between magnetostrophic waves and g and r modes. Due to the strong shear in angular velocity (resulted from helioseismology data) in the overshoot layer, r waves could be easily excited by the Kelvin–Helmholtz instability. Resonant interactions with g and magnetostrophic waves having a possible role in the dynamo mechanism could give rise to magneto-cyclonic vortices, similar to cyclogenetic zones determined by the baroclinic instability in the Earth's atmosphere. While cyclonic motions might help magnetic field concentrations at preferred longitudes, vertical motions developed in the vortex core may influence the dynamics of flux tubes.

Starting from the ideal MHD equations written in spherical coordinates, after scale analysis and applying asymptotic approximations, a quasimagnetostrophic model able to describe quantitatively the above scenario, would be derived in the sequel. The procedure is similar to that used by Pedlosky (1979) for a consistent and elegant derivation of quasigeostrophic model for large-scale atmospheric flow.

The MHD equations in (r, θ, ϕ) (radial distance, latitude and longitude) coordinates read

$$\frac{du}{dt} + \frac{uw}{r} - \frac{uv}{r}\tan\theta - 2\Omega v \sin\theta + 2\Omega w \cos\theta = -\frac{1}{\rho r \cos\theta}\frac{\partial}{\partial\phi}(p + \frac{B^2}{2\mu})$$
$$+ \frac{B_\phi}{\mu\rho r \cos\theta}\frac{\partial B_\phi}{\partial\phi} + \frac{B_\theta}{\mu\rho r}\frac{\partial B_\phi}{\partial\theta} + \frac{B_r}{\mu\rho}\frac{\partial B_\phi}{\partial r} - \frac{B_\phi B_\theta}{\mu\rho r}\tan\theta + \frac{B_\phi B_r}{\mu\rho r}, \quad (1)$$

$$\frac{dv}{dt} + \frac{vw}{r} + \frac{u^2}{r}\tan\theta + 2\Omega u \sin\theta = -\frac{1}{\rho r}\frac{\partial}{\partial\theta}(p + \frac{B^2}{2\mu})$$
$$+ \frac{B_\phi}{\mu\rho r \cos\theta}\frac{\partial B_\theta}{\partial\phi} + \frac{B_\theta}{\mu\rho r}\frac{\partial B_\theta}{\partial\theta} + \frac{B_r}{\mu\rho}\frac{\partial B_\theta}{\partial r} + \frac{B_r B_\theta}{\mu\rho r} + \frac{B_\phi^2}{\mu\rho r}\tan\theta, \quad (2)$$

$$\frac{dw}{dt} - \frac{u^2 + v^2}{r} - 2\Omega u \cos\theta = -\frac{1}{\rho}\frac{\partial}{\partial r}(p + \frac{B^2}{2\mu}) - g$$
$$+ \frac{B_\phi}{\mu\rho r \cos\theta}\frac{\partial B_r}{\partial\phi} + \frac{B_\theta}{\mu\rho r}\frac{\partial B_r}{\partial\theta} + \frac{B_r}{\mu\rho}\frac{\partial B_r}{\partial r} - \frac{B_\phi^2 + B_\theta^2}{\mu\rho r}, \quad (3)$$

$$\frac{dB_\phi}{dt} = \frac{B_\theta}{r}\frac{\partial u}{\partial\theta} - \frac{B_\phi}{r}\frac{\partial v}{\partial\theta} + B_r\frac{\partial u}{\partial r} - B_\phi\frac{\partial w}{\partial r}$$
$$- \frac{u}{r}B_\theta\tan\theta + B_r\frac{u}{r} - 2B_\phi\frac{w}{r} + \frac{v}{r}B_\theta\tan\theta, \quad (4)$$

$$\frac{dB_\theta}{dt} = \frac{1}{r\cos\theta}(B_\phi\frac{\partial v}{\partial\phi} - B_\theta\frac{\partial u}{\partial\phi}) + B_r\frac{\partial v}{\partial r} - B_\theta\frac{\partial w}{\partial r}$$
$$+ \frac{u}{r}B_\phi\tan\theta + B_r\frac{v}{r} - 2B_\theta\frac{w}{r} + \frac{v}{r}B_\theta\tan\theta, \quad (5)$$

$$\frac{dB_r}{dt} = \frac{1}{r\cos\theta}(B_\phi\frac{\partial w}{\partial\phi} - B_r\frac{\partial u}{\partial\phi}) + \frac{1}{r}(B_\theta\frac{\partial w}{\partial\theta} - B_r\frac{\partial v}{\partial\theta})$$
$$- \frac{uB_\phi - vB_\theta}{r} - 2B_r\frac{w}{r} + \frac{v}{r}B_r\tan\theta. \quad (6)$$

$$0 = \frac{d\rho}{dt} + \rho \nabla \cdot \mathbf{v}, \tag{7}$$

$$0 = c_V \frac{dT}{dt} + RT\nabla \cdot \mathbf{v}, \tag{8}$$

$$p = \rho RT, \tag{9}$$

where (B_r, B_θ, B_ϕ) is the magnetic induction and $\mathbf{v} = (u, v, w)$ the velocity; ρ is the density, T the temperature, c_V the specific heat, and $R = R^*/\mu^*$, where R^* is the universal gas constant and μ^* the molar mass of plasma. The angular velocity of the Sun is Ω. The material derivative is

$$\frac{d}{dt} = \frac{\partial}{\partial t} + \frac{u}{r \cos \theta} \frac{\partial}{\partial \phi} + \frac{v}{r} \frac{\partial}{\partial \theta} + w \frac{\partial}{\partial r} \tag{10}$$

and the velocity divergence is

$$\nabla \cdot \mathbf{v} = \frac{1}{r \cos \theta} \frac{\partial u}{\partial \phi} + \frac{1}{r} \frac{\partial v}{\partial \theta} + \frac{\partial w}{\partial r} + \frac{2w}{r} - \frac{v}{r} \tan \theta. \tag{11}$$

New coordinates are introduced:

$$x = \phi r \cos \theta_0, \qquad y = (\theta - \theta_0) r, \qquad z = r - r_0,$$

where θ_0 is the central latitude for the domain of interest and r_0 the radius of the spherical shell considered.

Fifteen dimensional parameters define the flow: the characteristic horizontal scale L, the characteristic horizontal velocity U, the mean pressure at the bottom of the overshoot layer P, the characteristic horizontal magnetic field M, the thickness of the layer d, the gravitational acceleration g, the Coriolis parameter $f_0 = 2\Omega \sin \theta_0$, the magnetic permeability μ, and R, r_0, c_V, x, y, z, and t. By the π theorem, one may form ten dimensionless parameters: $t' = tU/L$, $x' = x/L$, $y' = y/L$, $z' = z/d$, $\lambda = L/r_0$, the aspect ratio $\delta = d/L$, the Rossby number $Ro = U/f_0 L$, the plasma beta $\beta_p = 2\mu P/M^2$, the squared Alfvén–Mach number $Ma = U^2 \mu P/M^2 gd$, and $\nu = R/c_V = \gamma - 1$, where γ is the adiabatic exponent. The nine dimensionless functions are now: $u' = u/U$, $v' = v/U$, $w' = w/\delta U$, $B'_\phi = B_\phi/M$, $B'_\theta = B_\theta/M$, $B'_r = B_r/\delta M$, $p' = p/P$, $T' = RT/gd$ and $\rho' = \rho gd/P$. The resulting dimensionless system is represented by equations (10a)–(14) in the paper by Ghizaru (1992).

Asymptotic expansions of the form

$$X \sim X_0 + \epsilon X_1 + \epsilon^2 X_2 + \epsilon^3 X_3 + \dots \tag{12}$$

are used for each unknown function, except

$$w \sim \epsilon^2 w_0 + \epsilon^3 w_1 + \dots$$

Trigonometric functions are Taylor expanded around θ_0. It is also assumed that $T_0 = 1$ and p_0, ρ_0, p_1, ρ_1 and T_1 are time-independent.

3 QUASIMAGNETOSTROPHIC APPROXIMATION

Assuming that $d = 5 \times 10^5$m, $L = 10^8$m, $U = 30\,\mathrm{m\,s^{-1}}$, $f_0 = 3 \times 10^{-6}\mathrm{s^{-1}}$ and $\gamma = 5/3$, the order of magnitude of the dimensionless parameters as ϵ goes to zero is: $Ro = O(\epsilon)$, $\nu = O(1)$, $\delta = O(\epsilon^2)$ and $\lambda = O(\epsilon)$. It is also assumed that $\beta_p = O(\epsilon^{-2})$ and $Ma = O(\epsilon)$. Then, the quasimagnetostrophic equations will be obtained at fourth order in the asymptotic expansion. The hydrodynamic part of the equations is obtained in the same form as by Vamos & Georgescu (1990).

The first and second-order equations read

$$\frac{\partial p_0}{\partial x} = 0, \qquad \frac{\partial p_0}{\partial y} = 0, \qquad \rho_0 = -\frac{\partial p_0}{\partial z}, \qquad \frac{\partial u_0}{\partial x} + \frac{\partial v_0}{\partial y} = 0, \qquad p_0 = \rho_0, \tag{13}$$

and

$$\frac{\partial p_1}{\partial x} = 0, \qquad \frac{\partial p_1}{\partial y} = 0, \qquad \rho_1 = -\frac{\partial p_1}{\partial z},$$

$$\frac{\partial u_1}{\partial x} + \frac{\partial v_1}{\partial y} - y\frac{\partial v_0}{\partial y}\tan\theta_0 - v_0\tan\theta_0 = 0, \qquad p_1 = \rho_0 T_1 + \rho_1. \tag{14}$$

The third-order equations are

$$-\rho_0 v_0 = -\frac{\partial}{\partial x}(p_2 + B_0^2) + (\mathbf{B_0}\cdot\nabla)B_{\phi 0},$$

$$\rho_0 u_0 = -\frac{\partial}{\partial y}(p_2 + B_0^2) + (\mathbf{B_0}\cdot\nabla)B_{\theta 0},$$

$$\rho_2 = -\frac{\partial}{\partial z}(p_2 + B_0^2),$$

$$\frac{D_0 B_{\phi 0}}{Dt} = B_{\theta 0}\frac{\partial u_0}{\partial y} - B_{\phi 0}\frac{\partial v_0}{\partial y} + B_{r0}\frac{\partial u_0}{\partial z},$$

$$\frac{D_0 B_{\theta 0}}{Dt} = B_{\phi 0}\frac{\partial v_0}{\partial x} - B_{\theta 0}\frac{\partial u_0}{\partial x} + B_{r0}\frac{\partial v_0}{\partial z}, \tag{15}$$

$$B_{r0}v_0\tan\theta_0 = B_{\phi 0}u_0 + B_{\theta 0}v_0,$$

$$\frac{D_0 T_2}{Dt} + \frac{\partial u_2}{\partial x} + \frac{\partial v_2}{\partial y} + \frac{\partial w_0}{\partial z} - y\frac{\partial v_1}{\partial y}\tan\theta_0 - \frac{1}{2}y^2\frac{\partial v_0}{\partial y} - v_1\tan\theta_0 - yv_0 = 0,$$

$$\frac{D_0\rho_2}{Dt} + \rho_0\left(\frac{\partial u_2}{\partial x} + \frac{\partial v_2}{\partial y} - y\frac{\partial v_1}{\partial y}\tan\theta_0 - \frac{y^2}{2}\frac{\partial v_0}{\partial y} - v_1\tan\theta_0 - yv_0\right) + \frac{\partial}{\partial z}(\rho_0 w_0) = 0,$$

$$p_2 = \rho_2 + \rho_1 T_1 + \rho_0 T_2.$$

The first two equations represent the magnetostrophic balance. The fourth-order momentum and induction equations read

$$\rho_0\left(\frac{D_0 u_0}{Dt} - 2yv_0\tan^{-1}2\theta_0 - v_1\right) - \rho_1 v_0 = -\frac{\partial}{\partial x}\left(p_3 + 2\mathbf{B_0}\cdot\mathbf{B_1}\right)$$

$$+ (\mathbf{B_0} \cdot \nabla) B_{\phi 1} + (\mathbf{B_1} \cdot \nabla) B_{\phi 0} - \left(B_{\theta 0} \frac{\partial B_{\phi 0}}{\partial y} y + B_{r0} \frac{\partial B_{\phi 0}}{\partial z} y + B_{\phi 0} B_{\theta 0} \right) \tan \theta_0,$$

$$\rho_0 \left(\frac{D_0 v_0}{Dt} + 2 y u_0 \tan^{-1} 2 \theta_0 + u_1 \right) + \rho_1 u_0 = - \frac{\partial}{\partial y} \left(p_3 + 2\mathbf{B_0} \cdot \mathbf{B_1} \right)$$

$$+ (\mathbf{B_0} \cdot \nabla) B_{\theta 1} + (\mathbf{B_1} \cdot \nabla) B_{\theta 0} + B_{\phi 0}^2 \tan \theta_0$$

$$+ \left(\frac{\partial}{\partial y} (p_2 + B_0^2) - B_{\theta 0} \frac{\partial B_{\theta 0}}{\partial y} - B_{r0} \frac{\partial B_{\theta 0}}{\partial z} \right) y \tan \theta_0,$$

$$\frac{D_0 B_{\phi 1}}{Dt} + A_1 B_{\phi 0} - y v_0 \frac{\partial B_{\phi 0}}{\partial y} \tan \theta_0 = -y \left(B_{\theta 0} \frac{\partial u_0}{\partial y} - B_{\phi 0} \frac{\partial v_0}{\partial y} + B_{r0} \frac{\partial u_0}{\partial z} \right) \tan \theta_0$$

$$+ B_{\theta 0} \frac{\partial u_1}{\partial y} + B_{\theta 1} \frac{\partial u_0}{\partial y} - B_{\phi 0} \frac{\partial v_1}{\partial y} - B_{\phi 1} \frac{\partial v_0}{\partial y} \qquad (16)$$

$$+ B_{r0} \frac{\partial u_1}{\partial z} + B_{r1} \frac{\partial u_0}{\partial z} + (B_{\phi 0} v_0 - B_{\theta 0} u_0) \tan \theta_0,$$

$$\frac{D_0 B_{\theta 1}}{Dt} + A_1 B_{\theta 0} - y v_0 \frac{\partial B_{\theta 0}}{\partial y} \tan \theta_0 = B_{\phi 0} \frac{\partial v_1}{\partial x} + B_{\phi 1} \frac{\partial v_0}{\partial x} - B_{\theta 0} \frac{\partial u_1}{\partial x} - B_{\theta 1} \frac{\partial u_0}{\partial x}$$

$$+ (B_{\phi 0} u_0 + B_{\theta 0} v_0) \tan \theta_0 + B_{r0} \frac{\partial v_1}{\partial z} + B_{r1} \frac{\partial v_0}{\partial z},$$

$$\frac{D_0 B_{r0}}{Dt} = -B_{r0} \frac{\partial u_0}{\partial x} - B_{r0} \frac{\partial v_0}{\partial y} - B_{\phi 1} u_0 - B_{\phi 0} u_1 - B_{\theta 1} v_0 - B_{\theta 0} v_1$$

$$+ B_{r1} v_0 \tan \theta_0 + B_{r0} v_1 \tan \theta_0 + y B_{r0} v_0.$$

Here,

$$A_1 = u_1 \frac{\partial}{\partial x} + v_1 \frac{\partial}{\partial y} \qquad \text{and} \qquad \frac{D_0}{Dt} = \frac{\partial}{\partial t} + u_0 \frac{\partial}{\partial x} + v_0 \frac{\partial}{\partial y}.$$

Since the quasimagnetostrophic equations obtained generalize the quasi-geostrophic model for large-scale flows, there are hopes that they form a first step towards a suitable characterisation of the processes leading to the presence of active longitudes.

Acknowledgements
I would like to thank Dr Peter Fox for useful discussions on Solar Rossby waves and for suggestions made for future developments of these ideas.

REFERENCES
De Luca, E.E. & Gilman, P.A. 1991 The Solar dynamo. In *Solar Interior and Atmospheres* (ed. A.N. Cox, W.C. Livingston & M.S. Matthews), pp. 275–303. Tucson.

Ghizaru, M. 1992 Quasiheliostrophic and quasimagnetostrophic asymptotic approximations by scale analysis. *Romanian Astronomical Journal* (submitted).

Gilman, P.A. 1967 Stability of baroclinic flows in a zonal magnetic field: part I. *J. Atm. Sci.* **24**, 101–118.

Gilman, P.A. 1969a A Rossby-wave dynamo for the Sun, I. *Solar Phys.* **8**, 316–330.

Gilman, P.A. 1969b A Rossby-wave dynamo for the Sun, II. *Solar Phys.* **9**, 3–18.

Moffatt, H.K. 1978 *Magnetic Field Generation in Electrically Conducting Fluids.* Cambridge University Press.

Papaloizou, J. & Pringle, J.E. 1978 Non-radial oscillations of rotating stars and their relevance to the short-period oscillations of cataclysmic variables. *Mon. Not. R. Astron. Soc.* **182**, 423–442.

Parker, E.N. 1955 Hydromagnetic dynamo models. *Astrophys. J.* **122**, 293–314.

Pedlosky, J. 1979 *Geophysical Fluid Dynamics.* Springer.

Provost, J., Berthomieu, G. & Rocca, A. 1981 Low frequency oscillations of a slowly rotating star: quasi-toroidal modes. *Astron. Astrophys.* **94**, 126–133.

Saio, H. 1982 R-mode oscillations in uniformly rotating stars. *Astrophys. J.* **256**, 717–735.

Schmitt, D. 1984 Dynamo action of magnetostrophic waves. In *The Hydromagnetics of the Sun* (ed. T.D. Guyenne & J.J. Hunt), pp. 223–224. ESA SP-220, Noordwijkerhout.

Smeyers, P., Craeynest, D. & Martens, L. 1981 Rotational modes in a slowly and uniformly rotating star. *Astrophys. Space Sci.* **78**, 483–501.

Vamos, C. & Georgescu, R. 1990 Models of asymptotic approximation for synoptic-scale flow. *Z. Meteorol.* **40**, 14–20.

Ward, F. 1965 The general circulation of the Solar atmosphere and the maintenance of the equatorial acceleration. *Astrophys. J.* **141**, 534–547.

Wolff, C.L. 1974 Rigid and differential rotation driven by oscillations within the Sun. *Astrophys. J.* **194**, 489–498.

Wolff, C.L. 1984 Solar irradiance changes caused by g-modes and large scale convection. *Solar Phys.* **93**, 1–13.

Wolff, C.L. & Blizard, J.B. 1986 Properties of r-modes in the Sun. *Solar Phys.* **105**, 1–15.

Wolff, C.L. & Hickey, J.R. 1987 Multiperiodic irradiance changes caused by r-modes and g-modes. *Solar Phys.* **109**, 1–18.

Wolff, C.L. & Hoegy, W.R. 1989 Periodic Solar EUV flux monitored near Venus. *Solar Phys.* **123**, 7–20.

Simple Dynamical Fast Dynamos

A.D. GILBERT

Department of Applied Mathematics and Theoretical Physics
University of Cambridge, Silver St., Cambridge, CB3 9EW UK

N.F. OTANI

School of Electrical Engineering
Cornell University, Ithaca, NY 14853 USA

S. CHILDRESS

Courant Institute of Mathematical Sciences
New York University, New York, NY 10012 USA

Fast dynamo saturation is explored numerically using a simplified model. The magnetic field has many degrees of freedom and allows the generation of fine structure at large R_m. The velocity field is constrained, containing two Fourier modes and so eight degrees of freedom; the Lorentz force is projected onto these modes. Numerical simulations at varying R_m are discussed.

Fast dynamo instabilities are the subject of intense research (reviewed in Childress 1992), through numerical simulations and analytical studies of simple models. However little is known about how a fast dynamo instability might saturate and what the resulting spatial structure and temporal behaviour of the field might be. Does a fast dynamo saturate by suppressing the flow field until the effective magnetic Reynolds number is reduced to a value of order unity or by modifying transport effects of the flow (Vainshtein *et al.* 1993)? Is the saturated magnetic energy in equipartition with the kinetic energy and how is the magnetic energy distributed; in particular how much energy is stored in large-scale field components (Vainshtein & Cattaneo 1992)? Does the field contain the fine structure typical of kinematic fast dynamo instabilities and is it intermittent in time? The difficulty in answering these questions is that dynamo action only allows growth of 3-d magnetic fields and through the Lorentz force this leads to all the complexities of 3-d MHD turbulence. Numerical studies are computationally expensive and only moderate values

129

M.R.E. Proctor, P.C. Matthews & A.M. Rucklidge (eds.)
Theory of Solar and Planetary Dynamos, 129–136
©1993 Cambridge University Press.

of R_m have been achieved (see, for example, Gilman 1983, Glatzmaier 1985, Meneguzzi & Pouquet 1989, Nordlund *et al.* 1991 and Galanti *et al.* 1992).

Our approach is to introduce a constrained model in the spirit of the fast dynamo but which allows efficient computation. In the model the magnetic field is followed exactly, without any kind of truncation; however, the velocity field is constrained to contain only two Fourier modes $\exp(ix)$, $\exp(iy)$ (together with their complex conjugates) and so has eight real degrees of freedom. The Lorentz force is projected onto these modes. Informally one might think of the flow as being very viscous so that only the largest scales are excited significantly (although in the simulations below the viscosity $\nu = 1$). This limit is in fact inappropriate for the Sun, where the viscosity is much smaller than the magnetic diffusivity. Note also that the model does not allow the possibility of saturation by the generation of localised small-scale flows as can occur in flux ropes (Galloway, Proctor & Weiss 1978). The model is, however, intended only as a first step in exploring saturation mechanisms and further complications will be added in future research.

In the model the magnetic field obeys the induction equation

$$\partial_t \mathbf{B} = \nabla \times (\mathbf{U} \times \mathbf{B}) + \eta \nabla^2 \mathbf{B}, \qquad \nabla \cdot \mathbf{B} = 0, \qquad (1)$$

and the field is restricted to take the form

$$\mathbf{B}(x, y, z) = \mathbf{b}(x, y)e^{ikz} + \mathbf{b}^*(x, y)e^{-ikz}, \qquad (2)$$

where $\mathbf{b}(x, y)$ is a complex field. This form is preserved by the flow field, which is truncated to just two modes:

$$\mathbf{U} = \mathbf{u}(t)e^{ix} + \mathbf{u}^*(t)e^{-ix} + \mathbf{v}(t)e^{iy} + \mathbf{v}^*(t)e^{-iy} \qquad (3)$$

with complex $\mathbf{u} = (0, u_2, u_3)$ and $\mathbf{v} = (v_1, 0, v_3)$. These components are taken to obey the Navier–Stokes equation projected onto these modes:

$$(\partial_t + \nu)u_l = P_l + F_l, \qquad (\partial_t + \nu)v_m = Q_m + G_m \qquad (4)$$

for $l = 2, 3$ and $m = 1, 3$. Here F_l and G_m comprise the external force maintaining fluid motion, and P_l and Q_m are components of the Lorentz force projected onto the modes:

$$P_l = <ib_1 b_l e^{-ix}>, \qquad Q_m = <ib_2 b_m e^{-iy}>, \qquad (5)$$

the average $<\cdot>$ being taken over the 2π-periodicity of the flow. Note that energy is conserved by the induction and Lorentz force terms, despite the constrained nature of the flow.

The external force is given by specifying a 'basic flow', which is the flow the force maintains in the absence of magnetic field; the force is calculated from (4) with P_i and Q_j set to zero. We have used four basic flows:

$$\mathbf{U}_{CP} = \sqrt{3/2}(0, \sin(x + \cos t), \cos(x + \cos t))$$
$$+ \sqrt{3/2}(\cos(y + \sin t), 0, \sin(y + \sin t)),$$
$$\mathbf{U}_{LP} = \sqrt{3/2}(0, \sin(x + \cos t), \cos(x + \cos t))$$
$$+ \sqrt{3/2}(\cos(y + \cos t), 0, \sin(y + \cos t)),$$
$$\mathbf{U}_{PW\pm} = 2\cos^2 t(0, \sin x, \cos x) + 2\sin^2 t(\sin y, 0, \cos y),$$
$$\mathbf{U}_{PW+} = 2\cos^2 t(0, \sin x, \cos x) + 2\sin^2 t(\cos y, 0, \sin y).$$

The first two flows are the circularly polarized (CP) and linearly polarized (LP) cellular flows of Galloway & Proctor (1992) and for these we take $k = 0.57$ and 0.62 respectively. The second two flows are pulsed Beltrami waves (see Bayly & Childress 1988, 1989 and Otani 1992) and we take $k = 1$ for both of these. For the flow PW± the waves have opposite signs of helicity, while in PW+ the signs are the same. In the code the magnetic field is represented in Fourier space using between 49^2 and 97^2 modes; the induction term is calculated in Fourier space using the simple form of the flow field (3). The fields are evolved using the Adams–Bashforth scheme. Only the b_1 and b_2 components are time-stepped; b_3 is calculated using $\nabla \cdot \mathbf{B} = 0$ when needed to evaluate the Lorentz force (5). A viscosity of $\nu = 1$ is taken in all runs.

The equations have been integrated up to $t = 500\pi$, which is 250 turnover times, starting with a weak field $\mathbf{b}(x, y) = $ const. We follow the magnetic and kinetic energy densities, $E_M = <\frac{1}{2}\mathbf{B}^2>$ and $E_V = <\frac{1}{2}\mathbf{U}^2>$. Since these quantities fluctuate strongly over each period, we average them over every period $(2m\pi, 2(m + 1)\pi)$ to obtain smoothed energy densities $\overline{E}_M$ and $\overline{E}_V$. Figure 1 shows the evolution of $\overline{E}_M$ in the four flows for different values of η^{-1} from 100 to 2000; the curves are separated by adding a constant to $\overline{E}_M$ and in each graph the diffusivity decreases from the lowest to the highest curve. After a short kinematic regime in which $\overline{E}_M$ grows exponentially, saturation occurs. In the saturated state there is a variety of behaviour: chaos in LP (b) and PW± (c), chaos with very long transients in in CP (a) and quiescent evolution with intermittent bursts in PW+ (d).

Figure 2 shows statistical information. We define $\langle \cdot \rangle$ to be the average of a quantity from $t = 100\pi$ to 500π. We also define an effective magnetic Reynolds number $R_m^{EFF} = \sqrt{2\overline{E}_V/\eta}$; in the kinematic regime $\overline{E}_V = 3/2$ for all four flows and $R_m = R_m^{EFF} = \sqrt{3}/\eta$. Figure 2 shows (a) $\langle E_M \rangle$ and (b) $\langle E_V \rangle$ as a function of R_m for each run. The magnetic energy is roughly constant for flows CP, PW±, PW+, but increases with R_m for the LP flow. The kinetic

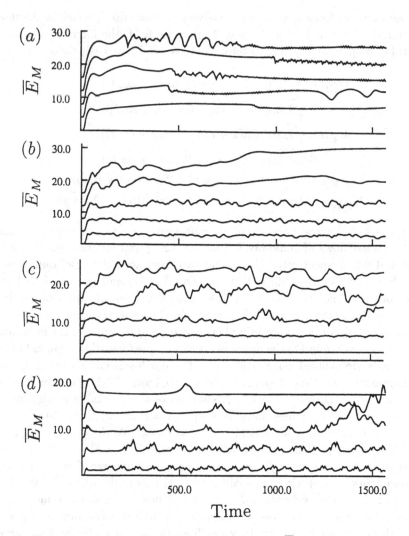

Figure 1. Evolution of the smoothed magnetic energy density $\overline{E}_M$ as a function of time for the flows (a) CP, (b) LP, (c) PW± and (d) PW+. Curves are given for $\eta^{-1} = 100, 200,$ 500, 1000 and 2000 and are separated by adding 0, 4, 8, 12 and 16 respectively to $\overline{E}_M$.

energy is suppressed, but not greatly, for all flows except PW±, where it rises above the kinematic value of 1.5 for $\eta^{-1} = 1000$ and 2000. Figure 2(c) shows $\langle R_m^{\mathrm{EFF}} \rangle$ plotted against R_m. $\langle R_m^{\mathrm{EFF}} \rangle$ increases with R_m, indicating that saturation is not achieved by reducing R_m^{EFF} to values of order unity. In fact $\langle R_m^{\mathrm{EFF}} \rangle / R_m$ varies little with η for each flow (Vainshtein *et al.* 1993). Finally Figure 2(d) shows the ratio of average magnetic energy to average kinetic energy; there is no clear pattern. Flows CP and LP show notable deviations

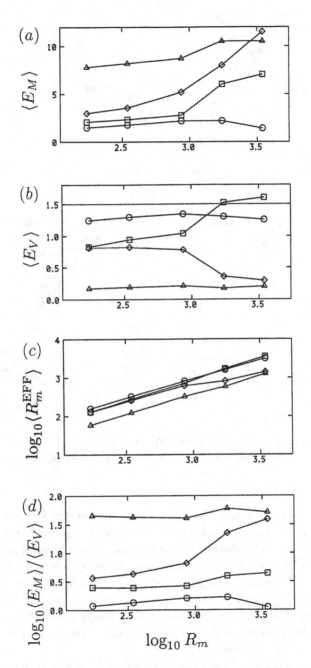

Figure 2. Average quantities as a function of $\log_{10} R_m$ for CP (triangles), LP (diamonds), PW± (squares) & PW+ (circles): (a) $\langle E_M \rangle$, (b) $\langle E_V \rangle$, (c) $\log_{10}\langle R_m^{\mathrm{EFF}} \rangle$ and (d) $\log_{10}\langle E_M \rangle/\langle E_V \rangle$.

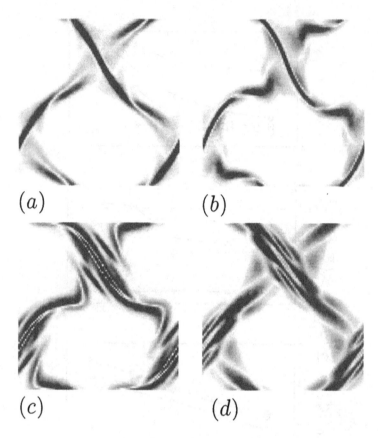

Figure 3.   Boundary layer burst for the PW+ flow with $\eta^{-1} = 1000$. The quantity $\sqrt{B_x^2 + B_y^2}$ at $z = 0$ is plotted on a grey scale for (a) $t = 119\pi$ (b) 121π (c) 122π and (d) 126π.

from equipartition; this may result from using a truncated flow field.

Pictures of the magnetic fields generally show disorganised spatial structure; there is little resemblance to the kinematic eigenfunction and the dynamical fields have less fine-scale structure. An exception is the flow PW+, which has the intermittent bursts noted above. Consider $\eta^{-1} = 1000$; after the initial transient the field is organised in boundary layers which evolve extremely slowly (Figure 3(a)), and then suddenly become unstable and fold up (b,c). The folds then slowly diffuse away (d) and the field eventually regains a quasi-steady state similar to (a).

Finally we return to discuss the distribution of energy in kinematic dynamos; Vainshtein & Cattaneo (1992) have argued that for the kinematic dynamo problem $\gamma(R_m) = \lim_{t\to\infty} <|\mathbf{b}|^2>/|<\mathbf{b}>|^2$ is proportional to R_m^n

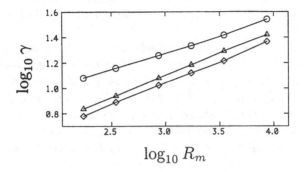

Figure 4. Plot of $\log_{10} \gamma(R_m)$ against $\log_{10} R_m$ for CP (triangles), LP (diamonds) and PW+ (circles).

for some constant n as $R_m \to \infty$. The constant n measures the efficiency of generating large-scale field, being 1 for (inefficient) 2-d folding and 0 for the STF model which has perfect reinforcement of flux and uniform stretching. It is important for questions of saturation, because small-scale components of field exert a mean Lorentz force which may lead to saturation when the large scale field $<\mathbf{b}>$ is small, this problem being more serious the larger n is (Vainshtein & Cattaneo 1992). Figure 4 shows $\log_{10} \gamma$ as a function of $\log_{10} R_m$ obtained from kinematic simulations. There is clear linear behaviour with $n \simeq 0.35$ (CP), 0.32 (LP) and 0.27 (PW+); γ converges very slowly for PW±, presumably because of eigenfunction degeneracy, and so is not shown. The small values of n imply that folding is relatively efficient in these models.

In conclusion we emphasize that all the dynamical results have been obtained in a simplified model in which the flow field is allowed very few degrees of freedom, the magnetic field has a simple z-dependence and which is driven by deterministic body forces. Topics for future research include using a random body force and allowing the flow field many degrees of freedom (while remaining two-dimensional); such a constrained MHD dynamo model would allow the possibility of saturation at small scales.

Acknowledgements
We thank B.J. Bayly, D.J. Galloway, I. Klapper, A.M. Soward and S.I. Vainshtein for useful comments. This work commenced during a visit by A.D.G. to the Courant Institute of New York University, supported by N.S.F. grant DMS-8922676, and was completed at the Isaac Newton Institute of the University of Cambridge. A.D.G. is supported financially by Gonville & Caius College, Cambridge.

REFERENCES

Bayly, B.J. & Childress, S. 1988 Construction of fast dynamos using unsteady flows and maps in three dimensions. *Geophys. Astrophys. Fluid Dynam.* **44**, 211–240.

Bayly, B.J. & Childress, S. 1989 Unsteady dynamo effects at large magnetic Reynolds number. *Geophys. Astrophys. Fluid Dynam.* **49**, 23–43.

Childress, S. 1992 Fast dynamo theory. In *Topological Aspects of the Dynamics of Fluids and Plasmas* (ed. H.K. Moffatt, G.M. Zaslavsky, P. Comte & M. Tabor), pp. 111–147. Kluwer Academic Publishers.

Galanti, B., Sulem, P.L. & Pouquet, A. 1992 Linear and non-linear dynamos associated with ABC flows. *Geophys. Astrophys. Fluid Dynam.* **66**, 183–208.

Galloway, D.J. & Proctor, M.R.E. 1992 Numerical calculations of fast dynamos for smooth velocity fields with realistic diffusion. *Nature* **356**, 691–693.

Galloway, D.J., Proctor, M.R.E. & Weiss, N.O. 1978 Magnetic flux ropes and convection. *J. Fluid Dyn.* **87**, 243–261.

Gilman, P.A. 1983 Dynamically consistent nonlinear dynamos driven by convection in a rotating spherical shell. II. Dynamos with cycles and strong feedbacks. *Astrophys. J. Suppl.* **53**, 243–268.

Glatzmaier, G.A. 1985 Numerical simulations of stellar convective dynamos. II. Field propagation in the convection zone. *Astrophys. J.* **192**, 300–307.

Meneguzzi, M. & Pouquet, A. 1989 Turbulent dynamos driven by convection. *J. Fluid Dyn.* **205**, 297–318.

Nordlund, A., Brandenburg, A., Jennings, R., Rieutord, M., Ruokolainen, J., Stein, R.F. & Tuominen, I. 1991 Dynamo action in stratified convection with overshoot. *Astrophys. J.* **392**, 647–652.

Otani, N.F. 1992 A fast magnetohydrodynamic dynamo in two-dimensional time-dependent flows. *J. Fluid Mech.* (submitted).

Vainshtein, S.I. & Cattaneo, F. 1992 Nonlinear restrictions on dynamo action, *Astrophys. J.* **393**, 165–171.

Vainshtein, S.I., Tao, L., Cattaneo, F. & Rosner, R. 1993 Turbulent magnetic transport effects and their relation to magnetic field intermittency. In this volume.

A Numerical Study of Dynamos in Spherical Shells with Conducting Boundaries

W. HIRSCHING & F.H. BUSSE

Institute of Physics
University of Bayreuth
D-8580 Bayreuth, Germany

The problem of the generation of magnetic fields by convection in rotating spherical shells is considered in the case when the boundaries of the fluid shell exhibit a finite electrical conductivity. This problem is of geophysical interest because Lorentz forces acting in the boundaries provide a mechanical coupling that was not included in previous computations by Zhang & Busse (1988, 1989). The vanishing torques between fluid shell and boundaries determine the relative rotation between the three regions of the problem. But the finite conductivity does not seem to improve the numerical convergence for dynamo solutions.

---

1 INTRODUCTION

The mathematical difficulties in deriving solutions for growing magnetic fields in spherical geometries have long puzzled dynamo theoreticians. In contrast to the solutions of the kinematic dynamo problem found by Roberts (1970, 1972) and others in the case of periodic velocity fields in infinitely extended electrically conducting fluids, dynamo action often seems to disappear as soon as insulating boundaries are introduced. Motivated by this observation Bullard & Gubbins (1977) have investigated kinematic dynamos in a spherical domain of constant conductivity with insulating exterior for velocity fields with alternating signs as a function of radius. As expected the critical magnetic Reynolds number decreases significantly as the number of sign changes increases and the limit of a periodic velocity field is approached. When the radial velocity component does not change sign, a dynamo solution was not obtained unless an outer shellular region of finite conductivity was introduced.

137

M.R.E. Proctor, P.C. Matthews & A.M. Rucklidge (eds.)
Theory of Solar and Planetary Dynamos, 137–144
©1993 Cambridge University Press.

In the present paper a similar approach is used in the case of the physical dynamo problem of magnetic field generation by buoyancy driven convection in rotating spherical shells. Although the numerical convergence of the dynamo solutions obtained by Zhang & Busse (1988, 1989, referred to in the following by ZB88, ZB89) appeared to be satisfactory at the relatively low truncation levels used for the Galerkin expansion, more detailed studies at higher truncation levels have revealed a tendency of a diverging critical magnetic Prandtl number. It is the purpose of this paper to demonstrate this tendency in a particular case and to investigate the possibility of a more satisfactory convergence in the case when conducting layers are added outside the boundaries of the spherical fluid shell.

Regions of finite conductivity enclosing the fluid outer core of the Earth are physically realistic and play an important role in the electromagnetic coupling between core and mantle. Because viscous stresses are usually neglected at the boundaries because of the small value of the viscosity of the fluid core, Lorentz forces in a conducting mantle provide the only mechanical coupling between core and mantle unless deviations from a spherical boundary are taken into account. In the model of ZB89 the relative motion between core and mantle could not be determined because of the assumption of an insulating exterior. The analysis of the present paper resolves this indeterminacy through the addition of conducting layers inside and outside the fluid shell.

In accordance with the above remarks the analysis of the following sections is divided into two parts. After the formulation of the mathematical problem in section 2 the kinematic problem of the growth of infinitesimal magnetic fields is considered in section 3. The second part is described in section 4 and deals with the full magnetohydrodynamic dynamo problem in which the Lorentz forces in the equation of motion are taken into account. A discussion of the convergence problem and an outlook to future work are given in the concluding section.

2 MATHEMATICAL FORMULATION

We consider a rotating self-gravitating spherical fluid shell of thickness d which permits a motionless state as a solution of the basic equations. The temperature distribution of the static state corresponds to a uniform distribution of heat sources within the fluid and the enclosed inner spherical core, $T_s = T_o - \beta d^2 r^2 / 2$ where r is the dimensionless distance from the center. The gravity force is given by $\mathbf{g} = -\gamma d\mathbf{r}$. Using d as the length scale, d^2/ν as the time scale, $\beta d^2 \nu / \kappa$ as the scale of the temperature, and $\nu(\mu\rho)^{1/2}d$ as the scale of the magnetic flux density we obtain dimensionless equations for the dependent variables of the problem. The kinematic viscosity ν, the thermal diffusivity κ, the magnetic permeability μ, and the density ρ have been used in the definitions of the scales. Since both the velocity field $\mathbf{u}$ and the magnetic field $\mathbf{B}$ are solenoidal vector fields it is convenient to use the general representation in terms of poloidal and toroidal components

$$\mathbf{u} = \nabla \times (\nabla \times \mathbf{r}v) + \nabla \times \mathbf{r}w, \qquad \mathbf{B} = \nabla \times (\nabla \times \mathbf{r}h) + \nabla \times \mathbf{r}g. \qquad (1)$$

Using a spherical system of coordinates (r, θ, φ) the basic equations for the scalar fields v, w, h, g and for the deviation Θ from the static temperature distribution can be written in the form (ZB88, ZB89)

$$[(\nabla^2 - \frac{\partial}{\partial t})L_2 + \tau\frac{\partial}{\partial \phi}]\nabla^2 v + \tau Q w - RL_2\Theta = -\mathbf{r}\cdot\nabla\times(\nabla\times(\mathbf{u}\cdot\nabla\mathbf{u} - \mathbf{B}\cdot\nabla\mathbf{B})),$$
$$(2a)$$

$$[(\nabla^2 - \frac{\partial}{\partial t})L_2 + \tau\frac{\partial}{\partial \phi}]w - \tau Q v = \mathbf{r}\cdot\nabla\times(\mathbf{u}\cdot\nabla\mathbf{u} - \mathbf{B}\cdot\nabla\mathbf{B}), \qquad (2b)$$

$$\nabla^2\Theta + L_2 v = P(\frac{\partial}{\partial t} + \mathbf{u}\cdot\nabla)\Theta, \qquad (2c)$$

$$\nabla^2 L_2 h = Pm[\frac{\partial}{\partial t}L_2 h - \mathbf{r}\cdot\nabla\times(\mathbf{u}\times\mathbf{B})], \qquad (2d)$$

$$\nabla^2 L_2 g = Pm[\frac{\partial}{\partial t}L_2 g - \mathbf{r}\cdot\nabla\times(\nabla\times(\mathbf{u}\times\mathbf{B}))], \qquad (2e)$$

where the definitions

$$L_2 \equiv -r^2\nabla^2 + \frac{\partial}{\partial r}r^2\frac{\partial}{\partial r}, \quad Q = r\cos\theta\nabla^2 - (L_2 + r\frac{\partial}{\partial r})(\cos\theta\frac{\partial}{\partial r} - \frac{\sin\theta}{r}\frac{\partial}{\partial\theta})$$

have been used and the Rayleigh number R, the rotation parameter τ, the Prandtl number P and magnetic Prandtl number Pm have been introduced:

$$R = \frac{\alpha\gamma\beta d^6}{\nu\kappa}, \quad \tau = \frac{2\Omega d^2}{\nu}, \quad P = \frac{\nu}{\kappa}, \quad Pm = \frac{\nu}{\lambda}.$$

The formulation of the mathematical problems differs from that of ZB88, ZB89 only in that the magnetic boundary conditions are applied at spherical surfaces beyond the boundaries of the fluid shell,

$$g = h - h^{(e)} = \frac{\partial}{\partial r}(h - h^{(e)}) = 0 \text{ at } r = \hat{r}_i \equiv \frac{\eta}{1 - \eta} - \delta \text{ and } r = \hat{r}_0 \equiv \frac{1}{1 - \eta} + \delta,$$
$$(3)$$

where η is the radius ratio of the fluid shell and where $h^{(e)}$ refers to the potential fields in the regions $r < \hat{r}_i$ and $r > \hat{r}_o$. As in ZB88, ZB89 stress-free conditions apply at the boundaries of the fluid shell:

$$v = \frac{\partial^2 v}{\partial r^2} = \frac{\partial}{\partial r}(\frac{w}{r}) = \Theta = 0 \quad \text{at} \quad r = r_i \equiv \frac{\eta}{1 - \eta} \quad \text{and} \quad r = r_0 \equiv \frac{1}{1 - \eta}.$$
$$(4)$$

For simplicity we have assumed that the conducting layers of thickness δ have the same magnetic diffusivity λ as the fluid. Insulating spaces are assumed for $r < \hat{r}_i$ and for $r > \hat{r}_0$. This choice minimizes the number of parameters of the problem, but is not very realistic from a geophysical point of view. We expect, however, that qualitative aspects will not change much if different diffusivities are chosen for fluid and boundaries or if different thicknesses of the inner and outer conducting layers are assumed.

For the solution of (2) we employ the same Galerkin method that has been used in ZB88 and ZB89 with the exception that r_i is replaced by $\hat{r}_i$ in the case of the magnetic variables h, g. As examples we give the expressions for w and g:

$$w = r \sum_{\nu,n,\hat{l}} c_{\nu ni} \exp\{i\nu m(\varphi - ct)\} P_{\hat{l}}^{|\nu|m}(\cos\vartheta)\cos(n-1)\pi(r - r_i) \quad (5a)$$

$$g = \sum_{\nu,n,l} g_{\nu nl} \exp\{i\nu m(\varphi - ct)\} P_l^{|\nu|m}(\cos\vartheta)\sin[n\pi(r - \hat{r}_i)(1 + 2\delta)^{-1}] \quad (5b)$$

where the index n runs through all positive integers, ν runs through all integers and $\hat{l}$ and l run through odd and even integers, respectively. The latter choice corresponds to dipolar magnetic fields. The complex coefficients $c_{\nu ni}$ and $g_{\nu nl}$ must satisfy the usual relations to ensure that expressions (5) are real.

After the representations of the form (5) have been introduced into equations (2) and these equations have been projected onto the expansion functions, a system of nonlinear algebraic equations for the unknown coefficients $c_{\nu ni}, g_{\nu nl}$ etc. is obtained. After this system has been truncated through the neglect of all coefficients and corresponding equations whose subscripts satisfy

$$2n + \hat{l} - |\nu|m + 2|\nu| > 2N_T + 3, \quad (6)$$

it can be solved by the Newton–Raphson method. The truncation procedure (6) is applied for l in place of $\hat{l}$ as well.

Besides this nonlinear magnetohydrodynamic dynamo problem a simpler kinematic dynamo problem can be solved for a given solution v, w, Θ with $h \equiv g \equiv 0$. In this case only equations $(2d, e)$ are considered and the common factor $\exp\{\sigma t\}$ is added in the representations for h and g. Equations $(2d, e)$ thus lead to linear eigenvalue problems with the growth rate σ as eigenvalue. This latter problem will be considered in the next section.

3 GROWTH RATES OF MAGNETIC FIELDS GENERATED BY CONVECTION

For a given convection solution which has been obtained for the truncation parameter $N_T = 3$, the linear equations $(2d, e)$ have been solved for different values of the truncation parameter applied to the representation for h and g.

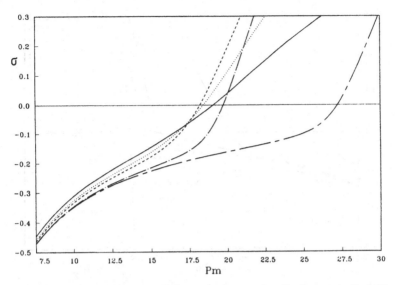

Figure 1. The growth rate $\sigma = \sigma_r$ of the strongest growing dipolar magnetic field as a function of the magnetic Prandtl number Pm for $R = 3100, P = 0.1, \eta = 0.4, m = 2, \tau = 7.66 \cdot 10^2, \delta = 0$. The curves correspond to $N_B = 5$ (solid), 6 (dotted), 7 (dashed), 8 (dash-dotted), and 9 (long-dash – short-dash).

In order to avoid misunderstandings we shall refer to this latter truncation parameter by N_B instead of N_T. In Figures 1 and 2 the real part σ_r of the growth rate σ has been plotted for different values of N_B as a function of the magnetic Prandtl number Pm. In Figure 1 a particular case with $\delta = 0$ listed in Table 2 of ZB88 has been used and extended to higher values of N_B. The results indicate a reasonable numerical convergence at $N_B = 6$ or 7. But the real growth rate σ obtained for $N_B = 9$ deviates by a large amount as soon as it exceeds a value of about -0.2. Because the magnetic Reynolds number based on the convection velocity exceeds the order unity, flux expulsion occurs which tends to compress magnetic flux into boundary layers. These boundary layers can be resolved numerically only with sufficiently high values of N_B. As these boundary layers become resolved, the magnetic flux available for the interaction with the velocity field diminishes. A higher magnetic Prandtl number is thus required for dynamo action leading to a negative feedback mechanism since the magnetic Reynolds number is proportional to Pm.

It was expected that the thin current layers corresponding to the magnetic boundary layers could be avoided in part in the case when the insulating boundaries are replaced by conducting layers of thickness δ. In Figure 2 growth rates σ_r are plotted for a case with $\delta = 0.05$. In contrast to the case of $\delta = 0$ in which the strongest growing magnetic fields corresponded to real values of σ, the imaginary parts of the eigenvalues σ with the largest real

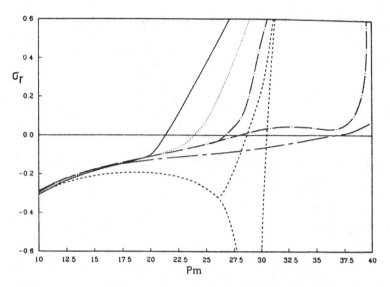

Figure 2. The same case as Figure 1 except for the addition of static conducting boundary layers of thickness $\delta = 0.05$. The values of N_B corresponding to the different curves have been increased by one. In the case of bifurcating curves the upper one corresponds to a finite imaginary part of σ.

part do not always vanish. No apparent convergence for intermediate values of N_B is noticeable in this case owing to the discontinuities in the velocity field caused by the fact that v and w are vanishing in the conducting layers. But the convergence properties also do not seem to be improved for the highest values of N_B shown in the figure.

4 CORE–MANTLE COUPLING

The introduction of an electrically conducting boundary provides the opportunity to resolve an indeterminacy in previous dynamo computations. In the analysis of ZB89 the rigid rotation component of the velocity field given by the coefficient c_{011} in (5a) remained undetermined since no torque was acting between fluid shell and boundaries. In the case of finitely conducting boundaries the relative rotation between boundary and the fluid can be determined through the condition that the torque exerted on the boundary vanishes. This relative rotation is finite, in general, and may be reduced by viscous and topographic coupling torques.

Assuming as before the frame of reference defined by $c_{011} = 0$ in the fluid core we find for the angular rotation rates $\omega_{o,i} = c_{011}^{o,i}/\sqrt{3}$ of outer (mantle) and inner (inner core) boundaries the following relationships:

$$\int_{r_0}^{\hat{r}_0} \int_0^{2\pi} \int_{-1}^{+1} \mathbf{k} \times \mathbf{r} \cdot ((\nabla \times \mathbf{B}) \times \mathbf{B}) d\cos\vartheta \, d\varphi \, r^2 dr - f(\hat{r}_0^5 - r_0^5) 4\pi\omega_o/5\sqrt{2} = 0, \quad (7a)$$

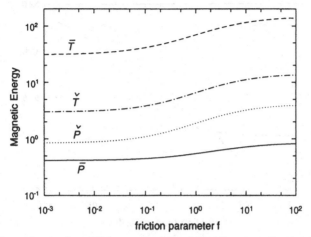

Figure 3. Magnetic energies of the mean and the fluctuating components of the toroidal $(\bar{T}, \check{T})$ and of the poloidal magnetic field $(\bar{P}, \check{P})$ as function of the friction parameter f. The parameter values are those of Figure 1 except that $N_B = 7$ and $Pm = 40$ have been chosen.

$$\int_{\check{r}_i}^{r_i} \int_0^{2\pi} \int_{-1}^{+1} \mathbf{k} \times \mathbf{r} \cdot ((\nabla \times \mathbf{B}) \times \mathbf{B}) d\cos\vartheta \, d\varphi r^2 dr - f(r_i^5 - \hat{r}_i^5) 4\pi \omega_i / 5\sqrt{2} = 0, \quad (7b)$$

where a friction parameter f has been introduced which parametrizes processes that tend to reduce the differential rotation between core and mantle and between inner and outer core to zero in the absence of Lorentz forces. In the limit of large f the differential rotations ω_o and ω_i tend to zero and only the effect of the static conducting layer is felt by the magnetic field which has been discussed in the preceding section for the case of vanishing magnetic field strength. As the friction parameter f decreases, an increasing shear between fluid and boundaries develops which in general leads to a decrease of the magnetic energy for a fixed value of Pm. This influence is seen in Figure 3 and the corresponding variations of ω_o and ω_i are shown in Figure 4. The drift c of the convection rolls, also seen in the figure, exhibits relatively small variations. Since both c and ω_o have the same sign, the drift of the rolls relative to the mantle is reduced. While this effect is rather small in the case of Figure 4, larger values for ω_o in the limit of small f have been obtained in other cases. We always find that ω_o and ω_i have opposite signs.

5 CONCLUDING REMARKS

Although the numerical convergence of the convection driven dynamo solutions discussed here and in the previous work ZB88, ZB89 is doubtful, most qualitative features of the nonlinear interaction between magnetic fields and convection are expected to persist in the case of more complicated realistic dynamos. Preliminary work (Hirsching 1992) has shown that aperiodic time

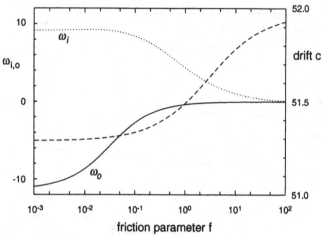

Figure 4. The rotation rates ω_o and ω_i of the inner core and of the mantle, respectively. The drift rate c of the convection rolls is given by the dashed curve and corresponds to the right ordinate.

dependent dynamo solutions can be obtained at higher Rayleigh numbers but at significantly lower magnetic Reynolds numbers which appear to avoid the problem of flux expulsion. It is most likely that the time averaged properties of these solutions will exhibit, for example, similar relationships for the differential rotations of inner core and mantle discussed in the preceding section.

The research reported in this paper was supported by the Deutsche Forschungsgemeinschaft under Grant Bu589/1. We are grateful to Dr. K. Zhang for providing solutions with which the computer code could be checked.

REFERENCES

Bullard, E.C. & Gubbins, D. 1977 Generation of magnetic fields by fluid motions of global scale. *Geophys. Astrophys. Fluid Dynam.* **8**, 43–56.

Hirsching, W. 1992 Konvektion in Kugeln und das Problem der Magnetfelderzeugung. Dissertation, University of Bayreuth.

Roberts, G.O. 1970 Spatially periodic dynamos. *Phil. Trans. R. Soc. Lond.* A **226**, 335–558.

Roberts, G.O. 1972 Dynamo action of fluid motions with two dimensional periodicity. *Phil. Trans. R. Soc. Lond.* A **271**, 411–454.

Zhang, K. & Busse, F.H. 1988 Finite amplitude convection and magnetic field generation in a rotating spherical shell. *Geophys. Astrophys. Fluid Dynam.* **44**, 33–53.

Zhang, K. & Busse, F.H. 1989 Convection driven magnetohydrodynamic dynamos in rotating spherical shells. *Geophys. Astrophys. Fluid Dynam.* **49**, 97–116.

Non-axisymmetric Shear Layers in a Rotating Spherical Shell

R. HOLLERBACH

Department of Mathematics
University of Exeter
Exeter, EX4 4QE UK

M.R.E. PROCTOR

Department of Applied Mathematics and Theoretical Physics
University of Cambridge, Silver St., Cambridge, CB3 9EW UK

It is pointed out that in general non-axisymmetric solutions of
the forced momentum equation in a rotating spherical shell are
singular in the inviscid limit, with all three components of the
flow discontinuous across the cylinder tangent to the inner core
and parallel to the axis of rotation. A constraint on the forcing
is derived which must be satisfied if the inviscid solution is to be
non-singular. It is suggested that the Lorentz force will evolve to
satisfy this constraint, thereby eliminating the need for the viscous
shear layer that would otherwise resolve the singularity.

It is generally recognized that the Earth's magnetic field is maintained against
Ohmic decay by fluid motions in its liquid iron outer core. Obtaining solutions
of the momentum equation in a rotating spherical shell is thus of considerable geophysical interest. In this work we focus on the role of viscosity in
obtaining sensible non-axisymmetric solutions, which we believe has been in-
sufficiently considered in the past. That non-axisymmetric fluid motions must
play an essential part in maintaining the magnetic field follows from Cowling's
theorem, which states that a purely axisymmetric flow cannot maintain an
axisymmetric field.

So, consider the inertia-less momentum equation

$$2\hat{\mathbf{k}} \times \mathbf{u} = -\nabla p + \epsilon \nabla^2 \mathbf{u} + \mathbf{f}, \tag{1}$$

where $\mathbf{u}$ is the fluid flow to be determined, p is the pressure, and ϵ is the
(small) Ekman number. For now we simply take $\mathbf{f}$ to be some given forcing,
having the azimuthal dependence $\exp(im\phi)$, which factor will henceforth be

145

M.R.E. Proctor, P.C. Matthews & A.M. Rucklidge (eds.)
Theory of Solar and Planetary Dynamos, 145–152
©1993 Cambridge University Press.

implicit throughout. Since (1) is linear, we can restrict attention to a single azimuthal mode without loss of generality.

The neglect of inertia in (1) considerably alters the character of the equation, changing it from parabolic to elliptic. Nevertheless, subject to suitable boundary conditions, it continues to be the case that (1) has a unique, sensible solution for any forcing $\mathbf{f}$. However, if one attempts to neglect viscosity as well, that is no longer the case: taking the curl of the inviscid momentum equation

$$2\hat{\mathbf{k}} \times \mathbf{u} = -\nabla p + \mathbf{f}, \tag{2}$$

and using the incompressibility condition $\nabla \cdot \mathbf{u} = 0$, one obtains

$$-2\frac{\partial}{\partial z}\mathbf{u} = \nabla \times \mathbf{f}, \tag{3}$$

$$\mathbf{u} = -\frac{1}{2}\int^z \nabla \times \mathbf{f}\, dz' + \mathbf{c}(s). \tag{4}$$

At this point one must be careful that any component-by-component integration of (4) is done in cylindrical coordinates (z, s, ϕ), to ensure that one's unit vectors in fact remain constant over the path of integration. Letting subscripts denote the indicated components, two boundary conditions on appropriate linear combinations of u_z and u_s will turn out to determine c_z and c_s. To determine c_ϕ, consider $\nabla \cdot \mathbf{u} = 0$, which implies

$$\frac{\partial}{\partial z}c_z(s) + \frac{1}{s}\frac{\partial}{\partial s}\big(sc_s(s)\big) + \frac{im}{s}c_\phi(s) = \frac{1}{2}\,\nabla \cdot \int^z \nabla \times \mathbf{f}\, dz', \tag{5}$$

which determines c_ϕ if $m \neq 0$. If $m = 0$, c_ϕ is undetermined, and there is instead the solvability condition (Taylor 1963)

$$\frac{1}{s}\frac{\partial}{\partial s}\big(sc_s(s)\big) = \frac{1}{2}\,\nabla \cdot \int^z \nabla \times \mathbf{f}\, dz'. \tag{6}$$

For the axisymmetric case $m = 0$, $\mathbf{f}$ must be such that c_s, which has already been determined by the boundary conditions, also just happens to satisfy (6). If it does not there is no solution at all, and if it does, the solution is still only determined to within an arbitrary geostrophic flow $c_\phi(s)$.

In contrast, for the non-axisymmetric cases $m \neq 0$ there appear to be no such difficulties, since we have explicitly demonstrated how to compute a unique solution for any $\mathbf{f}$. Nevertheless, in a spherical shell there is a considerable difficulty, since the unique solution just computed will not necessarily be sensible. To see why, we consider the two boundary conditions that need to be imposed at each cylindrical radius s to determine c_z and c_s. At the outer boundary the no-normal-flow boundary condition becomes

$$zu_z + su_s = 0 \qquad \text{at} \quad z = (r_o^2 - s^2)^{1/2}, \tag{7}$$

and this applies for all $s \leq r_o$. At the inner boundary the no-normal-flow boundary condition becomes

$$zu_z + su_s = 0 \qquad \text{at } z = (r_i^2 - s^2)^{1/2}, \qquad (8a)$$

but this applies only for $s \leq r_i$. For $r_i \leq s \leq r_o$ we take instead the symmetry condition

$$u_z = 0 \qquad \text{at } z = 0, \qquad (8b)$$

appropriate for solutions having u_z antisymmetric and u_s and u_ϕ symmetric about the equator. (This requires of course that f satisfies the same symmetry conditions.) The point now is that $(8a)$ and $(8b)$ do not join smoothly at $s = r_i$, $z = 0$, where $(8a)$ becomes $u_s = 0$ instead of $u_z = 0$. This discontinuity in the applied boundary conditions will then induce a discontinuity in the constants of integration c_z and c_s, and according to (4) this discontinuity will manifest itself in the flow for all z on the tangent cylinder at $s = r_i$. Since (3) is now hyperbolic, any discontinuity on the boundary is transmitted undiminished along the characteristics into the interior.

If one really believes the inviscid limit to be appropriate, the only resolution to this singularity is to assume that f satisfies some condition that ensures that c_z and c_s just happen to be continuous. After all, for *some* f it must surely be the case that both u_z and u_s are zero at $s = r_i$, $z = 0$, and if that is the case then $(8a)$ and $(8b)$ will join somewhat more smoothly. So, according to (4), if u_z and u_s are both to be zero at $s = r_i$, $z = 0$, they must take the form

$$u_z = -\frac{1}{2} \int_0^z [\nabla \times \mathbf{f}]_z \, dz' \qquad \text{at } s = r_i, \qquad (9a)$$

$$u_s = -\frac{1}{2} \int_0^z [\nabla \times \mathbf{f}]_s \, dz' \qquad \text{at } s = r_i. \qquad (9b)$$

Since this flow must still satisfy the outer boundary condition (7), that gives us the constraint on f

$$z \int_0^z [\nabla \times \mathbf{f}]_z \, dz' + s \int_0^z [\nabla \times \mathbf{f}]_s \, dz' = 0, \qquad (10)$$

which must be satisfied at $s = r_i$, $z = (r_o^2 - r_i^2)^{1/2}$. If (10) is not satisfied, the solution (4), although well-defined and unique, will have a very severe singularity across the tangent cylinder at $s = r_i$, whereas if it is satisfied the singularity will be considerably weaker.

The reason the singularity is not eliminated entirely is that although we have ensured that c_z and c_s are continuous across $s = r_i$, we have not ensured that dc_z/ds and dc_s/ds are, and so according to (5) c_ϕ is still discontinuous.

One could derive additional constraints similar to (10), although of ever-increasing complexity and presumably ever-decreasing relevance, which will ensure that this singularity is progressively eliminated.

Alternatively, if one insists on obtaining solutions for forcings $\mathbf{f}$ that do not satisfy (10), one must invoke some small viscosity to resolve the singularity. A direct spectral solution of (1) explicitly resolving the boundary layer structure has been developed. Due to limitations on space, we merely present some examples here, leaving the numerical details to be presented elsewhere.

Consider first $\mathbf{f}_1 = [-4isz\hat{\mathbf{e}}_z + 2is^2\hat{\mathbf{e}}_s]\exp(i\phi)$, for which the inviscid solution is

$$u_z = -sz + c_z(s),$$

$$u_s = -z^2 + c_s(s),$$

$$u_\phi = -i(z^2 + s^2) + c_\phi(s).$$

From (7) and (8), the constants of integration turn out to be: for $s \leq r_i$,

$$c_z = 2s[(r_o^2 - s^2)^{1/2} + (r_i^2 - s^2)^{1/2}],$$

$$c_s = -2(r_o^2 - s^2)^{1/2}(r_i^2 - s^2)^{1/2},$$

and for $s \geq r_i$,

$$c_z = 0, \qquad c_s = 2(r_o^2 - s^2), \qquad c_\phi = 2i(r_o^2 - 3s^2).$$

Figure 1 shows the singularity in the inviscid solution, and also how the viscous solution at $\epsilon = 10^{-5}$ resolves the singularity.

Consider next $\mathbf{f}_2 = [-4is^3z\hat{\mathbf{e}}_z + 6is^2z^2\hat{\mathbf{e}}_s]\exp(i\phi)$, for which the inviscid solution is

$$u_z = -sz^3 + c_z(s),$$

$$u_s = -s^2z^2 + c_s(s),$$

$$u_\phi = -6is^2z^2 + c_\phi(s).$$

From (7) and (8), the constants of integration turn out to be: for $s \leq r_i$,

$$c_z = s[(r_o^2 - s^2)^{1/2} + (r_i^2 - s^2)^{1/2}](r_o^2 + r_i^2 - s^2),$$

$$c_s = -(r_o^2 - s^2)^{1/2}(r_i^2 - s^2)^{1/2}(r_o^2 + r_i^2 - s^2) - (r_o^2 - s^2)(r_i^2 - s^2),$$

and for $s \geq r_i$,

$$c_z = 0, \qquad c_s = r_o^2(r_o^2 - s^2), \qquad c_\phi = ir_o^2(r_o^2 - 3s^2).$$

Figure 2 shows the singularity in the inviscid solution, and also how the viscous solution at $\epsilon = 10^{-5}$ resolves the singularity.

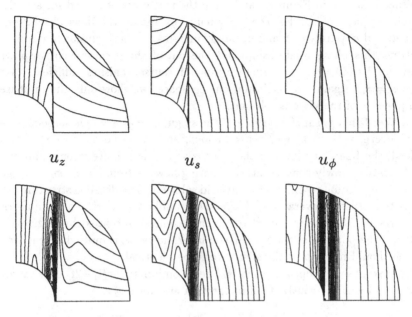

Figure 1. The inviscid (top) and viscous (bottom) solutions corresponding to $\mathbf{f}_1$. Contour intervals 0.2 for u_z, 0.4 for u_s, 2 for u_ϕ.

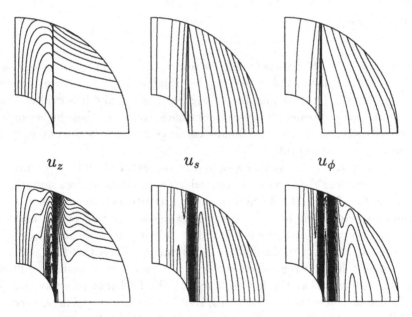

Figure 2. The inviscid (top) and viscous (bottom) solutions corresponding to $\mathbf{f}_2$. Contour intervals 0.2 for u_z, 0.4 for u_s, 2 for u_ϕ.

What is shown in Figures 1 and 2 are the real parts of u_z and u_s, and the imaginary parts of u_ϕ. For these particular examples, the inviscid solutions have u_z and u_s purely real, and u_ϕ purely imaginary. The inclusion of viscosity destroys this phase relationship, but the out-of-phase component displays a similar boundary layer structure, and of course vanishes sufficiently far from $s = r_i$. Again due to limitations on space, we show only the in-phase components of the viscous solutions.

Neither $\mathbf{f}_1$ nor $\mathbf{f}_2$ satisfies (10), and the singularity at $s = r_i$ is consequently very severe, with c_z and c_s discontinuous, and $c_\phi \to \infty$ as $s \to r_i^-$. And indeed, the boundary layer in the viscous solutions is quite strong, and evidently rather slowly convergent, considering how far from $s = r_i$ one must go before the solution resembles the inviscid solution. The detailed structure appears to consist of several nested layers, not unlike the classical axisymmetric Stewartson (1966) layer, considered in the geophysical context by Ruzmaikin (1989). The purpose of the Stewartson layer, however, is quite different, and so one should not expect the structure to be identical.

However, the appropriately chosen linear combination $r_o^2 \mathbf{f}_1 - 2\mathbf{f}_2$ does satisfy (10), and for it the constants of integration are: for $s \leq r_i$,

$$c_z = 2s(s^2 - r_i^2)[(r_o^2 - s^2)^{1/2} + (r_i^2 - s^2)^{1/2}],$$

$$c_s = 2(r_o^2 - s^2)^{1/2}(r_i^2 - s^2)^{3/2} + 2(r_o^2 - s^2)(r_i^2 - s^2),$$

and for $s \geq r_i$,

$$c_z = c_s = c_\phi = 0.$$

The singularity is thus considerably less severe, with c_z and c_s continuous but still not differentiable, and $c_\phi \to -4i(r_o^2 - r_i^2)r_i^2$ as $s \to r_i^-$. Figure 3 shows the appropriate linear combination of Figures 1 and 2, and it is evident that that portion of the boundary layer responsible for resolving these higher-order singularities is rather weak and rapidly convergent. Even the discontinuity in c_ϕ causes no great difficulty.

Returning now to the proper geophysical context, $\mathbf{f}$ should not simply be taken to be some given, fixed forcing, which either does or does not satisfy (10). Instead, $\mathbf{f}$ is itself a dynamic variable, consisting of the buoyancy force ultimately responsible for driving the fluid motion, as well as the Lorentz force exerted by the magnetic field. And considering the stiffening effect that this Lorentz force has on the fluid, it seems rather unlikely that the magnetic field would tolerate strong shear layers in the interior of a conducting fluid. So, we postulate that the flow will distort the field into precisely such a configuration where the constraint (10) is satisfied, much as Malkus & Proctor (1975) postulated that the flow will distort the field into precisely such a configuration where the Taylor's constraint (6) is satisfied.

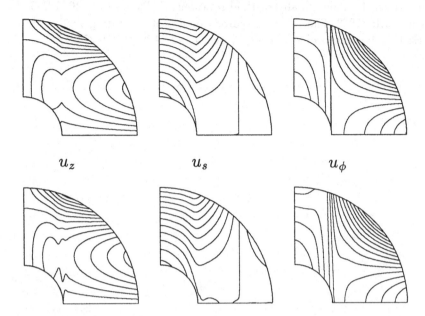

u_z $\qquad\qquad$ u_s $\qquad\qquad$ u_ϕ

Figure 3. The linear combination r_o^2 times Figure 1 minus 2 times Figure 2. Contour intervals 0.2 for u_z, 0.4 for u_s, 1 for u_ϕ.

The need to satisfy (10) could then significantly influence the pattern of magnetoconvection in the vicinity of the inner core tangent cylinder. Of course, the constraint (10) is weaker than Taylor's constraint, in the sense that it must be satisfied only at $s = r_i$ rather than at all s. On the other hand, Taylor's constraint applies only to the axisymmetric mode $m = 0$, whereas the constraint (10) applies to *all* non-axisymmetric modes $m \neq 0$.

Finally, the alternative, that the Lorentz force does *not* adjust to satisfy (10), would be equally significant, since one would then be forced to conclude that viscous shear layers do exist in the Earth's core, and that viscosity, however small, always influences the pattern of magnetoconvection. Either way, this issue clearly deserves further attention.

We thank Andrew Soward, Chris Jones, and David Fearn for valuable discussions. This work was funded by the Science and Engineering Research Council under grant number GR/E93251.

REFERENCES

Malkus, W.V.R. & Proctor, M.R.E. 1975 The macrodynamics of α-effect dynamos in rotating fluids. *J. Fluid Mech.* **67**, 417–443.

Ruzmaikin, A.A. 1989 A large-scale flow in the Earth's core. *Geomag. Aeron.* **29**, 299–303.

Stewartson, K. 1966 On almost rigid rotations. *J. Fluid Mech.* **26**, 131–144.

Taylor, J.B. 1963 The magnetohydrodynamics of a rotating fluid and the Earth's dynamo problem. *Proc R. Soc. Lond. A* **274**, 274–283.

Testing for Dynamo Action

D.W. HUGHES

Department of Applied Mathematical Studies
The University
Leeds, LS2 9JT UK

It is important to determine whether a cosmical magnetic field is
a consequence of dynamo action or, alternatively, is a slowly decaying fossil field. Similarly, in numerical simulations of magnetohydrodynamic turbulence we should like to distinguish between a
dynamo-generated magnetic field and one that is simply decaying, albeit slowly. Here certain criteria are presented that must
be satisfied before any positive claims can be made for dynamo
action.

1 INTRODUCTION

Given the existence of a naturally occurring magnetic field, be it astrophysical
or geophysical, it is natural to ask whether the field is generated by dynamo
action or if instead it is a fossil field, trapped in the body since its formation.
In certain contexts it is possible to give a definitive answer. For example, the
Ohmic diffusion time of the Earth's core is of the order of 10^4 years whereas
paleomagnetic records show that the magnetic field of the Earth has existed
for 10^9 years. Consequently, since the field has been maintained for so many
Ohmic decay times it must be generated by some sort of dynamo process. For
astrophysical bodies on the other hand, for which typically the Ohmic time
is comparable to the lifetime of the body itself, it is not so straightforward to
assert that a field is dynamo-generated. Of course, there may be other factors
suggesting the origin of the field, but simply on the basis of the Ohmic decay
time the issue often cannot be decided. What we would like therefore is a
test to distinguish between these two possibilities.

M.R.E. Proctor, P.C. Matthews & A.M. Rucklidge (eds.)
Theory of Solar and Planetary Dynamos, 153–159
©1993 Cambridge University Press.

A major difficulty in devising such a test is that we know very little about either the flow or the magnetic field in astrophysical bodies; even for the Sun our knowledge is rather limited. Thus we shall focus here on the closely related problem of deciding whether dynamo action is present in a numerical MHD simulation, where all the data are known. (This may be thought of as analysing a star, for example, for which we have a detailed knowledge of its interior behaviour.) We shall be concerned with dynamos driven by turbulent velocity fields, with no added extras such as prescribed α-effects. The standard procedure in such calculations is to inject a large-scale weak field into a statistically stationary turbulent flow and then to watch the field evolve. In terms of such a model the question we would like to answer is, 'For how long do we need to study the evolving field before we can make any positive claims of dynamo action?'

The essence of a dynamo is that there is a balance between induction and dissipation; if dissipation wins then there is no dynamo. One would thus certainly have confidence that a dynamo was operative if the field persisted for several Ohmic decay times, based on the large-scale field. However for a high-resolution simulation with a large magnetic Reynolds number R_m this time is immense and it is impractical to integrate for so long. In a turbulent flow, however, the timescale of interest is the effective turbulent diffusion time, which may be much smaller than the Ohmic time τ_η. In the next section we show, by consideration of one particular two-dimensional flow, how the effective diffusivity can be affected by the strength of the magnetic field. In section 3 we consider both steady and cyclic dynamos and provide the time for which integration must be performed in order to say anything about dynamo action. In section 4 we discuss briefly some of the many unresolved difficulties associated with the turbulent advection and diffusion of magnetic fields. Sections 2 and 3 are based on the paper by Cattaneo, Hughes & Weiss (1991), in which considerably more details may be found, together with a discussion of various numerical simulations that purport to exhibit dynamo action.

2 TURBULENT DIFFUSION OF MAGNETIC FIELDS

Understanding the behaviour of a magnetic field in a turbulent fluid is a problem of long standing, dating back to the work of Batchelor (1950) and Biermann & Schlüter (1951). The problem is considerably more complicated than the (already difficult) one of the turbulent advection and diffusion of a scalar; not only are there geometric complications arising from the vector nature of the magnetic field but, more crucially, the magnetic field is dynamic, in that it can react back on the velocity field through the Lorentz force. In order to illustrate one way in which a dynamic magnetic field may affect the process of turbulent diffusion we shall consider below a simulation of the relatively simple case of two-dimensional turbulence.

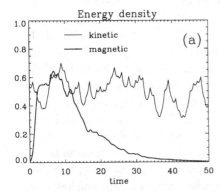
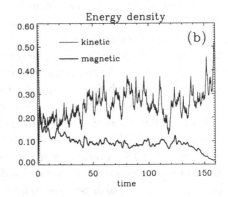

Figure 1. Time histories of the kinetic and magnetic energy densities in a two-dimensional turbulent flow. (a) Purely kinematic evolution: the scale for the magnetic energy is arbitrary. (b) Dynamic evolution.

2.1 Two-dimensional turbulence

It is well known that in two dimensions dynamo action is impossible and hence any magnetic field must eventually decay; nevertheless it is instructive to examine the timescales involved in the decay process. Cattaneo & Vainshtein (1991) simulated numerically the evolution of an initially large-scale magnetic field in randomly-forced two-dimensional turbulence, the magnetic field being in the plane of the fluid motions. The time histories of the kinetic and magnetic energy densities are shown in Figure 1. The evolution of Figure 1(a) is purely kinematic; that of Figure 1(b) is dynamic (i.e. the Lorentz force is included) but the initial large-scale field is weak, the magnetic energy being just 3% of the kinetic energy.

In both cases the magnetic energy increases rapidly as the field is amplified by turbulent motion without significant dissipative losses. In the kinematic evolution $\langle B^2 \rangle$ remains quasi-stationary for roughly twelve turnover times before decaying exponentially. The timescale for the decay is the classical turbulent diffusion time τ_e. Suppose that initially the field varies on a scale L and that the turbulence has a characteristic velocity U and characteristic lengthscale l. Then assuming an eddy diffusivity of the form $\eta_t \sim Ul$ gives

$$\tau_e \sim \frac{L^2}{\eta_t} \sim \frac{L^2}{Ul} = \left(\frac{L}{l}\right)^2 \tau_0, \tag{1}$$

where $\tau_0 = l/U$ is the eddy turnover time. This may also be expressed as $\tau_e = \tau_\eta / R_m$, where $R_m = Ul/\eta$ is the magnetic Reynolds number and $\tau_\eta = L^2/\eta$ is the Ohmic decay time.

The dynamic evolution, portrayed in Figure 1(b), is drastically different. Whereas in the kinematic evolution there is no impediment to the formation of the small scales necessary for rapid diffusion, in the dynamic evolution the

Lorentz force can become sufficiently strong as to inhibit severely the formation of small scales and hence to prolong the lifetime of the field. Furthermore this may occur even for a weak large-scale field, since the distortion of the field can lead to a strong (equipartition) field on the small scales. From Figure 1(b) we see that the magnetic energy remains quasi-stationary for about 130 turnover times before, eventually, the back reaction of the field on the flow ceases to be important and the field decays rapidly as in the kinematic case.

If we were not aware of anti-dynamo theorems forbidding the possibility of two-dimensional dynamos then the dynamic evolution of Figure 1(b), shown only up to $t \approx 130$, would look like a promising candidate for dynamo action. For three-dimensional flows, for which dynamos can of course exist, care must be taken to distinguish between genuine dynamo action on the one hand and slow decay of the field, due to inhibited diffusion, on the other.

3 CRITERIA FOR DYNAMO ACTION

In order to establish the existence of a *kinematic* dynamo it is necessary to show that the magnetic field exhibits exponential (possibly oscillatory) growth for times long compared to the turbulent diffusion time of the large-scale field. Hence assuming, very simply, a turbulent decay time of the form (1), we conclude that numerical integration should be performed for times $t \gg (L/l)^2 \tau_0$; only if exponential growth of the field persists for such times can we be sure that the flow is acting as a kinematic dynamo.

In dynamic simulations (Lorentz force included) the possibility of inhibited diffusion, as illustrated above for a two-dimensional flow, introduces a further complication. It is instructive to consider separately the cases of steady and cyclic behaviour.

3.1 Steady dynamos

Following the introduction of a weak, large-scale magnetic field into a turbulent flow we expect the magnetic energy to grow initially, due to stretching by the chaotic particle paths, before dynamical effects come into play and saturation occurs. It is then important to determine whether the field is subsequently maintained by dynamo action or whether it decays, possibly on a long timescale. Consequently, what we need to determine is the effective decay time of the magnetic field.

Let us define the timescale $T_0(t)$ as the ratio of the average magnetic energy to the average rate of dissipation:

$$T_0(t) = \frac{\langle B^2 \rangle}{\eta \langle |\nabla \times \mathbf{B}| \rangle^2} . \tag{2}$$

This would be the required timescale of decay if *all* of the magnetic energy were contained in the small scales of $O(l)$. However, a weak large-scale field, while making no significant contribution to (2), will decay on a much longer timescale. Arguing as in expression (1), we may therefore deduce that the decay time for a large-scale field with lengthscale L is

$$T_e = \frac{L^2}{l^2} T_0 \,. \tag{3}$$

Although T_0 is a function of t, an appropriate mean value may be obtained by averaging over a few turnover times once the field has evolved to a quasi-stationary state. The term $(L/l)^2$ in (3) is essentially a safety factor; although it may be reduced from a detailed knowledge of the field structure, in general, when such information is unavailable, it must be retained. Here l is the scale of the dominant helical eddies; in forced turbulence it is obviously well-determined, but for flows with more freedom, such as those driven by convection, the estimation of l may be more tricky.

In the discussion above we have implicitly assumed that the flow is turbulent everywhere; however there is also the possibility, often realised, that the flow becomes segregated into fairly distinct turbulent and quiescent regions. If this occurs due to the Lorentz force becoming sufficiently strong locally as to suppress the fluid motion then the timescale T_e, given by (3), is still appropriate. Alternatively, the stratification may be such as to inhibit motions in certain regions, as occurs in penetrative convection; T_e must then be evaluated separately for each layer, giving the appropriate timescale for the whole system as T_{max}, the maximum value of T_e over all the layers.

It should be stressed that the above timescales are the *minimum* times for which a magnetic field must be studied before any claim can be made for dynamo action. However, for both linear and nonlinear models, there is a transition from dynamo action to decay as R_m is decreased and so, near criticality, the time-scale becomes arbitrarily large. This situation may be recognised by performing the simulations for a range of R_m.

3.2 Cyclic dynamos

If the magnetic field is cyclic then the relevant timescale is no longer T_e, and may be shorter. Cyclic behaviour is typically the result of rotating convection in a sphere (Gilman 1983; Glatzmaier 1985) and so in this subsection it is natural to adopt the notation of spherical polar coordinates. As a first step we separate the cyclic component, with a mean period P_{cyc}, from any steady field (in the Sun, for example, this would involve the removal of any steady field within the radiative zone). After averaging the magnetic field azimuthally so as to produce a mean field $\bar{\mathbf{B}}(r, \theta, t)$ we average over a time $t_0 \gg P_{cyc}$ to obtain

$$\mathbf{B}_1(r, \theta) = \frac{1}{t_0} \int_t^{t+t_0} \bar{\mathbf{B}} dt\,, \qquad B_2(r, \theta) = \left(\frac{1}{t_0} \int_t^{t+t_0} |\bar{\mathbf{B}}|^2 dt \right)^{1/2}.$$

We now exclude from consideration those regions with a steady field, where $|\mathbf{B}_1| \approx B_2$, and focus on the region of cyclic activity, where $B_2 \gg |\mathbf{B}_1|$.

There are two possible mechanisms for producing cyclic fields. One is dynamo action, where both poloidal and toroidal fields reverse every half-cycle; the other is an oscillation, where a reversing toroidal field is generated by shearing a non-reversing poloidal component. In both cases we envisage that the differential rotation will produce azimuthal fields that are locally much stronger than poloidal fields. If there is a significant steady poloidal field B_{1p} in the cyclic region, so that

$$B_{1p} \gtrsim B_2/(\Delta \Omega P_{cyc}), \qquad (4)$$

where $\Delta \Omega$ is the variation in angular velocity in the region, then the cyclic behaviour is due to an oscillator. On the other hand, if

$$B_2 \gg \Delta \Omega P_{cyc} B_{1p}, \qquad (5)$$

then the poloidal field, averaged over many cycles, is small and the cyclic field is due to dynamo action. Thus, to distinguish between oscillators and dynamos, it is necessary to integrate for many cycles and then to decide which of criteria (4) or (5) is appropriate.

4 DISCUSSION

We have provided the timescales for which integration must be performed before any positive claim should be made for dynamo action. The criteria for kinematic and cyclic dynamos are fairly straightforward to implement; more care is needed in evaluating the timescale for steady, dynamic dynamos. The beauty of the criteria of section 3 is that they do not rely on a detailed knowledge of the diffusion processes; for steady dynamos the influence of the magnetic field on the dissipation will simply be reflected in the denominator of expression (2). However, that is not to say that there are not fundamental unanswered questions concerning the turbulent diffusion of magnetic fields.

The simulations of Cattaneo & Vainshtein (1991), discussed in section 2, illustrate clearly how the diffusion of a magnetic field can be suppressed by the action of the Lorentz force on small scales. The question remains though whether such behaviour is generic, even for two-dimensional flows. In the Cattaneo & Vainshtein calculation the turbulence is randomly forced over a small band of wavenumbers, and is uncorrelated with the magnetic field; it is certainly an interesting possibility that the nature of the diffusion will be different in convectively driven turbulence, for example, which may possess large-scale flows, and in which there is a greater correlation between the velocity and magnetic fields.

Tied in with the question of the inhibition of diffusion is that of the spectrum of the magnetic field; in particular, is equipartition reached on all scales (as proposed by Biermann & Schlüter 1951), or just on the small scales (as suggested by Batchelor 1950, by pursuing an analogy with the vorticity spectrum)? In the calculations of Cattaneo & Vainshtein, equipartition is reached on the small scales, with the large-scale field remaining weak. Vainshtein & Cattaneo (1992) have argued, on the basis of these calculations, that a dynamo will therefore only be able to produce a large-scale field of order $R_m^{-1/2}$ times equipartition value. Since, typically, R_m is immense in stellar convection zones this would imply that only extremely feeble large-scale fields could result from dynamo action. However, it is not clear that their rather provocative conclusion is correct. Their argument is based on the assumption that the magnetic field fills the turbulent region, whereas there is also the alternative possibility that the field will be intermittent. For steady convective flows it is well-known that flux expulsion leads to localised regions of strong field (see, for example, Galloway & Weiss 1981); whether this phenomenon persists, even in part, in fully turbulent convection has yet to be determined. Observations of the Solar surface certainly suggest that elements of strong field are brought together by a large-scale flow to form a large-scale field. Furthermore, in addition to small-scale turbulent motions, there is the possibility of a mean flow, such as differential rotation, producing a strong large-scale field by shearing a weak field. In conclusion, there are still many issues to resolve concerning the interaction of magnetic fields with turbulent velocity fields; at the moment it is dangerous to make sweeping statements about the diffusion of the field or the nature of the field produced by a dynamo.

It is a pleasure to thank Professor Nigel Weiss for many interesting discussions.

REFERENCES

Batchelor, G.K. 1950 On the spontaneous magnetic field in a conducting liquid in turbulent motion. *Proc R. Soc. Lond. A* **201**, 405–416.

Biermann, L. & Schlüter, A. 1951 Cosmic radiation and cosmic magnetic fields. II. Origin of cosmic magnetic fields. *Phys. Rev.* **82**, 863–868.

Cattaneo, F. & Vainshtein, S.I. 1991 Suppression of turbulent transport by a weak magnetic field. *Astrophys. J.* **376**, L21–L24.

Cattaneo, F., Hughes, D.W. & Weiss, N.O. 1991 What is a stellar dynamo? *Mon. Not. R. Astron. Soc.* **253**, 479–484.

Galloway, D.J. & Weiss, N.O. 1981 Convection and magnetic fields in stars. *Astrophys. J.* **243**, 945–953.

Gilman, P.A. 1983 Dynamically consistent nonlinear dynamos driven by convection in a rotating spherical shell. II. Dynamos with cycles and strong feedbacks. *Astrophys. J. Suppl.* **53**, 243–268.

Glatzmaier, G.A. 1985 Numerical simulations of stellar convective dynamos. II. Field propagation in the convection zone. *Astrophys. J.* **291**, 300–307.

Vainshtein, S.I. & Cattaneo, F. 1992 Nonlinear restrictions on dynamo action. *Astrophys. J.* **393**, 165–171.

Alpha-quenching in Cylindrical Magnetoconvection

C.A. JONES

Department of Mathematics
University of Exeter
Exeter, EX4 4QE UK

D.J. GALLOWAY

School of Mathematics and Statistics
University of Sydney
Sydney, NSW 2006 Australia

The α-effect is calculated for the case of axisymmetric magnetoconvection in a cylinder with an imposed swirl. A fixed total flux of magnetic field is imposed at the cylinder endwalls, and the azimuthal motion generates azimuthal field. The α-effect can be compared with asymptotic kinematic results of Childress (1979), and the nonlinear effects of the Lorentz force on the magnitude of α are computed.

---

1 INTRODUCTION

The problem of axisymmetric magnetoconvection in a cylinder with an azimuthal magnetic field has been studied recently (Jones & Galloway 1992), henceforth referred to as JG. In that paper, the conditions for the formation of twisted flux tubes were considered, along with some of the dynamical properties. The particular model considered is that of a cylinder with its axis parallel to gravity containing Boussinesq fluid and heated from below. A fixed amount of vertical magnetic flux is imposed through the cylinder. The azimuthal flux is generated here by imposing an azimuthal swirl velocity on the curved boundary. The meridional flow generated by the convection then distributes angular momentum throughout the cylinder. This then generates azimuthal magnetic field through the ω-effect.

The corresponding kinematic problem, in which the fluid velocity and the angular momentum are prescribed, and the meridional and azimuthal magnetic fields are calculated, was solved in the large R_M limit by Childress

161

M.R.E. Proctor, P.C. Matthews & A.M. Rucklidge (eds.)
Theory of Solar and Planetary Dynamos, 161–170
©1993 Cambridge University Press.

(1979), henceforth referred to as C79; see also Childress & Soward (1986). Childress pointed out that in this cylindrical geometry there is an α-effect, in the sense that if the magnetic field $\mathbf{B} = \mathbf{B}_0 + \mathbf{B}'$, where $\mathbf{B}_0 = B_0\hat{\mathbf{z}}$ is the initially imposed uniform magnetic field, the horizontal average of the e.m.f. in the z-direction, $\langle \mathbf{u} \times \mathbf{B}' \rangle_z$, is non-zero. We can therefore define the α_{33}-component of the α-tensor by the relation

$$\alpha_{33}B_0 = \langle \mathbf{u} \times \mathbf{B}' \rangle_z. \tag{1}$$

Since the dominant contribution to the e.m.f. comes from the boundary layers, for which Childress was able to find analytic solutions, the value of α_{33} could be explicitly evaluated. It was therefore possible to note the remarkable fact that α_{33} is finite in the limit $R_M \to \infty$, whereas in two dimensions the corresponding effect gives $\alpha \to 0$ as $R_M \to \infty$.

The purpose of this paper is to investigate how this α-effect varies in a fully convective model where Lorentz forces can be taken into account. It is often supposed that Lorentz forces act on convection in such a way as to reduce the helicity and hence suppress the α-effect. This process is known as α-quenching, and has traditionally been modelled (see e.g. Stix 1972; Jepps 1975; Weiss, Cattaneo & Jones 1984) by assuming a form

$$\alpha = \alpha_0/(1 + \lambda|\mathbf{B}|^2) \tag{2}$$

or similar. Recently Vainshtein & Cattaneo (1992) have suggested that at the very large magnetic Reynolds numbers in many astrophysical bodies, Lorentz forces may be so efficient at preventing small scale motion that the effective values of the turbulent diffusivity and the α-effect are so low as to cast doubt on the viability of the traditional mean-field dynamo model. It may be argued that our cylindrical convection model is not particularly representative of the situation on the Sun, because there the main function of the α-effect is to turn toroidal field into poloidal field, so that the 'given' toroidal field is perpendicular to gravity, whereas in our model the imposed mean field is parallel to gravity. Nevertheless, this cylindrical model is amenable to computational exploration, as it is axisymmetric, and also can be analysed to some extent by asymptotic methods.

2 EQUATIONS OF THE MODEL

The equations we need to describe the convection are the Navier–Stokes equation and the induction equation, together with various supplementary relations defining magnetic potential, streamfunction and current. The viscous, thermal, and magnetic diffusivities are ν, κ, and η respectively. We give the equations in dimensionless form, with d^2/κ as the unit of time, d, the height of the cylinder, as the unit of length, B_0 as the unit of magnetic field, and

ΔT as the unit of temperature, T. Since the fluid is incompressible, and $\nabla \cdot \mathbf{B} = 0$, we can let

$$\mathbf{u} = (-\frac{1}{r}\frac{\partial \psi}{\partial z}, \frac{h}{r}, \frac{1}{r}\frac{\partial \psi}{\partial r}) \quad \text{and} \quad \mathbf{B} = (-\frac{1}{r}\frac{\partial \chi}{\partial z}, rb, \frac{1}{r}\frac{\partial \chi}{\partial r})$$

where ψ is the streamfunction, h is the specific angular momentum, χ is the poloidal magnetic potential and b is a quantity conserved by axisymmetric motion if there is no diffusion. The current is $\mathbf{j} = \nabla \times \mathbf{B} = (j_r, rJ, j_z)$. The potential vorticity $\Omega = \omega_\theta / r$, where ω_θ is the θ-component of the vorticity, $\nabla \times \mathbf{u}$. Then

$$\Omega = -\nabla \cdot (\frac{1}{r^2}\nabla \psi), \tag{3}$$

and the poloidal part of the velocity distribution is given by

$$\frac{\partial \Omega}{\partial t} + \mathbf{u} \cdot \nabla \Omega = \frac{\partial}{\partial z}(\frac{h^2}{r^4}) + \frac{Q\sigma}{q}\left(\mathbf{B} \cdot \nabla J - \frac{\partial}{\partial z}b^2\right) - R\sigma\frac{\partial T}{\partial r} + \sigma\nabla \cdot (\frac{1}{r^2}\nabla(r^2\Omega)). \tag{4}$$

Here the Chandrasekhar number, Q, the Rayleigh number, R, the Prandtl number, σ, and the Roberts number q are given by

$$Q = \frac{B_0^2 d^2}{\mu \rho \eta \nu}, \qquad R = \frac{g\alpha\Delta T d^3}{\kappa \nu}, \qquad \sigma = \frac{\nu}{\kappa}, \qquad q = \frac{\kappa}{\eta},$$

μ being the permeability, ρ the density, α the coefficient of expansion and g the local gravity. Note that h is measured in units of κ. The angular momentum is determined by

$$\frac{\partial h}{\partial t} + \mathbf{u} \cdot \nabla h = \frac{Q\sigma}{q}\mathbf{B} \cdot \nabla(r^2 b) + \sigma\nabla \cdot (r^2\nabla(\frac{h}{r^2})). \tag{5}$$

The toroidal field is governed by the θ component of the induction equation,

$$\frac{\partial b}{\partial t} + \mathbf{u} \cdot \nabla b = \mathbf{B} \cdot \nabla(\frac{h}{r^2}) + \frac{1}{q}\nabla \cdot (\frac{1}{r^2}\nabla(r^2 b)), \tag{6}$$

while the poloidal part is governed by

$$\frac{\partial \chi}{\partial t} + \mathbf{u} \cdot \nabla \chi = \frac{1}{q}\nabla \cdot (r^2\nabla(\frac{\chi}{r^2})). \tag{7}$$

The final equation we need is the temperature equation,

$$\frac{\partial T}{\partial t} + \mathbf{u} \cdot \nabla T = \nabla^2 T. \tag{8}$$

The boundary conditions are

$$\psi = \Omega = \frac{\partial \chi}{\partial z} = h = \frac{\partial b}{\partial z} = 0 \quad \text{on} \quad z = 0, 1 \tag{9}$$

and

$$\psi = \Omega = b = 0, \quad \chi = 0.5A^2 \quad \text{and} \quad h = h_0 \quad \text{on} \quad r = A. \tag{10}$$

Note that $\chi = 0.5r^2$ gives a uniform vertical field of strength one in the static state; also, in this paper all results presented are for the numerical value $h_0 = 10$.

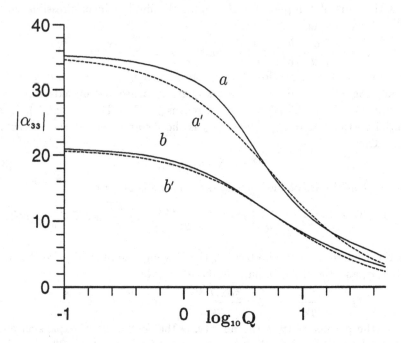

Figure 1. $|\alpha_{33}|$ as a function of Chandrasekhar number Q with $Ra = 20{,}000, q = 5, h_0 = 10, A = 1$. Case (a) is $\sigma = 0.2$, case (b) is $\sigma = 1$. The corresponding dashed curves are generated by the formula given in the text.

3 NUMERICAL RESULTS

The equations and boundary conditions (3)–(10) were integrated forward in time using a finite difference scheme described in JG. A small initial seed field was imposed, and the equations integrated until a steady state was achieved. The problem has a symmetry between solutions with fluid rising or sinking at the axis; the seed field was chosen so that the solution with fluid rising at the axis was selected. As noted in JG, for certain ranges of h_0, the imposed azimuthal swirl, the final state is oscillatory. We have not investigated the α-effect in these circumstances. When a steady state is found, the numerical evaluation of α_{33} from equation (1) is straightforward.

There is a large parameter space associated with this problem, which we have not explored in detail. In Figure 1, we show $|\alpha_{33}|$ as a function of Q. In case (a) the Prandtl number is 0.2 and $q = 5$, so the magnetic Reynolds number R_M and the fluid Reynolds number Re are equal. If R_M is defined using the maximum vertical velocity and the cylinder height, $R_M = 482.5$ at $Q = 0.1$ and $R_M = 374.5$ at $Q = 50.0$. In case (b) $\sigma = 1$ and $q = 5$ so $R_M = 5Re$ and $R_M = 553.7$ at $Q = 0.1$ and $R_M = 402.5$ at $Q = 50$. We see that $|\alpha_{33}|$ is quenched as Q increases; for comparison we have plotted dashed

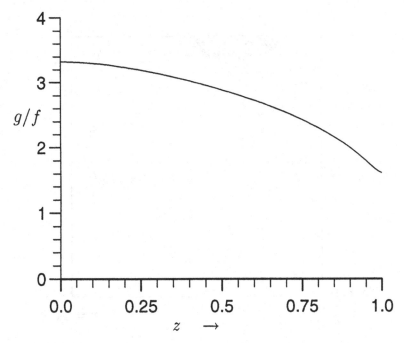

Figure 2. $g(z)/f(z)$ as a function of z.

curves of the formula $|\alpha_{33}| = C_1/(1+C_2Q)$, which is often used in α-quenching models. Here C_1 has been chosen so that $|\alpha_{33}|$ is exact as $Q \to 0$, and C_2 is chosen so that $|\alpha_{33}|$ is exact when $|\alpha_{33}| = 0.5C_1$. This simple formula gives a reasonable representation of α-quenching in this model. It should be noted that this does not necessarily imply that other models will behave similarly.

We have also compared our results to the asymptotic estimates at large R_M given in C79. The main difficulty in making the comparison is that C79 assumes that the angular momentum h is a function of ψ near the axis. In our calculations,

$$\psi \sim 0.5r^2 f(z) \quad \text{and} \quad h \sim 0.5r^2 g(z)$$

near the axis. Although both f and g satisfy $f = g = 0$ on $z = 0, 1$, the two functions are not multiples of each other as can be seen from Figure 2 where $g(z)/f(z)$ is plotted. To make the comparison with C79 we therefore took the integral of g/f and used this as the estimate of the quantity $dh/d\psi$ at $r = 0$ required in the C79 formula (2.4). This procedure is not exact in any asymptotic limit, but it nevertheless gives reasonable agreement with the C79 picture. In the case (a) in Figure 1 at $Q = 0.1$ we get $\alpha_{33} = -32.6$ whereas the C79 formula gives -35.1 for the case where fluid rises at the axis. In case (b) for the same value of Q we get $\alpha_{33} = -18.8$ while the C79 formula gives -20.1.

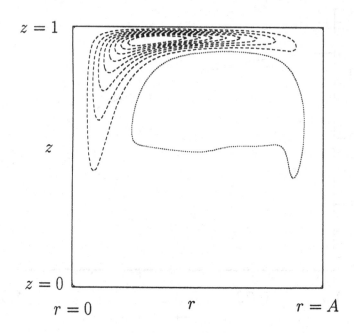

Figure 3. Contours of $(\mathbf{u} \times \mathbf{B'})_z$ on a vertical section of the cylinder. Here and in Figure 4 continuous, dotted and dashed contours are greater than, equal to, or less than 0 respectively.

In Figure 3, the distribution of $(\mathbf{u} \times \mathbf{B'})_z$, not just its integral, is shown. The dominant contribution comes from the top boundary, confirming the C79 asymptotic results, where it was shown that although the formula depends on conditions at the axis, the dominant contribution comes from the horizontal boundary layers. Finally, in Figure 4 we show various fields associated with the case $Q = 0.1$ with the Figure 1 case (a) parameter values. This shows the structure of the full solution at moderately large R_M.

4 NONLINEAR LIMITING OF α_{33} AT LARGE R_M AND Re

It was shown by Galloway *et al.* (1978) that when there is no azimuthal magnetic field or velocity, the Lorentz force starts to limit the kinematic amplification of field when $Q \sim 1/\ln R_M$ (see their equation 3.18). Kinematic concentration gives a maximum vertical field strength $B_{max} \sim R_M$ and a central rope of thickness $\sim R_M^{-1/2}$. Clearly, the value of α_{33} will be affected for $Q \sim 1/\ln R_M$ or greater, as a change in the central meridional flux rope will affect α_{33}. However, it is not clear *a priori* whether smaller values of Q can reduce α_{33} through the new nonlinearities involving the azimuthal flow and field. Throughout, we make use of the kinematic asymptotic estimates

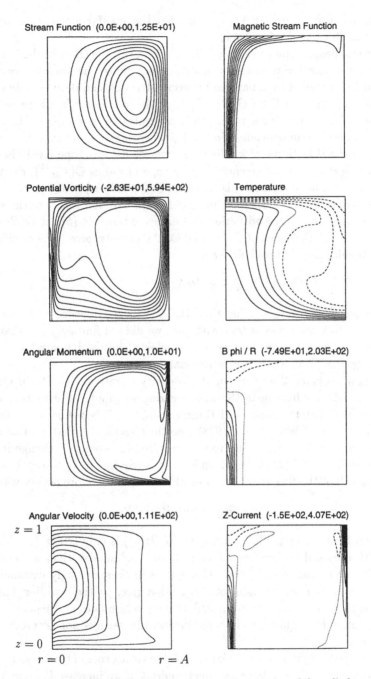

Figure 4. Contours of various fields are shown on a vertical section of the cylinder, so that the left edge is the axis of the cylinder. Contours are equally spaced and the maximum and minimum values of all quantities are shown.

of C79. We also assume that $\nu \geq \eta$, so that the angular momentum rope is at least as thick as the flux rope.

First we consider the term $Q\sigma q^{-1}(\mathbf{B} \cdot \nabla)r^2 b$ in (5), the effect of the Lorentz force on the angular momentum distribution. Since the vortex core near the axis is controlled by a balance between advection and diffusion, its characteristic thickness will be $O(Re^{-1/2})$, and the value of h/r^2 there will be $O(h_0 Re)$. In the kinematic regime, b is concentrated in a rope of thickness $O(R_M^{-1/2})$ and has magnitude $O(\sigma^{-1}h_0 R_M)$, bearing in mind that h_0 is non-dimensionalised in terms of κ. Since the meridional flux rope itself also has strength $O(R_M)$ and is concentrated in a rope of radius $O(R_M^{-1/2})$, the term $Q\sigma q^{-1}\mathbf{B} \cdot \nabla r^2 b$ in equation (5) is therefore $O(Qq^{-1}R_M h_0)$ near the axis; note that the contributions from $B_r \partial/\partial r$ and $B_z \partial/\partial z$ are necessarily of the same order of magnitude. The advection and diffusion terms in (5) are $O(Re\,\sigma h_0)$ near the axis, so the value of Q required for the Lorentz force to be significant in the angular momentum balance is

$$Q \sim q\sigma Re R_M^{-1} = 1.$$

This is formally a larger value than the $O(1/\ln R_M)$ needed to affect the meridional flux rope. Consistent with this, we did not find evidence that the angular momentum near the flux rope is affected by the Lorentz force until Q is large enough to affect the meridional flux rope.

We now consider the terms in the vorticity equation. The term $Q\sigma \mathbf{B} \cdot \nabla J/q$ is $O(Q\sigma q^{-1}R_M^3)$ near the axis, bearing in mind the definition of J and the fact that the rope is of thickness $R_M^{-1/2}$. This seems large, but it was noted by Galloway $et\ al.$ (1978) (see their section 3) that a value of Ω of $O(R_M^2 q^{-1}/\ln R_M)$ in the flux rope is required to produce a change in the vertical velocity of $O(R_M q^{-1})$, which is what is required to significantly affect the strength of the flux rope. Dynamical limitation therefore occurs when

$$Q\sigma q^{-1}R_M^3 \sim \sigma q^{-1}R_M^3/\ln R_M,$$

i.e. when $Q \sim 1/\ln R_M$. The term $Q\sigma q^{-1}\partial/\partial z(b^2)$ is only $O(Q\sigma^{-1}q^{-1}R_M^2 h_0^2)$ near the axis, and is therefore a factor of $R_M h_0^{-2}\sigma^2$ smaller than the Lorentz force from the meridional field. This term will therefore only dynamically limit the flow at a smaller value of Q than that given by the meridional flux if $h_0^2\sigma^{-2}$ is large, but then the term $\partial/\partial z(h^2/r^4)$, which does not depend on Q, will normally be important. In consequence, this term is not generally very significant.

The term $\partial/\partial z(h^2/r^4)$ is associated with the vortex core of thickness $Re^{-1/2}$. This term certainly can have an effect: indeed, if we increase h_0 from 10 to 30 with the Figure 4 parameter values, the solution is oscillatory, not steady. Using an analogous argument to that given in Galloway $et\ al.$ (1978), the term $\partial/\partial z(h^2/r^4)$ will significantly affect the vertical velocity when

$$h_0^2 \sim \sigma q^{-1} R_M / \ln Re.$$

In the interior, the term is dynamically important when $h_0^2 \sim Re^2$ which will normally be at a larger value of h_0. So the dynamical influence of this term will first be felt at the axis if h_0 is gradually increased. Note that h_0 is measured in diffusive units; the swirl velocities can therefore be small compared to the convective velocities and yet still have a significant effect.

5 CONCLUSIONS
The α-effect mechanism can be investigated in the context of cylindrical magnetoconvection, and the effect of the Lorentz force can be evaluated. We find that at moderate or large magnetic and fluid Reynolds numbers, the most important nonlinearity is that due to the Lorentz force produced by the meridional field on the meridional flow. This suppresses α in a straightforward manner.

Vainshtein & Cattaneo (1992) relate the field-strength at which the α-effect is quenched to the equipartition field strength, B_{eq}, defined as

$$B_{eq} = (\mu\rho)^{1/2}\langle u^2\rangle^{1/2}.$$

As noted by Brandenburg *et al.* (1993), equation (2) with $\lambda = 1/B_{eq}^2$ has been used by many authors. Vainshtein & Cattaneo claim that at large R_M the fluctuating field of magnitude B' which is generated by the concentration of flux into ropes will be much larger than the mean field B_0. Together with the expectation that α-quenching will occur when $B' \sim B_{eq}$, this implies a B_0 much less than B_{eq} (by a factor with R_m to some power in the denominator).

In our model, if we identify B' with the peak value of the field in the flux rope then $B' \sim B_0 R_M$ as we leave the kinematic regime and α-quenching starts to operate. Since $Q \sim O(1/\ln R_M)$ is the condition for this to happen, $B_0^2 d^2/\mu\rho\eta\nu \sim 1/\ln R_M$, which implies

$$\frac{B_0^2}{2\mu} \sim \frac{\rho u^2}{2\ln R_M Re R_M}.$$

So the ratio

$$\frac{B_0}{B_{eq}} \sim (\ln R_M)^{-1/2} Re^{-1/2} R_M^{-1/2}$$

at which α-quenching occurs is much less than unity. Our model therefore predicts neither the 'traditional' value B_{eq} nor the Vainshtein & Cattaneo value of B_0 for the onset of α-quenching. Our results are much closer to (and less than!) the Vainshtein–Cattaneo value (see e.g. Figure 9 of Brandenburg *et al.* 1993).

The magnetic energy in our model is dominated by the contribution from the rope, and is of order $0.5B_0^2 R_M V$, where V is the cylinder volume; the kinetic energy is of order $0.5\rho u^2 V$. (Because of the logarithmic singularity, the velocity remains unaffected far from the axis as the Lorentz force becomes significant in the flux rope.) The ratio of total magnetic to kinetic energy is therefore $\sim Q Re^{-1} \sim (\ln R_M)^{-1} Re^{-1} \ll 1$, so that the onset of α-quenching is not associated with equipartition of total energy.

Although our model has the advantage that the α-effect can be computed relatively easily for a wide range of parameters, it is not clear that our α-quenching results can be directly applied to the Solar dynamo, and our geometry means we can only calculate the component α_{33}. Brandenburg *et al.* (1993) advance a number of reasons why the ratio B'/B_0 should be less than the very large value of R_M in the Sun. Our work adds to the evidence that α-quenching is, unfortunately, a very model-dependent phenomenon.

REFERENCES

Brandenburg, A., Krause, F., Nordlund, A., Ruzmaikin, A., Stein, R.F. & Tuominen, I. 1993 On the magnetic fluctuations produced by a large scale magnetic field. *Astrophys. J.* (submitted).

Childress, S. 1979 Alpha-effect in flux ropes and sheets. *Phys. Earth Plan. Int.* **20**, 172–180.

Childress, S & Soward, A.M. 1986 Analytic theory of dynamos. *Adv. Space Res.* **6**, 7–18.

Galloway, D.J., Proctor, M.R.E. & Weiss, N.O. 1978 Magnetic flux ropes and convection. *J. Fluid Mech.* **87**, 243–261.

Jepps, S.A. 1975 Numerical models of hydromagnetic dynamos. *J. Fluid Mech.* **67**, 625–646.

Jones, C.A. & Galloway, D.J. 1992 Axisymmetric magnetoconvection in a twisted field. Preprint.

Stix, M. 1972 Non-linear dynamo waves. *Astron. Astrophys.* **20**, 9–12.

Vainshtein, S.I. & Cattaneo, F. 1992 Nonlinear restrictions on dynamo action. *Astrophys. J.* **393**, 165–171.

Weiss, N.O., Cattaneo, F. & Jones, C.A. 1984 Periodic and aperiodic dynamo waves. *Geophys. Astrophys. Fluid Dynam.* **30**, 305–341.

On the Stretching of Line Elements in Fluids: an Approach from Differential Geometry

T. KAMBE & Y. HATTORI

Dept. of Physics, University of Tokyo
Hongo, Bunkyo-ku, Tokyo 113, Japan

V. ZEITLIN

Observatoire de Nice
BP 229, 06304 Nice Cedex 4, France

The rate of stretching of line elements is studied for an incompressible ideal fluid, based on the frame of differential geometry of a group of diffeomorphisms. Riemannian curvature is closely connected with the time evolution of distance between two mappings of fluid particles. Exponential stretching of line elements in time is considered in the context of negative curvature in turbulent flows. The corresponding two-dimensional MHD problem of a perfectly conducting fluid with the current perpendicular to the plane of motion is also investigated. Simultaneous concentration of vortex and magnetic tubes is presented first.

1 INTRODUCTION

Stretching of line elements in fluids with or without conductivity is studied from various points of view. Firstly, simultaneous concentration of vortex and magnetic tubes is considered in section 2 by presenting an exact solution of the axisymmetric MHD equation for a viscous, incompressible, conducting fluid. This solution (Kambe 1985) tends to a stationary state that results from complete balance of convection, diffusion and stretching. Sections 3 and 4 are concerned with mathematical formulation based on Riemannian differential geometry, and global (mean) stretching of line elements is considered.

The general form of the Riemannian curvature tensor for any Lie group was derived by Arnold (1966), where explicit formulae for T² (two-torus) were given. Explicit expressions for diffeomorphism curvatures on Tⁿ (and even on any locally flat manifold) were described by Lukatskii (1981). Recently, Nakamura *et al.* (1992) considered the curvature form on T³ corresponding to three-dimensional motion of an ideal fluid with periodic boundary conditions

M.R.E. Proctor, P.C. Matthews & A.M. Rucklidge (eds.)
Theory of Solar and Planetary Dynamos, 171–179
©1993 Cambridge University Press.

in a cubic space. An immediate consequence is the property that the section curvature in the space of ABC diffeomorphisms is a *negative* constant for all the sections (Kambe *et al.* 1992). In section 4, some characteristic properties in the two-dimensional MHD problem (Zeitlin 1992) are considered beside the problem of line-element stretching. It is suggested that the geodesics of the problem are related to charge transport in addition to the diffeomorphism.

2 CONCENTRATION OF VORTEX AND MAGNETIC TUBES

Motion of a viscous, incompressible, conducting fluid is governed by the vorticity equation

$$\partial_t \omega + (\mathbf{v} \cdot \nabla)\omega = (\omega \cdot \nabla)\mathbf{v} + \nu\Delta\omega + \eta\nabla \times [(\mathbf{B} \cdot \nabla)\mathbf{B}], \qquad (1)$$

and the induction equation

$$\partial_t \mathbf{B} + (\mathbf{v} \cdot \nabla)\mathbf{B} = (\mathbf{B} \cdot \nabla)\mathbf{v} + \lambda\Delta\mathbf{B}, \qquad (2)$$

which are supplemented by the solenoidal conditions,

$$\nabla \cdot \mathbf{v} = 0, \qquad \nabla \cdot \mathbf{B} = 0, \qquad (3)$$

where $\mathbf{v}$ is the velocity, $\omega = \nabla \times \mathbf{v}$ is the vorticity, $\mathbf{B}$ the magnetic induction, and $\eta = 1/4\pi\rho\mu$ with ρ being the uniform density. It is assumed that the kinematic viscosity ν, the magnetic diffusivity $\lambda = 1/4\pi\mu\sigma$, and the magnetic permeability are all constant, with σ being the electrical conductivity.

Suppose that the total velocity field is composed of an axisymmetric irrotational part $\mathbf{v}_i$ and a rotational part $\mathbf{v}_r$ with a unidirectional vorticity ω: $\mathbf{v} = \mathbf{v}_i + \mathbf{v}_r$. In cylindrical coordinates (r, θ, z), the irrotational part is represented as $\mathbf{v}_i = (-\alpha r, 0, 2\alpha z)$, where α is a positive constant, while the rotational part is simply given by $\mathbf{v}_r = (0, u_\theta(r, t), 0)$. Hence the vorticity ω has only the z-component $\omega(r, t)$. We seek a solution in which the magnetic field $\mathbf{B}$ is also aligned with the vorticity ω. Thus $\omega = (0, 0, \omega(r, t))$ and $\mathbf{B} = (0, 0, b(r, t))$, where $\omega = r^{-1}\partial(ru_\theta)/\partial r$. From (1) and (2), one obtains the equations for $\omega(r, t)$ and $b(r, t)$:

$$\partial_t \omega = \alpha r\partial_r \omega + 2\alpha\omega + \nu\Delta\omega \quad \text{and} \quad \partial_t b = \alpha r\partial_r b + 2\alpha b + \lambda\Delta b, \qquad (4)$$

which are decoupled and almost identical in form, the only difference being the diffusivity. The three terms on the right hand side of both equations represent the rate of change of ω and b due to the convective transport, stretching and diffusion, respectively.

Introducing a scaling factor $A(t) \equiv e^{\alpha t}$ ($\alpha > 0$) and normalizing r, ω and b by A^{-1}, A^2 and A^2, respectively, one obtains the axisymmetric, z-independent, diffusion equations for the normalized variables and the transformed time τ

defined by $d\tau = A^2(t)dt$. The axisymmetric diffusion equation is immediately solved for arbitrary initial distribution, and the solution is represented in a convolution integral in terms of the Green's function and the initial profile.

It is remarkable to find (Kambe 1985) that a stationary state of ω and b is approached simultaneously from arbitrary initial states as a result of the complete balance of inward convection, outward diffusion and axial stretching of both ω and $\mathbf{B}$. The final stationary state is represented by

$$\omega(r, \infty) = \frac{\Gamma}{\pi l^2} e^{-r^2/l^2}, \qquad b(r, \infty) = \frac{\Phi}{\pi L^2} e^{-r^2/L^2}, \qquad (5)$$

where

$$\Gamma = \int_0^\infty \omega(r, 0) 2\pi r\, dr \quad \text{(total vorticity)}, \qquad l = (2\nu/\alpha)^{1/2},$$

$$\Phi = \int_0^\infty b(r, 0) 2\pi r\, dr \quad \text{(total flux)}, \qquad L = (2\lambda/\alpha)^{1/2}.$$

This is a state of simultaneous concentration of vortex and magnetic tubes at the same position.

3 RIEMANNIAN CURVATURE / RATE OF STRETCHING

3.1 Geodesic trajectories and Riemannian curvature
Motion of a fluid particle starting from any point $x \in M$ (domain of the flow) with a given initial condition is considered to be a *mapping* $g_t(x)$ with a continuous time parameter t, carrying the fluid particles from $x \in M$ at $t = 0$ to $y = g_t(x)$ at time t. The mapping $g_t : M \to M$ is an auto-diffeomorphism from the initial particle configuration to that at t. This diffeomorphism is volume preserving if the velocity field is divergence-free. The volume preserving diffeomorphisms from M to itself form an infinite dimensional group, denoted as $D_v(M)$. The flow g_t can be shown to be a geodesic curve on the group D_v (Arnold 1966, 1978; Nakamura *et al.* 1992; Kambe *et al.* 1992), and is generated by the velocity field $v(x, t)$ in the way $dg_t/dt = v(g_t(x), t)$ with $g_0(x) = x$. The velocity v is governed by Euler's equation of motion,

$$\partial_t v + (v \cdot \nabla)v = -\nabla p \qquad (6)$$

with the initial condition $v(x, 0) = v_0(x)$ and the prescribed boundary (periodicity) condition. The flow g_t is also written as $g_t(x) = g(x, t; v_0)$.

Now we study mean separation of two particle configurations which start from a common initial position but with different velocity fields v_a and v_b. The L^2-distance of the two mappings of the particle configuration is defined as

$$d(v_a, v_b; t) = \left(\int_{T^2} |g(x, t; v_a) - g(x, t; v_b)|^2 \, d^2x \right)^{\frac{1}{2}}. \qquad (7)$$

Obviously we have $d(v_a, v_b; 0) = 0$ by the assumed initial state. We try to find how the mean distance d grows in the initial stage as the time t increases from 0 and ask what is the implication.

Suppose that the two initial fields have a slight difference, i.e. $v_a = v_o + \epsilon v_p$ and $v_b = v_o - \epsilon v_p$, for a small parameter ϵ. Now we expand the integral expression of the distance d into the Taylor series with respect to t:

$$d(v_a, v_b; t) = 2\epsilon|v_p|(t - \frac{1}{6}|v_o|^2 \sin^2 \theta K(v_o, v_p)t^3) + O(\epsilon^2 t, \epsilon t^5) \qquad (8)$$

(Hattori 1990), where $\theta = \cos^{-1}[(v_o, v_p)/|v_o||v_p|]$ and the factor $K(v_o, v_p)$ is the Riemannian curvature of the section spanned by the vectors (velocity fields) v_o and v_p in the tangent space $T_e D_v$ (Arnold 1978; Kambe *et al.* 1992). When the curvature K is negative, the mean particle separation d is enhanced with respect to the linear growth of the first term.

To illustrate this, we consider a simple case of two-dimensional (2-d) steady flow fields v_a and v_b given by

$$v_o = 2\Re[e_k + e_l], \qquad v_p = 2\Re[i(k \cdot p)e_k + i(l \cdot p)e_l],$$

where k, l and p are 2-d vectors given by $k = (k_1, k_2) = (a, b)$, $l = (a, -b)$, $p = (1, 0)$, and $e_k = (ik_2, -ik_1)e^{ik \cdot x}$ represents a 2-d solenoidal velocity field given by the stream function $e^{ik \cdot x}$, and similarly for e_l. From the definition of curvature we find

$$K(v_o, v_p) = \frac{2}{(2\pi)^2} \frac{a^2 b^2(a^2 - 2b^2)}{(a^2 + b^2)^2} \qquad (9)$$

(Hattori 1990). Hence K is negative or positive according as a^2 is less or greater than $2b^2$, respectively.

Numerical computations were carried out to illustrate the above properties for the same parameters. The cubic growth of d in addition to the linear one, represented by (8), is confirmed in Figure 1. One case of positive curvature $(a, b) = (4, 1)$ is shown in Figure 2(a), while a negative case of $(a, b) = (1, 4)$ is shown in Figure 2(b).

3.2 Curvature in the space of ABC diffeomorphisms (3-d)

The mathematical theory in the previous section can be applied to the ABC flow, which is represented as a linear combination of three Beltrami waves,

$$U_{ABC} = A[(i, 1, 0) e^{iz} + (-i, 1, 0) e^{-iz}] + B[(0, i, 1) e^{ix} + (0, -i, 1) e^{-ix}]$$
$$+ C[(1, 0, i) e^{iy} + (1, 0, -i) e^{-iy}], \qquad (10)$$

where A, B and C are real. We consider another ABC flow $U_{A'B'C'}$ with $(A', B', C') \neq (A, B, C)$. It can be shown (Kambe *et al.* 1992; Nakamura *et al.* 1992) that the curvature $K(U_{ABC}, U_{A'B'C'})$ is equal to a negative constant $-(4\pi)^{-3}$ independent of (A, B, C) and (A', B', C'). Thus any two-dimensional section of the ABC diffeomorphisms is analogous to that of a pseudo-sphere.

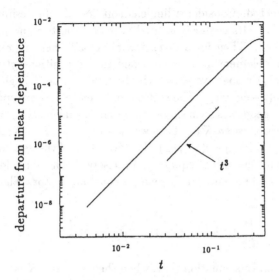

Figure 1. Growth of the distance minus the linear term: $d - 2\epsilon|v_p|t$.

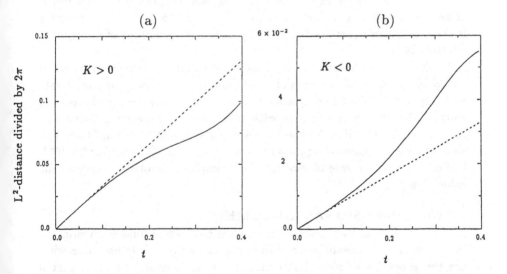

Figure 2. Time evolution of $d(t)$ in the two cases: (a) positive curvature with $(a, b) = (4, 1)$; (b) negative curvature with $(a, b) = (1, 4)$.

3.3 Exponential rate of stretching in turbulence (2-d)

Next, we consider mean evolution of all line elements in turbulent flow, that is, the problem of evolution of average magnitude of the line elements which are initially of the same size: infinitesimally small, uniformly distributed and isotropically directed. Consider a line element $2\delta a$ of infinitesimal length at each point x on T^2. Its center is located at x and the two end points are at $x + \delta a$ and $x - \delta a$. Then for a given initial velocity field $v_o(x)$, both ends of the initial line element at x are mapped to the positions $g(x + \delta a, t; v_o)$ and $g(x - \delta a, t; v_o)$ respectively, and the line element at t is given by $\delta l = g(x+\delta a, t; v_o) - g(x - \delta a, t; v_o)$. The motions of its endpoints are approximately described by two geodesics with slightly different initial fields $v_+ = v_o + v_p$ and $v_- = v_o - v_p$ where $v_p = \delta a \cdot \nabla v_o$. Thus we have $\delta l(t) = 2\delta a + \delta X(t) + O(|\delta a|^2)$, where $\delta X(t) = g(x, t; v_+) - g(x, t; v_-)$. The evolution of the geodesic variation δX is described by the Jacobi equation. Precisely the Jacobi field is given by $W = \partial \delta X / \partial \epsilon |_{\epsilon=0}$ where $\delta a = \epsilon a$, a being a constant vector. The norm $|W|^2$ is governed by

$$\frac{d^2}{dt^2}|W|^2 = -2K(W, \dot{g}_t)|W|^2 + 2|\frac{d}{dt}W|^2. \tag{11}$$

It is expected from this equation that, when the curvature K is negative, $|W|$ grows exponentially, at least initially. In fact, if K is a negative constant, we immediately obtain the exponential growth $|W(t)| \sim \exp\sqrt{-K}t$ asymptotically. In turbulence we expect the form, $\log[|\delta X(t)|/|\delta X(0)|] = \gamma t + \text{const}$, at an initial stage. A theoretical estimate gives the value $\gamma = \frac{1}{4}C\bar{\omega}$ where $\bar{\omega}$ is the root mean square of the vorticity and C is a constant of order unity (Hattori 1990).

A computer simulation solving the 2-d Euler equation numerically is carried out to find evidence of exponential growth and try to estimate the growth rate γ. Integration of the Euler equation is made by means of the pseudo-spectral method with 128^2 Fourier modes, with an initial energy spectrum distributed only in a circular shell in the Fourier space. Evolution of 32×32 or 64×64 line elements are traced, starting from the same initial size of 2×10^{-3} or 2×10^{-4}, to find the average rate of growth. The computer simulation suggests the value $C \approx 1.5$.

4 TWO-DIMENSIONAL IDEAL MHD

Mean stretching of line elements in an ideal conducting fluid is investigated in the context of the diffeomorphism of fluid motion. It is not clear what are the geodesics of ideal MHD motion. It is interesting to find that a two-dimensional MHD problem is analogous to a 3-d hydrodynamics independent of a third coordinate. We consider a two-dimensional problem in which the velocity $\mathbf{v}$ and the magnetic field $\mathbf{B}$ are represented in the form, $\mathbf{v} = (\partial_y \psi, -\partial_x \psi, 0)$, $\mathbf{B} = (\partial_y a, -\partial_x a, 0)$, where $\psi(x, y, t)$ and $a(x, y, t)$ are the stream function and magnetic potential respectively. Thus the solenoidal conditions (3) are satisfied automatically. The MHD equations for an ideal conducting fluid are given by (1) and (2) with setting $\nu = 0$ and $\lambda = 0$, where the vorticity ω has only the z component $\omega = -\Delta\psi$. The z-component of the vorticity equation (1) reduces to (with an appropriate normalization)

$$\partial_t\omega + (\mathbf{v}\cdot\nabla)\omega = \mathbf{B}\cdot\nabla j \tag{12}$$

where the electric current $\mathbf{j}$ is given by the Ampere's law, $\mathbf{j} = \nabla\times\mathbf{B} = (0,0,j)$ with $j = -\Delta a$, having only one non-vanishing component. The induction equation (2) can be integrated once, yielding the equation for the magnetic potential,

$$\partial_t a + (\mathbf{v}\cdot\nabla)a = 0. \tag{13}$$

This system of 2-d MHD equations (12) and (13) has three integral invariants which are quadratic or bilinear with respect to ψ and a:

$$E = \frac{1}{2}\int(|\mathbf{v}|^2 + |\mathbf{B}|^2)\,d^2x, \quad H = \int\mathbf{v}\cdot\mathbf{B}\,d^2x, \quad A = \int a^2\,d^2x. \tag{14}$$

Equation (13) describes the property that the magnetic potential $a(x,y,t)$ is merely transported by the fluid velocity $\mathbf{v}$. Correspondingly, the integral of any function of a, i.e. $\int f(a)\,d^2x$, is conserved. On the other hand, in (12) the vorticity ω changes during the convective transport due to non-uniformity of the current j in the direction of $\mathbf{B}$. Nevertheless, an amount of vorticity contained in between any pair of equipotential lines is conserved, that is $\int\omega g(a)\,d^2x$ is conserved, where g is another arbitrary function. These two families of integrals are the *Casimir* invariants for 2-d MHD.

A computer simulation has been carried out with 85^2 active Fourier modes, and the growth of average length $\langle\delta l\rangle$ of 40×40 line elements in a turbulent field is examined numerically. Figure 3 shows evidence of exponential growth of $\langle\delta l\rangle$, after the initial transient interval, and the influence of the magnetic field.

The equations (12) and (13) are rewritten in the form

$$\dot\omega + J(\omega,\psi) + J(a,j) = 0, \quad \dot a + J(a,\psi) = 0, \tag{15}$$

where J denotes the Jacobian. The infinitesimal group transformation (the Lie algebra) which preserves the above-mentioned Casimirs is given by the following transformation of the fields ω and a (Zeitlin 1992),

$$\delta\omega = J(\chi,\omega) + J(\sigma,a), \quad \delta a = J(\chi,a). \tag{16}$$

Here χ and σ are two independent functions, namely two infinitesimal parameters of the transformation. The energy (kinetic + magnetic) is chosen as a metric on the algebra:

$$E = -\frac{1}{2}\int_D d^2x(\omega\Delta^{-1}\omega + a\Delta a). \tag{17}$$

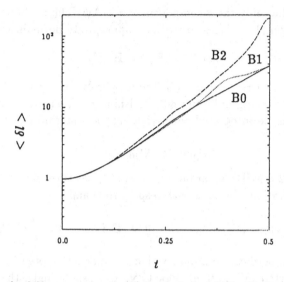

Figure 3. Growth of the average length $\langle \delta l \rangle$ in a turbulent field with three different initial conditions. B0: $E = E_K$, $A = 0$; B1: $E_M/E_K = 0.2$, $|H|/E = 0.8$; B2: $E_M/E_K = 5.0$, $|H|/E = 0.8$, where E_K and E_M are the kinetic and magnetic energy.

Here Δ^{-1} denotes a Green's function for the Laplacian and periodic boundary conditions are assumed for simplicity with zero mean-values for ω and a. We see that the stream function $\psi = -\Delta^{-1}\omega$ and the current $j = -\Delta a$ transform according to the following formulae:

$$\delta\psi = J(\chi,\psi), \qquad \delta j = J(\chi,j) + J(\sigma,\psi). \tag{18}$$

Now if we want to consider (15) as a geodesic equation on the group manifold, the question arises what is the meaning of the coordinate related to the σ-transformation in (16) and (18). As to the χ-transformation, its meaning is clear: it is just an area-preserving change of variables and the corresponding group coordinate is a Lagrangian coordinate of the fluid particle on the plane. To answer the posed question, we shall use the following realization of the group. If one takes a Lie algebra of 3-d divergenceless vector fields (a Lie algebra of 3-d volume-preserving diffeomorphisms) which is given by their commutator, $[\mathbf{U}, \mathbf{V}] = \mathbf{U} \cdot \nabla\mathbf{V} - \mathbf{V} \cdot \nabla\mathbf{U}$, $\nabla = (\partial_x, \partial_y, \partial_z)$, and restricts these fields to the forms

$$\mathbf{V} = (\mathbf{v}(x,y), W(x,y)), \quad \text{div}\,\mathbf{v} = 0, \quad \mathbf{v} = (v_1, v_2),$$

a straightforward calculation shows that one gets a subalgebra of the initial algebra, which is isomorphic to the algebra of the infinitesimal transformations (18) where j corresponds to W and ψ is a stream function for $\mathbf{v}$.

Returning to 2-d MHD we see that, as the group manifold is the same, it may be described by the same coordinates, the shift in the new (with respect to Lagrangian coordinates on the plane) coordinate is generated by the current j in the same way as the shift in z is generated by vertical velocity W. Hence, this new coordinate is related to a charge transport in the system. Moreover, although the coordinates are the same, the Riemannian structure and, as a consequence, the behaviour of geodesics are different.

REFERENCES

Arnold, V.I. 1966 Sur la geometric differentielle des groupes de Lie de dimension infinie et ses applications a l'hydrodynamique des fluides parfaits. *Ann. Inst. Fourier* (Grenoble) **16**, 319–361.

Arnold, V.I. 1978 *Mathematical Methods of Classical Mechanics*. Springer.

Hattori, Y. 1990. *Master's Thesis*, Dept. of Physics, University of Tokyo.

Kambe, T. 1985 A simultaneous concentration of a vortex and a magnetic field in magnetohydrodynamic flows. *Phys. Fluids* **28**, 2321–2323.

Kambe, T., Nakamura, F. & Hattori, Y. 1992 Kinematical instability and line-stretching in relation to the geodesics of fluid motion. In *Topological Aspects of the Dynamics of Fluids and Plasmas* (ed. H.K. Moffatt, G.M. Zaslavsky, P. Comte & M. Tabor), pp. 493–504. Kluwer Academic Publishers.

Lukatsukii, A.M. 1981 On the curvature of the group of measure-preserving diffeomorphisms of an n-dimensional torus. *Russ. Math. Surv.* **36** (2), 179–180.

Nakamura, F., Hattori, Y. & Kambe, T. 1992 Geodesics and curvature of a group of diffeomorphisms and motion of an ideal fluid. *J. Phys. A: Math. Gen.* **25**, L45–L50.

Zeitlin, V. 1992 On the structure of phase-space, Hamiltonian variables and statistical approach to the description of two-dimensional hydrodynamics and magnetohydrodynamics. *J. Phys. A: Math. Gen.* **25**, L171–L175.

Instabilities of Tidally and Precessionally Induced Flows

R.R. KERSWELL

Dept. of Mathematics and Statistics
University of Newcastle upon Tyne
Newcastle upon Tyne, NE1 7RU UK

In this paper, we examine the stability of tidally and precessionally strained rotating flows. A tidally ('elliptically') distorted uniform vortex has recently been shown to be linearly unstable to three-dimensional disturbances. Here we illustrate how such instability manifests itself within a tidally-distorted, rotating spheroid of fluid and comment on possible relevance to the Earth's outer core. Preliminary results are also presented which indicate that Poincaré's solution for the flow within a precessing spheroid is similarly unstable.

---

1 INTRODUCTION

To a good approximation, the Earth's outer core is a fluid-filled, oblate spheroidal shell which rotates daily about its axis. Due to the misalignment of the Earth's equatorial bulge and its orbit, the Earth is forced to precess once every 25,800 years by the Sun and orbiting Moon. Additionally, both exert gravitational (tidal) strains throughout the Earth which rotate, all but a lag, with the source body. In this paper, we discuss separately the effect of tidal distortion and precession upon a uniformly rotating fluid in an oblate spheroid. Both processes modify the basic rotation by superimposing a constant straining upon the flow. For tidal distortion, this straining is in the plane of motion and gives rise to elliptical streamlines. The strain due to precession is directed out of the plane of motion and consists of both an elliptical distortion *and* a shearing of the closed streamlines so that the line joining their centres is no longer perpendicular to the plane of motion.

181

M.R.E. Proctor, P.C. Matthews & A.M. Rucklidge (eds.)
Theory of Solar and Planetary Dynamos, 181–188
©1993 Cambridge University Press.

For the case of tidal distortion, instability to three-dimensional disturbances has recently been established by Pierrehumbert (1986) and Bayly
(1986) in the unbounded domain (see also Craik 1989 and Waleffe 1990)
and by Waleffe (1989) in a distorted cylinder (see Malkus 1989, Gledzer &
Ponomarev 1992 and references therein for experimental studies). In section 2
we illustrate the 'elliptical' instability in a distorted oblate spheroid and discuss briefly possible relevance for the Earth's outer core. Section 3 contains
a derivation (in the unbounded domain) that precessional flow is *also* unstable. This instability has also been isolated in the bounded, oblate spheroidal
geometry but will not be detailed here (see Kerswell 1992b).

2 TIDAL (ELLIPTICAL) FLOW
We consider the two-dimensional flow

$$
\mathbf{U} = \begin{bmatrix} 0 & -\sqrt{\frac{1+\beta}{1-\beta}} & 0 \\ \sqrt{\frac{1-\beta}{1+\beta}} & 0 & 0 \\ 0 & 0 & 0 \end{bmatrix} \boldsymbol{x} \tag{1}
$$

within the elliptically-distorted spheroidal container

$$
\frac{x^2}{1+\beta} + \frac{y^2}{1-\beta} + \frac{z^2}{c^2} = 1. \tag{2}
$$

Here c measures the container's oblateness and has an inferred value of $c \approx
1 - \frac{1}{400}$ for the Earth's outer core. The parameter β is a measure of the
tidal distortion of the underlying rotation through the eccentricity of the
streamlines

$$
\frac{x^2}{1+\beta} + \frac{y^2}{1-\beta} = \left(1 - \frac{z^2}{c^2}\right).
$$

The motions of the Moon and Sun relative to the Earth mean that the two
tidally-induced ellipticities rotate at different rates about the Earth's axis,
alternately reinforcing and opposing each other. Compared to the daily
timescale of the Earth's rotation, the time dependence of the resultant ellipticity is slow, fluctuating between, say, $\beta = 3$ to 7×10^{-8}. As a result,
in what follows, the ellipticity of the Earth's core will be assumed constant
in magnitude at about $\beta = 5 \times 10^{-8}$ and fixed in space. Nothing is lost in
assuming that the orbiting Moon is in fact stationary.

A linear stability analysis of the basic state (1) suffers from the complication
that the steady state is *not* axisymmetric. However, this difficulty may be
shifted from the basic state into the governing equations by use of the elliptico-
polar coordinate transformation, which maps the elliptical streamlines into
circular ones. Bayly was the first to realise the relevance of this coordinate

system to elliptical flow and Waleffe (1989) the first to use it in his bounded domain of a distorted cylinder. The transformation is

$$x = s\sqrt{1 + \beta}\cos\phi; \quad y = s\sqrt{1 - \beta}\sin\phi$$

with the corresponding new velocity components u, v and w defined such that

$$\boldsymbol{u} = u_x\hat{\boldsymbol{x}} + u_y\hat{\boldsymbol{y}} + u_z\hat{\boldsymbol{z}} = u\tilde{\boldsymbol{s}} + v\tilde{\boldsymbol{\phi}} + w\hat{\boldsymbol{z}}.$$

The base vectors

$$\tilde{\boldsymbol{s}} = \sqrt{1 + \beta}\cos\phi\,\hat{\boldsymbol{x}} + \sqrt{1 - \beta}\sin\phi\,\hat{\boldsymbol{y}},$$

$$\tilde{\boldsymbol{\phi}} = -\sqrt{1 + \beta}\sin\phi\,\hat{\boldsymbol{x}} + \sqrt{1 - \beta}\cos\phi\,\hat{\boldsymbol{y}}$$

are *neither* orthogonal *nor* of unit length in this system, where the basic flow is now one of *uniform rotation* $\boldsymbol{u} = s\tilde{\boldsymbol{\phi}}$. The price of this simplification is paid in the new form of the momentum equation. With respect to a frame in which the underlying flow is at rest, the linearised momentum equation for a small disturbance is (in $[\tilde{\boldsymbol{s}}, \tilde{\boldsymbol{\phi}}, \hat{\boldsymbol{z}}]$ components) exactly

$$\frac{\partial \boldsymbol{u}}{\partial t} + 2\begin{bmatrix} -v \\ u \\ 0 \end{bmatrix} + \nabla p = \frac{1}{2}\beta\left[e^{2i(\phi+t)}\mathbf{N} + e^{-2i(\phi+t)}\mathbf{N}^*\right]\nabla p, \tag{3}$$

$$\nabla \cdot \boldsymbol{u} = 0, \tag{4}$$

$$\boldsymbol{u} \cdot \boldsymbol{n}|_{\partial V} = 0 \qquad \partial V : x^2 + y^2 + \frac{1 - \beta^2}{c^2}z^2 = 1, \tag{5}$$

where the pressure p, the axial velocity w and the z coordinate have been rescaled as follows

$$p \to (1 - \beta^2)p, \quad w \to w\sqrt{1 - \beta^2}, \quad z \to z\sqrt{1 - \beta^2},$$

$$\mathbf{N} = \begin{bmatrix} 1 & i & 0 \\ i & -1 & 0 \\ 0 & 0 & 0 \end{bmatrix} \quad \text{and} \quad \overline{\nabla} = \tilde{\boldsymbol{s}}\frac{\partial}{\partial s} + \tilde{\boldsymbol{\phi}}\frac{1}{s}\frac{\partial}{\partial \phi} + \hat{\boldsymbol{z}}\frac{\partial}{\partial z}$$

is a pseudo gradient operator. The new right hand side of (3) is a direct consequence of the non-orthogonality of the transformation. For $\beta = 0$, the system (3)–(5) describes the linear oscillations upon a uniformly rotating flow. These Poincaré modes are well known for the case of an oblate spheroid (Kudlick 1966). For $0 < \beta \ll 1$, the right hand side of (3) can be interpreted as a coupling operator between two such modes of the unstrained system. A disturbance solution consisting of two modes to leading order,

$$\boldsymbol{u} = A\mathbf{Q}_a e^{i\lambda_a t} + B\mathbf{Q}_b e^{i\lambda_b t} + \beta\boldsymbol{u}_1 + O(\beta^2),$$

may possess secular growth if $|\lambda_b - \lambda_a \pm 2| = O(\beta)$. We consider the case $\lambda_b = \lambda_a + 2 + \Delta\beta$ to be specific, and via a multiple time scale analysis obtain the parametric equations

$$\frac{\partial A}{\partial \tau} = \frac{1}{2}\langle \mathbf{Q}_a, e^{-2i\phi}\mathbf{N}^*\nabla p_b\rangle Be^{i\Delta\tau},$$

$$\frac{\partial B}{\partial \tau} = \frac{1}{2}\langle \mathbf{Q}_b, e^{2i\phi}\mathbf{N}\nabla p_a\rangle Ae^{-i\Delta\tau},$$

where τ is the slow time βt. Searching for solutions of the form $A \propto e^{(\sigma+\frac{1}{2}i\Delta)\tau}$ and $B \propto e^{(\sigma-\frac{1}{2}i\Delta)\tau}$ leads to the expression

$$\sigma^2 = \frac{1}{4}\left\{\langle \mathbf{Q}_a, e^{-2i\phi}\mathbf{N}^*\nabla p_b\rangle\langle \mathbf{Q}_b, e^{2i\phi}\mathbf{N}\nabla p_a\rangle - \Delta^2\right\}. \tag{6}$$

The product of matrix elements is real and positive definite (Kerswell 1992a) if the azimuthal wavenumbers satisfy $m_b = m_a + 2$ and both waves share the same symmetry in z, being either both symmetric or antisymmetric about the equator. Exponential growth of the two waves can then occur on the β time scale provided the detuning Δ is sufficiently small (given by (6)). A special subset of such resonances is defined by the conditions $\lambda_b = -\lambda_a = 1$ and $m_b = -m_a = 1$, i.e. a $\lambda = m = 1$ 'spinover' mode couples with its complex conjugate. Figure 1 shows the growth rate for a spinover resonance plotted against the particular resonant geometry c for the first 65 such couplings. Two features are of note. Firstly the upper bound of $\frac{9}{16}\beta$ on the growth rates is consistent with Waleffe's (1990) value for the unbounded domain, and the general distribution of resonant geometries. For $\beta = 5 \times 10^{-8}$, the instability has an interesting e-folding time of 97,000 years in the Earth, to be compared with the timescales of the geomagnetic secular variation at $O(10\text{--}1000)$ years and inter-reversal times which range from hundreds of thousands to millions of years.

Dissipation may be incorporated to leading order into this inviscid picture by consideration of the linear decay rates of the individual waves, ν, say. Instability will now occur only if

$$\sigma\beta > \sqrt{\nu_a\nu_b}. \tag{7}$$

In the Earth's outer core, such global modes experience both Ohmic and viscous damping with Ohmic effects appearing dominant (the magnetic Ekman number $E_m \sim 10^{-9}$ as opposed to the viscous Ekman number of $E \sim 10^{-15}$). However, due to the small relative size of the basic Alfvén speed to rotational velocity, the Ohmic and viscous decay rates are actually estimated to be of the same order and *comparable* to the elliptical growth rate. A calculation for the lowest spinover mode in a sphere with basic magnetic field

$$\mathbf{H} = \alpha r\hat{\phi} + \gamma\hat{z}$$

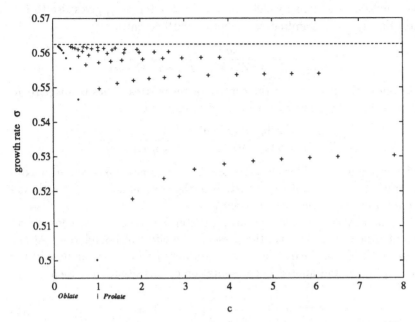

Figure 1. Growth rates for spinover resonances plotted against geometry c. Oblateness corresponds to $c < 1$ and prolateness $c > 1$. A dot is used to mark the initial point on the ridge.

leads to a total decay rate overpowering the excitation rate by a factor of 3 at $\alpha = 100$ gauss and $\gamma = 8$ gauss for an insulating mantle. This result is only slightly modified if the mantle is acknowledged to be conducting.

3 PRECESSIONALLY INDUCED FLOWS

Poincaré's (1910) solution for the core fluid response to the Earth's precession is

$$\boldsymbol{u} = \boldsymbol{w} \times \boldsymbol{r} + \nabla A$$

with

$$\boldsymbol{w} = \hat{\boldsymbol{z}} - \frac{2 + \eta}{\eta + 2(1 + \eta)\boldsymbol{\Omega} \cdot \hat{\boldsymbol{z}}} \hat{\boldsymbol{z}} \times (\hat{\boldsymbol{z}} \times \boldsymbol{\Omega})$$

and

$$A = \frac{\eta}{\eta + 2(1 + \eta)\boldsymbol{\Omega} \cdot \hat{\boldsymbol{z}}} (\boldsymbol{\Omega} \times \hat{\boldsymbol{z}} \cdot \boldsymbol{r})(\hat{\boldsymbol{z}} \cdot \boldsymbol{r})$$

in the container $\boldsymbol{r}^2 + \eta(\hat{\boldsymbol{z}} \cdot \boldsymbol{r})^2 = 1$ (see also Malkus 1993). The basic daily angular velocity, $\hat{\boldsymbol{z}}$, is recovered as the limiting solution when the precession rate $|\boldsymbol{\Omega}| \to 0$. The flow is two-dimensional in planes inclined at

$$\tan^{-1} \frac{2|\boldsymbol{\Omega} \times \hat{\boldsymbol{z}}|}{\eta + 2(1 + \eta)\boldsymbol{\Omega} \cdot \hat{\boldsymbol{z}}}$$

to the container's equator, which, due to the oblateness η, gives rise to two effects. Firstly the streamlines are *elliptical* with eccentricity

$$\beta = \frac{2\eta|\mathbf{\Omega} \times \hat{\mathbf{z}}|^2}{[\eta + 2(1 + \eta)\mathbf{\Omega} \cdot \hat{\mathbf{z}}]^2 + 2(2 + \eta)(\mathbf{\Omega} \times \hat{\mathbf{z}})^2}$$

and secondly the line joining the centres of these streamlines is *not* perpendicular to their plane but rather inclined at an angle

$$\tan^{-1} \frac{2\eta[\eta + 2(1 + \eta)\mathbf{\Omega} \cdot \hat{\mathbf{z}}]|\mathbf{\Omega} \times \hat{\mathbf{z}}|}{[\eta + 2(1 + \eta)\mathbf{\Omega} \cdot \hat{\mathbf{z}}]^2 + 4(1 + \eta)(\mathbf{\Omega} \times \hat{\mathbf{z}})^2}$$

to this perpendicular. This shearing of the streamlines is 5 orders of magnitude larger than the elliptical distortion of the streamlines in the Earth and it is upon this effect that we now focus.

As in the analysis of elliptical flow, a useful idealisation is to consider an unbounded flow, ignoring, for the present, the effect of boundaries. At the leading order of the shear, the streamlines may be taken as circular in the precessing frame and an exact representation of this situation is then

$$\mathbf{U} = \begin{bmatrix} 0 & -1 & 0 \\ 1 & 0 & -2\epsilon \\ 0 & 0 & 0 \end{bmatrix} \mathbf{x} = \mathbf{D} \cdot \mathbf{x} \tag{8}$$

in a rotating frame with

$$\mathbf{\Omega} = \begin{bmatrix} \epsilon \\ 0 \\ 0 \end{bmatrix}$$

and $\epsilon = |\mathbf{\Omega} \times \hat{\mathbf{z}}| \approx 4 \times 10^{-8}$. A Kelvin mode (see, for example, Greenspan 1968 or Bayly 1986)

$$[\mathbf{u}(\mathbf{x}, t), p(\mathbf{x}, t)] = [\hat{\mathbf{u}}(t), \hat{p}(t)]e^{i\mathbf{k}(t) \cdot \mathbf{x}}$$

is an exact, nonlinear, incompressible solution upon the basic state (8) if

$$\frac{d\mathbf{k}}{dt} + \mathbf{D}^t \cdot \mathbf{k} = 0 \tag{9}$$

and $$\frac{d\hat{\mathbf{u}}}{dt} + \mathbf{D} \cdot \hat{\mathbf{u}} + 2\mathbf{\Omega} \times \hat{\mathbf{u}} + ik\hat{p} = 0 \tag{10}$$

hold. The system (9) and (10) is in fact a Floquet problem for $\hat{\mathbf{u}}$ with only one free parameter: the angle between $\mathbf{k}$ and $\hat{\mathbf{z}}$. For all $\epsilon \neq 0$, there exists a band of angles in which the velocity amplitude $\hat{\mathbf{u}}$ grows exponentially. The Floquet problem is independent of $|\mathbf{k}|$ and hence this instability is *scale invariant*. The angle band of instability closes to a single angle of $\tan^{-1}\sqrt{15}$ as $\epsilon \to 0$. This may be verified analytically by use of the power statement

$$\frac{d}{dt}\left(\frac{1}{2}\hat{\mathbf{u}}^2\right) = 2\,\epsilon\,\hat{v}\hat{w} \tag{11}$$

obtained by taking the dot product of $\hat{u}$ and (10). For $\epsilon = 0$, the Kelvin mode with wave vector inclined at $\tan^{-1}\sqrt{15}$ to $\hat{z}$ is

$$k = \begin{bmatrix} -\sqrt{15}\sin t \\ \sqrt{15}\cos t \\ 1 \end{bmatrix} ; \quad \hat{u}_0 = \begin{bmatrix} \frac{3}{2}\cos\frac{3}{2}t + \frac{5}{2}\cos\frac{1}{2}t \\ \frac{3}{2}\sin\frac{3}{2}t + \frac{5}{2}\sin\frac{1}{2}t \\ \sqrt{15}\sin\frac{1}{2}t \end{bmatrix}. \tag{12}$$

When $0 < \epsilon \ll 1$ we can proceed perturbatively by use of the expansion

$$\hat{u} = [\hat{u}_0(t) + \epsilon\hat{u}_1(t) + O(\epsilon^2)]e^{\sigma\epsilon t}$$

in (11). Averaging over one period leads to the requirement that

$$\sigma = \frac{2\langle \hat{v}_0\hat{w}_0 \rangle}{\langle \hat{u}_0^2 \rangle} = \frac{5\sqrt{15}}{32}$$

to eliminate secularity in $\hat{u}_1$, i.e. *there is exponential growth on the ϵ timescale.* Rotating with the streamline, the underlying strain has a period 2π and the Kelvin mode in (12) has a period 4π; hence, as for elliptical flow, the strain excites a subharmonic instability. In the bounded domain, the instability occurs by the coupling of 2 Poincaré modes whose azimuthal wave numbers and frequencies differ by 1 (Kerswell 1992b).

The author thanks Professor Willem Malkus for many valuable discussions, and Professor Alar Toomre for indirectly stimulating this study of precession. The author also gratefully acknowledges the support of an Alfred P. Sloan Doctoral Dissertation Award and the National Science Foundation under grant ATM-8901473.

REFERENCES

Bayly, B.J. 1986 Three dimensional instability of elliptical flow. *Phys. Rev. Lett.* **57**, 2160–2163.

Craik, A.D.D. 1989 The stability of unbounded two- and three-dimensional flows subject to body forces: some exact solutions. *J. Fluid Mech.* **198**, 275–292.

Gledzer, E.B. & Ponomarev, V.M. 1992 Instability of bounded flows with elliptical streamlines. *J. Fluid Mech.* **240**, 1–30.

Greenspan, H.P. 1968 *The Theory of Rotating Fluids.* Cambridge University Press. Reprinted 1990 by Breukelen Press, Brookline.

Kerswell, R.R. 1992a Elliptical instabilities of stratified, hydromagnetic waves and the Earth's outer core. *Ph.D. Thesis*, M.I.T.

Kerswell, R.R. 1992b The instability of precessing flow. *Geophys. Astrophys. Fluid Dynam.* (submitted).

Kudlick, M.D. 1966 On transient motions in a contained, rotating fluid. *Ph.D. Thesis*, M.I.T.

Malkus, W.V.R. 1989 An experimental study of global instabilities due to tidal (elliptical) distortion of a rotating elastic cylinder. *Geophys. Astrophys. Fluid Dynam.* **30**, 123–134.

Malkus, W.V.R. 1993 Energy sources for planetary dynamos. In *Lectures on Solar and Planetary Dynamos* (ed. M.R.E. Proctor & A.D. Gilbert), Cambridge University Press.

Pierrehumbert, R.T. 1986 Universal short-wave instability of 2-dimensional eddies in a inviscid fluid. *Phys. Rev. Lett.* **57**, 2157–2159.

Poincaré, H. 1910 Sur la précession des corps deformable. *Bull. Astron.* **27**, 321.

Waleffe, F.A. 1989 On the 3D instability of a strained vortex and its relation to turbulence. *Ph.D. Thesis*, M.I.T.

Waleffe, F.A. 1990 On the three-dimensional instability of a strained vortex. *Phys. Fluids* **A2(1)**, 76–80.

Probability Distribution of Passive Scalars with Nonlinear Mean Gradient

Y. KIMURA

Isaac Newton Institute for Mathematical Sciences
University of Cambridge, 20 Clarkson Rd., Cambridge, CB3 0EH UK

Current address:
National Center for Atmospheric Research
P.O. Box 3000
Boulder, CO 80307, USA

Recently Pumir, Shraiman & Siggia (1991) proposed an idea that a nonlinear mean temperature is essential to produce exponential-like tails for probability density functions (PDFs) of temperature fluctuations in convection. In this paper, results of numerical simulations of the 3D random advection equation with a mean gradient term will be shown. Some theoretical analysis is given based on a transport equation without molecular diffusion. The simplified analysis can capture the characteristic shapes of PDFs well.

1 INTRODUCTION

The study of passive scalar advection provides fundamental understanding of various phenomena such as convection and mixing that are ubiquitous in nature. In particular, the probability distribution of amplitude and its spatial gradients are of vital importance in relation to recent active studies of non-Gaussian probability density functions (PDFs) endemic in turbulence.

Since Castaing et al. (1989) reported exponential-like tails on the PDF of temperature fluctuations in thermal convection at very high Rayleigh numbers, there has been increasing interest in the mechanism of the non-Gaussian tails on PDFs of amplitudes. In a recent paper, Pumir, Shraiman & Siggia (1991) have suggested that the non-Gaussian tails for an advected passive temperature field may be induced by the presence of a mean-temperature profile. A simple physical mechanism for this is proposed in the present paper. The resultant non-Gaussian statistics will be shown by numerical simulations and theoretical analysis for a transport equation without molecular

M.R.E. Proctor, P.C. Matthews & A.M. Rucklidge (eds.)
Theory of Solar and Planetary Dynamos, 189–193
©1993 Cambridge University Press.

diffusion. In this paper, the result on PDFs is summarized; other details will be presented elsewhere (Kimura & Kraichnan 1993).

2 NUMERICAL SIMULATION

Consider a temperature field $\theta(\mathbf{x}, t) = T(\mathbf{x}, t) + \psi(\mathbf{x}, t)$ where $T(\mathbf{x}, t)$ is an ensemble mean and $\psi(\mathbf{x}, t)$ is an zero-mean fluctuation part. Let $\theta(\mathbf{x}, t)$ obey

$$\frac{\partial \theta}{\partial t} + (\mathbf{u} \cdot \nabla)\theta = \kappa \Delta \theta, \tag{1}$$

where the advecting velocity $\mathbf{u}(\mathbf{x}, t)$ is solenoidal ($\nabla \cdot \mathbf{u} = 0$). To single out the effect of the nonlinear profile of the mean temperature, we prescribe the advecting velocity as a random Gaussian set satisfying the following energy spectrum:

$$E(k) = 16\sqrt{\frac{2}{\pi}} v_0^2 k_0^{-5} k^4 \exp(-2k^2/k_0^2). \tag{2}$$

Two types of time dependence of the velocity field are used. One is 'frozen' for which $\mathbf{u}(\mathbf{x}, t)$ is independent of time, and the other is 'refreshed' for which $\mathbf{u}(\mathbf{x}, t)$ is replaced by a new, statistically independent realization at the end of each time interval t_0. The advecting velocity can be approximated by a white noise when t_0 goes to zero, i.e. decorrelating rapidly.

The initial mean temperature $T(\mathbf{x}, 0)$ is taken as:

$$T(\mathbf{x}, 0) = \alpha z + \beta \sin z. \tag{3}$$

The parameters α and β are chosen as -1 and -0.75, respectively, which give maximum gradient at $z = 0$ and 2π, and minimum gradient at $z = \pi$. It is a model for the S-shape often observed for the mean temperature profile in convection. For the numerical simulation, a new variable $\theta'(\mathbf{x}, t) \equiv \theta(\mathbf{x}, t) - \alpha z$ is introduced which is periodic on the interval 2π with the initial condition (3) and obeys

$$\frac{\partial \theta'}{\partial t} + (\mathbf{u} \cdot \nabla)\theta' = -\alpha w + \kappa \Delta \theta', \tag{4}$$

where $\mathbf{u} = (u, v, w)$. Homogeneity in x- and y-directions is assumed and average is taken in a xy-plane ($\equiv \langle . \rangle_{xy}$). By definition

$$\theta'(\mathbf{x}, t) - \langle \theta'(\mathbf{x}, t) \rangle_{xy} = \langle \psi(\mathbf{x}, t) \rangle_{xy} \equiv \psi(z, t). \tag{5}$$

The simulations are performed in a 2π-periodic box with 64^3 grid points. In all runs, $k_0 = 4.767$ and $\kappa = 0.01$. The r.m.s. velocity is taken as $u_0 = 1.00$ for the frozen case and $u_0 = 2.24$ for the refreshed case. The refreshing time t_0 is 0.2, which is close to the eddy turnover time $\tau(\equiv (u_0 k_0)^{-1}) \approx 0.17$ for the frozen case. The higher value for the refreshed case is chosen so as to

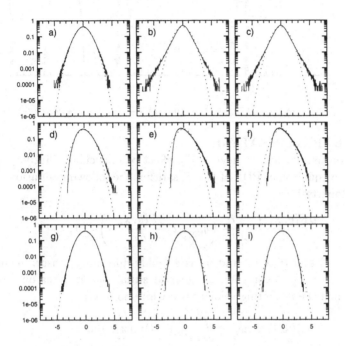

Figure 1. PDFs of $\psi(z,t)$ for the refreshed runs sampled in different planes (a,b,c: $z = \pi$, d,e,f: $z = \pi/2$, and g,h,i: $z = 0$) at different times (a,d,g: $t = 0.1$, b,e,h: $t = 0.5$, and c,f,i: $t = 1.0$). Ensemble averages were taken over 200 realizations.

compensate the decorrelation effect in the development of variance at each refreshment. The Peclét number $\equiv (\pi/2)u_0/k_0\kappa \sim 41$ for the frozen case.

Figure 1 shows the probabilty density functions of $\psi(z,t)$ for the refreshed runs at different z (π, $\pi/2$ and 0) and t (0.1, 0.5 and 1.0). The dotted lines denote Gaussian distributions. At $t = 0.1$ PDFs are close to Gaussian distributions in all planes. Then different features develop with time at different z, namely, exponential-like at $z = \pi$, skewed at $z = \pi/2$, and sub-Gaussian at $z = 0$. The exponential-like PDF in a mid-plane was found both in the experiment (Castaing *et al.* 1989) and in numerical simulations of Rayleigh–Bénard convection (Sirovich, Balachander & Maxey 1989; Kerr 1992). The skewed PDF has also been reported in these papers. The sub-Gaussian PDF seems special for the present simulation with the periodic boundary condition in z. The result from the frozen runs was almost identical with the refreshed ones.

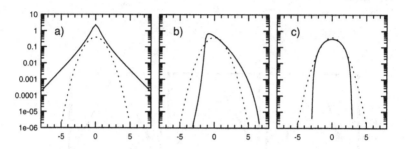

Figure 2. PDFs of $\psi(z,t)_{t=1.0}$ calculated from (9). (a) $z = \pi$, (b) $z = \pi/2$ and (c) $z = 0$.

3 THEORETICAL ANALYSIS

As a first approximation, we neglect the effect of molecular diffusivity κ in (1). Then the equation of PDF of $\theta(\mathbf{x},t)$ satisfies the following diffusion-type equation (Roberts 1961),

$$\frac{\partial P(\theta,\mathbf{x},t)}{\partial t} - \frac{\partial}{\partial x_i}\left(\eta_{ij}(\mathbf{x},t)\frac{\partial P(\theta,\mathbf{x},t)}{\partial x_j}\right) = 0\,, \tag{6}$$

where $P(\theta,\mathbf{x},t) \equiv P(\theta(\mathbf{x},t))$ and η_{ij} is an eddy-diffusivity tensor. Though η_{ij} has a complicated expression for general situations, it turns out to be simplified when the advecting velocity is a white-noise-type,

$$\eta_{ij}(\mathbf{x},t) = \eta\delta_{ij} \sim \int_{t-\delta t}^{t}\langle u_i(\mathbf{x},t)u_j(\mathbf{x},t')\rangle dt'\,, \tag{7}$$

where $\langle u_i(\mathbf{x},t)u_j(\mathbf{x},t')\rangle$ is the Lagrangian velocity autocorrelation function.

Under the above assumption, (6) becomes a pure diffusion equation, and the solution is

$$P(\theta,\mathbf{x},t) = \frac{1}{\sqrt{4\pi\eta t}}\int_{-\infty}^{\infty}e^{-|\mathbf{x}|^2/4\eta t}P(\theta,\mathbf{x},0)dx\,. \tag{8}$$

If temperature has a functional form $f(z)$ at $t = 0$, i.e. $\theta(\mathbf{x},0) = f(z)$, then $P(\theta,\mathbf{x},0) = \delta(\theta - f(z))$. With this initial condition (8) can be integrated analytically,

$$P(\theta,z,t) = \frac{-1}{\sqrt{4\pi\eta t}}\left(\frac{\partial f(z)}{\partial z}\right)_{z=z_0}^{-1}e^{-(z_0-z)^2/4\eta t}\,, \tag{9}$$

where z_0 is a solution of $\theta = f(z)$.

Substituting the temperature profile (3) into (9) and evaluating the derivative part by Newton's method, we obtained the PDFs ($t = 1.0$) at $z = \pi, \pi/2$ and 0 as in Figure 2. The eddy-diffusivity, η was estimated as

$$\eta \sim u_0^2\tau \approx 0.26\,. \tag{10}$$

Though the peak parts are exaggerated, each figure captures approximately the characteristic shapes of the PDF obtained by the numerical simulations, namely exponential-like $(z = \pi)$, skewed $(z = \pi/2)$ and sub-Gaussian $(z = 0)$.

Though the analysis by the zero-diffusivity transport equation is successful for reproducing the transient shapes of PDFs, the consideration of molecular diffusivity is essential to predict the long time development of PDFs. Modeling of the effect of molecular diffusivity is under way by the mapping closure technique developed by Kraichnan and co-workers (Chen, Chen & Kraichnan 1989).

Acknowledgements
The present work is a part of collaboration with Bob Kraichnan. The author expresses his cordial thanks to him. The author also thanks the organizers and the participants in the workshop on 'Dynamo Theory' at the Isaac Newton Institute for Mathematical Sciences in Cambridge for fruitful discussions and hospitality during my stay. NCAR is sponsored by the National Science Foundation.

REFERENCES
Castaing, B., Guneratne, G., Heslot, F., Kadanoff, L., Libchaber, A., Thomae, S., Wu, X.-Z., Zaleski, S. & Zanetti, G. 1989 Scaling of hard thermal turbulence in Rayleigh–Bénard convection. *J. Fluid Mech.* **204**, 1-30.

Chen, H., Chen, S., & Kraichnan, R.H. 1989 Probability distribution of a stochastically advected scalar field. *Phys. Rev. Lett.* **63**, 2657-2660.

Kerr, R.M. 1992 Rayleigh number scaling in numerical convection. (preprint).

Kimura, Y. & Kraichnan, R.H. 1993 Statistics of an advected passive scalar. *Phys. Fluids* (submitted).

Pumir, A., Shraiman, B. & Siggia, E.D. 1991 Exponential tails and random advection. *Phys. Rev. Lett.* **66**, 2984-2987.

Roberts, P.H. 1961 Analytical theory of turbulent diffusion. *J. Fluid Mech.* **11**, 257-283.

Sirovich, L., Balachandar, S. & Maxey, M.R. 1989 Simulations of turbulent convection. *Phys. Fluids* **A1**, 1911-1914.

Magnetic Fluctuations in Fast Dynamos

RUSSELL KULSRUD & STEPHEN ANDERSON*

Princeton Plasma Physics Lab
Princeton, NJ 08543 USA

*Current address:
Space Physics Directorate
Phillips Lab, Geophysics Directorate
Hanscom Air Force Base, MA 01731 USA

One theory for the origin of the galactic field is that it grows from a very weak seed field by fast dynamo action associated with turbulent motions. However, the dynamo also amplifies small scale fields faster than the large scale. In this paper we calculate the time evolution of the spectrum of small scale fields. We show that the magnetic turbulence reaches the resistive scale in a reasonably short time where some damping occurs. We also show that the damping is not strong enough to stop the exponential growth of the random turbulence which grows to equipartition with the turbulent power in a time short compared to the dynamo growth time for the large scale fields. Our conclusion is that a dynamo origin from a weak seed field is not plausible.

A hotly debated topic is the origin of the large scale galactic magnetic field. Originally, it was supposed by Fermi and others that the field had a primordial origin and was maintained against Ohmic decay by the large inductance of the galactic disk. (The time scale for Ohmic decay by ordinary Spitzer resistivity is extremely long, of order 10^{26} years.) However, there have been several objections to the primordial theory (Parker 1979). One objection is that turbulent resistivity is sufficiently large to destroy the field in a Hubble time. A second objection is that if it is not destroyed by turbulent resistivity, it can escape from the galactic disc by ambipolar diffusion. Probably the strongest objection has been that there seems no known way to produce a magnetic field in the early universe on a large enough scale and of sufficient strength to provide a primordial origin. However, there are counter arguments to these objections (Kulsrud 1990; Ratra 1992).

The second theory of the origin of the galactic field is that it is driven by a fast dynamo (Ruzmaikin, Shukurov & Sokoloff 1988). The dynamo must be

M.R.E. Proctor, P.C. Matthews & A.M. Rucklidge (eds.)
Theory of Solar and Planetary Dynamos, 195–201
©1993 Cambridge University Press.

fast compared to the very long resistive time to change the large scale field of the galaxy during its lifetime. The supposition of this theory is that there is an original very small (for example 10^{-17} gauss) magnetic field present when the galaxy was formed. Then the turbulent motions of the interstellar medium can amplify the mean field strength up to the present observed value of order several microgauss. This theory has a very plausible basis. The turbulence has a well defined origin in the acceleration of gas by supernova explosions and the winds and bubbles driven by hot stars. Moreover, the rotation of the galaxy gives this turbulence a helicity which leads to a finite value of the critical parameter for dynamo amplification, α. Further, employing reasonable estimates of the parameters of the turbulence one finds that this galactic dynamo actually leads to amplification of certain dynamo modes in a time short compared to the lifetime of the galaxy. Thus, a sufficient amplification of the initial magnetic field can be achieved to account for the present field. This theory represents one of the triumphs of fast dynamo theory in astrophysics.

However, the actual motion of the field lines during this amplification is very puzzling. If the interstellar plasma is taken as infinitely conducting, the field must be frozen into the plasma during the entire period of dynamo amplification. By a well known theorem the field strength divided by the plasma density remains proportional to its length during the entire period of amplification. Thus, assuming an amplification factor of 10^{11}, and assuming that there is no density change, the line of force must be increased in length by this same enormous factor. There is no room to put such a long line of force in the galactic disc unless it is very tangled. Thus, under the assumption of infinite conductivity, the random field component must be very large compared to the average field component. This problem is not usually addressed by the standard fast dynamo theory, which is only concerned with the mean field and not the fluctuations about the mean field. It has usually been assumed that any such fluctuations are propagated down to small scales and destroyed by resistivity (Ruzmaikin *et al.* 1988).

In order to address the question of how large magnetic fluctuations can become during fast dynamo amplification of the galactic field, and to judge the extent to which they are suppressed by resistivity, it is necessary to employ a mode coupling equation which gives the time evolution of the spectrum, $M(k)$ of magnetic fluctuations, where M is defined by

$$\frac{B^2}{4\pi\rho} = \int M(k)dk. \tag{1}$$

As a consequence of this equation, the very short scale magnetic fluctuations grow significantly and propagate to very short scales as has been suspected. Therefore, the only scales of hydrodynamic turbulence that affect these fluctuations are the shortest scales, for which one may neglect the helical part of the turbulence. Let us characterize the hydrodynamic turbulence by the spectrum $U(k)$, where the Fourier harmonics of the nonhelical part of the turbulent velocity satisfy a random phase approximation

$$\langle \mathbf{v}^*_{\mathbf{k}'}(t')\mathbf{v}_{\mathbf{k}}(t)\rangle = U(k) \left[\mathbf{I} - \hat{\mathbf{k}}\hat{\mathbf{k}}\right] \delta(\mathbf{k}' - \mathbf{k})\delta(t' - t). \qquad (2)$$

In the absence of resistivity, the mode coupling equation for $M(k)$ can be shown to be (Kulsrud & Anderson 1992)

$$\frac{\partial M}{\partial t} = \int K_m(k, k_0)M(k_0)dk_0 - 2\frac{\eta_T}{4\pi}k^2 M(k), \qquad (3)$$

where

$$K_m(k, k_0) = 2\pi k^4 \int \frac{\sin^3 \theta d\theta}{k_1^2}(k^2 + k_0^2 - kk_0 \cos \theta)U(k_1), \qquad (4)$$

$$k_1^2 = k^2 + k_0^2 - 2kk_0 \cos \theta, \qquad (5)$$

and

$$\frac{\eta_T}{4\pi} = \frac{1}{3} \int U(k_1)d\mathbf{k}_1. \qquad (6)$$

[This mode coupling equation is identical except for notation with the equation derived by Kraichnan & Nagarajan (1967).]

Integrate (3) over all k. The result is

$$\frac{\partial \mathcal{E}}{\partial t} = 2\gamma \mathcal{E}, \qquad (7)$$

where

$$\mathcal{E} = \frac{1}{2} \int M(k)dk, \qquad (8)$$

and

$$\gamma = \frac{1}{3} \int k^2 U(k)dk. \qquad (9)$$

The order of magnitude of γ is the rate of turnover of the smallest hydrodynamic eddy. To form an estimate of its size for the galactic dynamo we treat the interstellar turbulence as Kolmogoroff with a largest scale 100 parsecs and the random velocity on this scale 10 kilometers per second. We cut the turbulence off at the wave number $k = k_{max}$ by assuming a kinematic viscosity, due to neutral particles, of $\sim 10^{21}\text{cm}^2\,\text{s}^{-1}$. This leads to the cutoff length 10^{16}cm, and a value for γ of order the reciprocal of 10^4 years.

The implication of this result is that the fluctuations build up at an enormously rapid rate compared to the mean field dynamo rate. (The dynamo amplification time is typically hundreds of millions of years.) A cursory numerical estimate shows that all the magnetic energy builds up on scales shorter even than the smallest turbulent eddy and that it propagates towards shorter wave lengths. The critical question is whether the magnetic fluctuation energy is destroyed at the resistive scale faster than it is generated at longer scales. The exponential behavior of (7) is only valid in the absence of resistivity. However, resistivity is only important at extremely small length scales, smaller than $\lambda_R \approx \sqrt{\eta_S/\gamma}$ ($\approx 10^{10}$cm for the galactic example).

To settle the question concerning the fate of magnetic fluctuations, we must consider the mode coupling equation on scales smaller than the smallest hydrodynamic eddy scale. For $k \gg k_{max}$, one may expand the mode coupling equation (4), treating k_1 in it as small. The result is the partial differential equation for M in k and t,

$$\frac{\partial M}{\partial t} = \frac{\gamma}{5}[k^2\frac{\partial^2 M}{\partial k^2} - 2\frac{\partial M}{\partial k} + 6M] - 2\frac{k^2\eta_S}{4\pi}M \qquad k \gg k_{max}, \qquad (10)$$

where the last term is the normal Spitzer resistivity η_S term which becomes important for $k \approx 1/\lambda_R$. For k in the range $k_{max} \ll k \ll k_R$, we may still neglect it. If the time behavior of M is known at some k_{ref} one can solve (10) by a Green's function solution. Dropping resistivity the solution is

$$M(k,t) = \int_{-\infty}^{t} M(t', k_{ref})G(t - t', k/k_{ref})dt', \qquad (11)$$

where the Green's function is

$$G(k, \tau) = \sqrt{\frac{5}{4\pi}} \frac{k^{3/2}e^{(3/4)\gamma\tau}}{\gamma^{1/2}\tau^{3/2}} \ln k e^{-(5/4)\ln^2 k/\gamma\tau}. \qquad (12)$$

This solution is valid in the range $k_{ref} \ll k \ll k_R$.

This Green's function is seen to have the character of a spectrum increasing to a peak k_p as $k^{3/2}$, and then decaying exponentially in k. The integral of G over k increases as $e^{2\gamma t}$, consistent with (7).

Incidentally, this Green's function solution is what is required to complete the numerical solution of the mode coupling equation (3). Imagine solving (3) on the grid $k_{min} < k < k_{ref}$, where k_{ref} is a value k larger than k_{max}. Then the evolution of M for $k \approx k_{ref}$ involves M for $k < k_{ref}+k_{max}$, that is k's extended k_{max} beyond the grid. Thus, one can advance the Green's function solution and the solution on the numerical grid simultaneously. The numerical grid solution serves to advance $M(k_{ref})$ in time, and thus, to advance the Green's function solution of $M(k)$ for $k > k_{ref}$. At the same time, Green's function solution supplies the values of $M(k)$ over the range $k_{ref} < k < k_{ref} + k_{max}$, needed to advance the integral in (4).

When the numerical solution is carried out, it is found that on the numerical grid, M approaches a function which is a constant in time multiplied by a pure exponential factor $e^{3\gamma t/4}$. Assuming such a time behavior at k_{ref},

$$M(k_{ref}, t) = M_1 e^{3\gamma t/4}, \qquad (13)$$

the Green's function solution reduces to

$$M(k,t) = M_1 e^{(3/4)\gamma t} \left(\frac{k}{k_{ref}}\right)^{3/2} \left[1 - \mathrm{erf}\left(\ln\left(\frac{k}{k_{ref}}\right)\sqrt{\frac{5}{4\gamma t}}\right)\right], \quad (14)$$

where erf is the error function. Thus, $M(k,t)$ is a function that grows in time and spreads out in k-space at an exponentially fast rate. In fact, the integral of the expression over k approaches a constant times $e^{2\gamma t}$ consistent with (7). When k_p approaches k_R, the resistive term in (10) must be kept. The solution of (10) then is

$$M = M_1 \frac{k^{3/2}}{k_{ref}^{3/2}} \frac{K_0\left(\sqrt{\frac{5\eta_S}{\gamma}}k\right)}{K_0\left(\sqrt{\frac{5\eta_S}{\gamma}}k_{ref}\right)} e^{(3/4)\gamma t}, \quad (15)$$

where K_0 is the Bessel function of the second kind.

We see that the total fluctuation energy growth rate is reduced from 2γ to $(3/4)\gamma$ but that the fluctuation energy continues to grow exponentially. The flow of magnetic energy to the short scales where it is destroyed by resistivity is slower than the rate of amplification of this energy by stretching and shearing of the magnetic fluctuations by the turbulence. Therefore, the assumption, made in mean dynamo theory, that resistivity can suppress the magnetic fluctuations, is incorrect.

These results are all based on the kinematic assumption that the field is too weak to affect the turbulence in any way. This cannot continue to be the case because the power into magnetic field energy and into Joule heating $2\gamma\mathcal{E}$ grows exponentially on a very short time scale. Further, all this energy comes from the smallest hydrodynamic eddy and represents a drag on it. When the power into magnetic fluctuation energy and Joule heating becomes comparable to the total power into the hydrodynamic turbulence (which is also the power into the smallest eddy), the smallest eddy becomes suppressed, and the next larger eddy takes over. At this time, the value of γ is reduced to the turnover rate of the next larger eddy. (The turnover rate of an eddy of wave number k increases with k as $k^{2/3}$.) The magnetic fluctuations continue to grow, but at a slower rate. Next, this second eddy is suppressed. The turbulent cutoff scale progressively increases and the fluctuation energy continues to grow at a slower and slower rate so actually the growth during this nonkinematic stage is linear in time. Eventually, in a time of order a few million years the energy in magnetic fluctuations becomes comparable to the entire hydrodynamic energy. It is important to understand that this all happens during a time short compared to the dynamo amplification time.

[In actual fact, there is another nonlinear saturation mechanism that sets in even before the random magnetic energy spectrum reaches the resistive scale, k_R, unless the initial field is extremely weak, ($< 10^{-25}$ gauss). This mechanism is the ambipolar slippage of the ions through the neutrals. It is described by Kulsrud & Anderson (1992), who show that although the physics is different than that described above, the general conclusion that the turbulence saturates in a few million years, still holds. For simplicity, we ignore this effect in the present paper.]

It is difficult to calculate precisely what happens after the saturation of the turbulent hydrodynamic power occurs. The strong random fields will generate Alfvén wave motions, so that the turbulence will have the character of Alfvén wave turbulence, rather than hydrodynamic turbulence. Such motions cannot produce any further dynamo amplification of the mean field because of their high frequencies. The mean field strength will be saturated at a value very small compared to the random field. One must accept the result that the mean-field dynamo fails to produce a field at all similar to the currently observed galactic magnetic field. (In any event, the situation described above is far from that derived from the kinematic mean-field theory.)

One may say that the kinematic dynamo saturates not when the mean-field energy is comparable to the turbulent energy, but when the random magnetic fluctuation energy is comparable to the turbulent energy. Because γ is so large compared to the dynamo build-up rate of the mean field, the kinematic dynamo saturates in a time very short compared to the exponential build up time of the mean field.

The argument leading to this conclusion is bound up with the assumption of a very weak initial mean field. Such a field is unable to suppress the build-up of strong magnetic fluctuations. (This assumption is of course essential for a dynamo origin theory which starts with a weak seed field.) If the initial mean field were dynamically important, it would modify the turbulent motions in such a way as to prevent the catastrophic build-up of the magnetic fluctuation energy. This would be precisely the case if there were a substantial primordial field at the beginning of galactic evolution. The dynamo would no longer be kinematic, but could still modify the mean field substantially.

The conclusion one draws from this investigation of the magnetic fluctuations is that the success of the fast kinematic dynamo theory for the origin of our galactic field has to be critically reexamined. On the basis of this conclusion, the argument for the dynamo origin for the present galactic magnetic field, which assumes it arises by amplification from a very weak initial field, is no longer so strong compared to the argument for the primordial origin theory, that which assumes that it evolves from a strong initial magnetic field.

REFERENCES

Kraichnan, R. & Nagarajan, S. 1967 Growth of turbulent magnetic fields. *Phys. Fluids* **10**, 859–870.

Kulsrud, R.M. 1990 Galactic and intergalactic magnetic fields. In *IAU Symposium 140* (ed. R. Beck, P.P. Kronberg & Wielebinski), pp. 527–530. Netherlands.

Kulsrud, R.M. & Anderson, S.W. 1992 The spectrum of random magnetic fields in the mean dynamo theory of the galactic magnetic field. *Astrophys. J.* **396**, 606–630.

Parker, E.N. 1979 *Cosmical Magnetic Fields, Their Origin and Activity*. Clarendon, Oxford.

Ratra, B. 1992 Cosmological 'seed' field from inflation. *Astrophys. J.* **391**, L1–4.

Ruzmaikin, A.A., Shukurov, A.M. & Sokoloff, D.D. 1988 *Magnetic Fields of Galaxies*. Kluwer Academic Press, Dordrecht.

A Statistical Description of MHD Turbulence in Laboratory Plasmas

A. LAZARIAN

Department of Applied Mathematics and Theoretical Physics
University of Cambridge, Silver St., Cambridge, CB3 9EW UK

Turbulence plays a crucial role in dynamo processes. For example, turbulent diffusion is important for the existence of the Solar dynamo. Some turbulent phenomena may be studied with present-day measurement equipment. A number of relevant diagnostics are based on the interaction of an electromagnetic beam with plasma. Here we discuss the situation in which information on plasma properties is obtained by probing plasma with a plane polarized electromagnetic beam. It is shown that the problem of recovering statistical properties of turbulence from the line integrated data can be solved uniquely using a realistic model of plasma. Analytical expressions relating structure functions of both the random density field and random magnetic field to measured structure functions have been found. This information is of importance in studies of MHD turbulence.

---

1 STATISTICAL PROPERTIES OF PLASMA TURBULENCE

Recent measurements have shown the existence of fine-scale density structures in Tokamak plasmas (Cripwell & Costley 1991). There is also experimental evidence that the anomalous (i.e. greater than collisional) particle and energy transport may be in some circumstances due to particle drift motion caused by microturbulence. These facts make the investigation of the turbulence in Tokamaks very important. To describe the phenomena, it is useful to know statistics of random magnetic and density fields.

In this paper we discuss the statistical properties of plasma which can be studied with the so-called refractometry technique probing plasma with a plane polarized electromagnetic laser or microwave beam (Gill & Magyar 1987; Weisen et al. 1988). The interaction of an electromagnetic wave with plasma results in a phase delay of the beam and in Faraday rotation of its polarization plane. Turbulent motion in the plasma leads to fluctuations of these quantities. The measured phase delay of an electromagnetic wave is

203

M.R.E. Proctor, P.C. Matthews & A.M. Rucklidge (eds.)
Theory of Solar and Planetary Dynamos, 203–210
©1993 Cambridge University Press.

proportional to a line integral $\int n_e dl$ (where n_e is the electronic density). The measured Faraday rotation of the polarization plane is proportional to a line integral $\int n_e h_\parallel dl$ (where $h_\parallel$ is a component of the magnetic field parallel to the beam axis). Therefore the corresponding pointwise structure functions (e.g., the structure function of a random density field $d(\mathbf{r}, \mathbf{R}, t) = \langle (n(\mathbf{x}_1, t) - n(\mathbf{x}_2, t))^2 \rangle$, where the angular brackets $\langle ... \rangle$ denote the ensemble average, $\mathbf{r} = \mathbf{x}_1 - \mathbf{x}_2$, $\mathbf{R} = (\mathbf{x}_1 + \mathbf{x}_2)/2$, $n(\mathbf{x}, t)$ is a density at $(\mathbf{x}, t)$, and t is time) are not available directly. The problem we address here is to extract statistical information on pointwise functions from statistical information on the above line integrals. This problem is different from the one tomography deals with. We are interested in the statistical measure of the small-scale irregularities, and it can be shown that in this situation the inversion is not undetermined.

Previous attempts to solve this problem have been unsuccessful, in our view because a model of globally isotropic and homogeneous turbulence has been adopted (see Weisen *et al.* 1988). For example, Weisen *et al.* (1988) found that turbulence in plasma cross-sections varied along the electromagnetic beams which were used for probing the plasma. This contradicted their initial assumption that turbulence was globally homogeneous. Here the problem is treated using a much more realistic model where the the above assumption is dropped.

In this paper it is shown that the problem of expressing pointwise structure functions in terms of the structure functions of line integrals has a unique analytical solution. For further discussion we need to adopt a plasma model. In major Tokamaks (Gill & Magyar 1987), magnetic field pressure $B^2/8\pi$ is much greater than plasma pressure P, and plasma motion is dominated by the magnetic field. Therefore the small-scale turbulent motion of the plasma is likely to be axially symmetric (in the statistical sense), and the axis of symmetry lies in the direction of the magnetic field.

To simplify the presentation we restrict ourselves to a single arbitrarily chosen poloidal plane. Note that then both $\mathbf{r}$ and $\mathbf{R}$ lie in this plane. In view of this,

$$d_2(\mathbf{r}, \mathbf{R}, t) \approx N^2(\mathbf{R}, t) d_M(\mathbf{r}, \mathbf{R}, t) \qquad (1)$$

where $d_M(\mathbf{r}, \mathbf{R}, t) = \langle (d_\parallel(\mathbf{x}_1, t) - d_\parallel(\mathbf{x}_2, t))^2 \rangle$ is the structure function of the magnetic field, and $N(\mathbf{R}, t)$ is the mean density of plasma. Here, for the sake of simplicity, we ignore the vector character of the magnetic field. The full treatment of the problem will be given elsewhere (Lazarian, in preparation).

2 MEASURED STATISTICAL CHARACTERISTICS

An electromagnetic wave $u = u_0 \sin(kx - \omega t)$ passing through plasma in the direction l experiences a phase delay $\Phi(l, t) = (2\omega/c) \int \mu(x, l, t) dx$, where x is the coordinate along the l-axis, ω is the frequency of the signal, $\mu(x, l, t) =$

$\sqrt{1 - 4\pi n(x, 1, t)e^2/m_e\omega^2}$ is a dielectric permittivity (note that $\mu \approx 1 - 2\pi n(x, 1, t)e^2/m_e\omega^2$ for sufficiently high frequencies), $n_e(x, 1, t)$ is the density of electrons at a point $(x, 1)$, and m_e and e are the mass and the charge of an electron respectively (Ginzburg 1961).

The turbulence may deviate from axial symmetry on the largest scales, thus it is advantageous to filter out the low frequency component of Φ by introducing $\Psi(1, t, \Delta t) = \Phi(1, t_1) - \Phi(1, t_2)$, where $\Delta t = t_1 - t_2 \ll t_{pr}$, $t = (t_1 + t_2)/2$, and t_{pr} is the characteristic time of the process. Choosing the appropriate Δt, it may be possible to separate contributions from different spatial scales. Indeed, for Kolmogorov turbulence $v^2(k) \sim k^{-2/3}$. Thus the eddy rotation period for eddies of size $1/k$ is proportional to $1/kv(k) \sim k^{-2/3}$, which means that it is longer for larger eddies. Similar results are valid for Kraichnan magnetic turbulence (Kraichnan 1965). However, it cannot be claimed that they are universally valid (Bouchard *et al.* 1990). In other words, the filtration procedure is applicable only when the majority of particles obey classical transport laws. Note that $\Psi(l, t) \approx \kappa_1 \int \Delta n_e(x, 1, t)dx$, where $\kappa_1 = 4\pi e^2/cm_e\omega$, where $\Delta n_e(x, 1, t) = n_e(x, 1, t_1) - n_e(x, 1, t_2)$.

Statistical properties of the phase delay can be described by the structure function

$$D_1(l_1, l_2, t) = \langle(\Psi(l_1, t) - \Psi(l_2, t))^2\rangle. \qquad (2)$$

Considerations similar to the above ones are applicable to the Faraday rotation measure $\theta(1, t) = \kappa_2 N(1, t) \int h_\parallel(x, 1, t)dx$, where $\theta(1)$ is measured in radians, $\kappa_2 \approx 1.1 \times 10^6/\omega^2$ in c.g.s. units, and it has been assumed that in the adopted model the fluctuations of density δn_e are much less than the mean density N. The corresponding structure function

$$D_2(l_1, l_2, t) = \langle(\theta(l_1, t) - \theta(l_2, t))^2\rangle \qquad (3)$$

contains information on the statistics of magnetic field.

3 STATISTICS OF A RANDOM DENSITY FIELD AND OF A RANDOM MAGNETIC FIELD

In our simplified treatment statistical properties of D_1 and D_2 are similar, so that everywhere below we omit indices for D, d and κ and write

$$D_i(l_1, l_2, t) = 4\kappa_i^2 \langle (\int_0^{\delta/2} u_i(x, l_1, t)dx - \int_0^{\delta/2} u_i(x, l_2, t)dx)^2 \rangle \qquad (4)$$

where p is the distance between two parallel beams (see Figure 1), for D_1, $u_1(x, 1, t) = n(x, 1, t)$ and for D_2, $u_2(x, 1, t) = N(1, t) \int n_\parallel(x, 1, t)dx$; δ is the thickness of plasma ($\delta/2 \gg 1/\kappa$, where κ is a wavenumber under study). Using an equality

$$(\int_0^{\delta/2} \chi(y)dy)^2 = \int\int_0^{\delta/2} \chi(y_1)\chi(y_2)dy_1dy_2$$

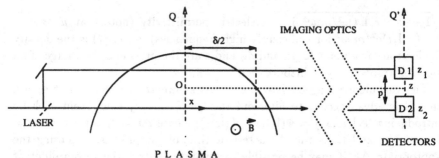

PLASMA

Figure 1. Schematic diagram of the refractometry correlation measurements. Q' is an image plane with the detectors D_1 and D_2 at positions z_1 and z_2 respectively, $p = z_1 - z_2$, $z = (z_1 + z_2)/2$, Q is the object plane coinciding with the plasma midplane, and x is zero at this plane.

and an elementary identity

$$(a - b)(c - d) = \frac{1}{2}[(a - d)^2 + (b - c)^2 - (a - c)^2 - (b - d)^2]$$

it is possible to obtain

$$\frac{\partial}{\partial p}D(p, z, t) = 8\kappa^2 p \int_p^{\delta/2} \frac{r\,dr}{\sqrt{r^2 - p^2}} \int_0^{\delta/2} \frac{1}{r}\frac{\partial}{\partial r}d(r, z, x, t)\,dx \qquad (5)$$

where r is the distance between correlating points ($r = \sqrt{(x_1 - x_2)^2 + p^2}$) and x is the distance along the beam ($x = (x_1 + x_2)/2$). Equation (5) is an Abel integral equation, which can be solved analytically to find (Gough 1984)

$$\int_0^{\delta/2} \frac{1}{r}\frac{\partial}{\partial r}d(r, z, x, t)\,dx = -\frac{1}{4\kappa^2\pi}\int_r^{\delta/2} \frac{dp}{\sqrt{p^2 - r^2}}\frac{\partial}{\partial p}(\frac{1}{p}\frac{\partial}{\partial p}D(p, z, t)). \qquad (6)$$

In general, Abel inversion has many practical advantages (Gough, private communication) and we are fortunate that it is applicable in our case. Transformations result in

$$F(r, z, t) = -\frac{1}{4\kappa^2\pi}\int_r^{\delta/2} \frac{dp}{\sqrt{p^2 - r^2}}\frac{\partial}{\partial p}D(p, z, t), \qquad (7a)$$

where

$$F(r, z, t) = \int_0^{\delta/2} \{d(r, z, x, t) - d(\delta/2, z, x, t)\}\,dx. \qquad (7b)$$

This result is different from the one obtained by Weisen *et al.* (1988, equation A.8). Here we have found an analytical solution for the inverse problem without assuming that turbulence is globally isotropic and homogeneous. Indeed, the turbulence characteristics in our model vary along the beam. The turbulence evolution with time can be traced by solving the inverse problem

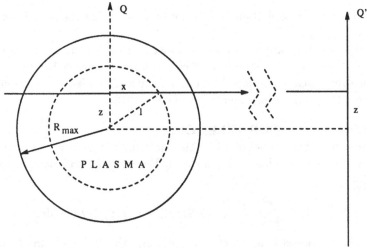

Figure 2. Schematic of a poloidal plane. R_{max} is the radius of the plasma cross-section. Since $p \ll z$, only one solid line is drawn to show the path of two separate beams on Figure 1.

for different t. Measuring turbulence inside different poloidal cross-sections, we can study spatial variations of statistical properties of the plasma distribution.

$F(r, z, t)$ corresponds to the difference of the structure functions of magnetic field and density averaged along the lines of sight. Statistics of this quantity are much more straightforwardly related to the statistics of turbulence than that of $D(p, z, t)$. For example, at a fixed r, F characterises the statistics of eddies with size r averaged along the beam, whereas D at a fixed p contains contributions from eddies of quite different sizes.

4 LOCAL TURBULENCE CHARACTERISTICS

In the previous section an analytic relationship has been found between $F(r, z, t)$ and measured statistics. If the plasma has an axially symmetric cross-section, it is possible to perform yet another analytical deconvolution to find the corresponding pointwise structure function d. Indeed, in this case structure functions depend only on two spatial coordinates r and l, e.g., $d(r, z, x, t) = d(r, \sqrt{z^2 + x^2}, t)$ (see Figure 2). Thus for $F(r, z, t)$ we can rewrite (7b) as

$$F(r, z, t) = \int_0^{R_{max}} \frac{l \, dl}{\sqrt{l^2 - z^2}} \{ d(r, l, t) - d(\delta/2, l, t) \}, \qquad (8)$$

where R_{max} is the radius of a plasma cylinder. This is an Abel equation similar to (6) and therefore

$$d(r, l, t) - d(\delta/2, l, t) = -\frac{1}{2\pi} \int_z^{R_{max}} \frac{dz}{\sqrt{z^2 - l^2}} \frac{\partial}{\partial z} F(r, z, t). \qquad (9)$$

The mean density distribution $N(l,t)$ can be found using the Abel inversion (Minerbo & Levy 1969).

For plasmas which are not axially symmetric, the problem of determining pointwise structure functions d from F can be solved numerically by adopting the algorithms already developed for calculating mean emissivity distribution from line integrated measurements (von Goller 1978).

6 SPECTRUM OF TURBULENCE.

The inverse problem can be solved for the spectrum of turbulence. For details the reader is referred to Lazarian (1992a,b). Here we just write down the final expression

$$\mathcal{E}(k,z,t) = \frac{\kappa^2}{8\pi\kappa^2} \int_0^{\delta/2} (D(\delta/2,z,t) - D(p,z,t)) J_0(kp) \mathrm{d}p, \qquad (10)$$

where J_0 is the Bessel function of zero order and the 3D spectrum of turbulence E is related to $\mathcal{E}$ by the equality

$$\mathcal{E}(k,z,t) = \int_0^{\delta/2} E(k,x,z,t)\mathrm{d}x. \qquad (11)$$

Unlike the Abel inversion for F (see section 3) the inversion for $\mathcal{E}$ does not require differentiation of measured quantities and provides a more reliable result when applied to data. Data processing based on the solution (10) is discussed elsewhere (Lazarian, in preparation). The spectrum $E(k,x,z,t)$ can be found by the second deconvolution described in the previous section.

7 DATA PROCESSING

The existence of analytical solutions $(7a)$ and (9) means that the inverse problem of finding statistics of density and magnetic field from the measured line integrals is not ill-posed. While processing real data one can use expressions $(7a)$ and (9) directly, but more sophisticated algorithms may be called for when the data is noisy.

Let us describe briefly inversion involving orthogonal polynomials. Consider an Abel equation

$$A(y) = 2 \int_y^a \frac{B(r)r}{\sqrt{r^2 - y^2}} \mathrm{d}r. \qquad (12)$$

Let the function $B(r)$ have an expansion

$$B(r) = \sum_m a_m f_m(r)$$

where $f_m(r)$ is a complete set of functions having the transform pairs

$$t_m(y) = 2 \int_y^a \frac{f_m(r)r}{\sqrt{r^2 - y^2}} \mathrm{d}r \qquad (13)$$

which are orthogonal with a norm P_m with respect to a scalar product

$$(g, h) = \int_0^a g(y)h(y)\mathrm{d}\alpha(y). \tag{14}$$

Then the coefficients a_m can be found (Gorenflo & Kovetz 1966). The coefficients a_m are given by

$$a_m = \frac{1}{P_m}(t_m, D) \tag{15}$$

where D is the known function, i.e., $D(p, z, t)$ for the solution $(7a)$ and $F(r, z, t)$ for the solution (9). Hermite polynomials and Genenbauer polynomials can be used as $f_m(r)$. A simpler technique has been suggested by Minerbo & Levy (1969), based on fitting the discrete function D with orthogonal polynomials. In this case finite sums and not integrals appear in (11). Using this approach, it is possible to show that if the standard deviation of $D(p)$ is σ, the standard deviation of $F(r)$ does not exceed 3σ.

8 DISCUSSION
It has been shown that pointwise statistical characteristics of plasma turbulence can be studied by processing line integral data from electromagnetic wave–plasma interaction. Statistical characteristics of random density and magnetic fields can be expressed analytically in terms of measured statistics. Our analysis shows that data processing may be performed using already existing algorithms in a manner which restricts the amplification of random noise. However, when two or more Abel transforms are used consecutively, as in section 5, an additional regularization might be called for (Gough 1985).

The technique described could be applicable to laboratory data obtained by probing plasma with an electromagnetic beam (see also Lazarian 1992c). A slight modification of the technique would make it suitable for processing images obtained with a pinhole camera while studying laser compressed self-emitting plasmas.

Acknowledgements
The author is greatly indebted to L. Fradkin for several helpful suggestions. An enlightening discussion with D. Gough is acknowledged. The research is supported by the Isaac Newton Scholarship.

REFERENCES
Bouchard, J.-P., Georges, A., Koplik, J., Provata, A. & Redner, S. 1990 Superdiffusion in random velocity fields. *Phys. Rev. Lett.* **64**, 2503–2506.
Cripwell, P. & Costley, A.E. 1991 Evidence for fine-scale density structures on JET under additional heated conditions. In *Proceedings of 19th EPS Conference, Berlin, 3–7 June* (in press).

Gill, R.D. & Magyar, G. 1987 Diagnostics. In *Tokamaks* (ed. J.A. Wesson), p. 239. Clarendon Press, Oxford.

Ginzburg, V.L. 1961 *Propagation of Electromagnetic Waves in Plasmas.* Gordon and Breach.

Gorenflo, R. & Kovetz, Y. 1966 Solution of the Abel-type integral equation in the presence of noise by quadratic programming. *Numer. Math.* **8**, 392-406.

Gough, D.O. 1984 On the rotation of the Sun. *Phil. Trans. R. Soc. Lond. A* **313**, 27-37.

Gough, D.O. 1985 Inverting helioseismic data *Solar Physics* **100**, 65-99.

Kraichnan, R.H. 1965 Inertial range spectrum of hydromagnetic turbulence *Phys. Fluids* **8**, 1385-1387.

Lazarian, A. 1992a Experimental study of turbulence in astrophysics. *Astron. Astrophys. Transactions* **3**, 33-51.

Lazarian, A. 1992b Study of turbulence using radiointerferometers. In Proceedings of ESA Colloquium *Targets for Space-Based Interferometry*, October 13-16, 1992, Côte d'Azur, France (in press).

Lazarian, A. 1992c Experimental study of 3D turbulence. Applied Scientific Research. An International Journal of Mechanical and Thermal Continua (in press).

Minerbo, G.N. & Levy, M.E. 1969 Inversion of Abel's integral equation by means of orthogonal polynomials. *SIAM J. Numer. Anal.* **6**, 598-614.

von Goeler, S. 1978. In *Diagnostics for Fusion Experiments*, (ed. E. Sindoni & C. Wharton). Pergamon Press, p. 79.

Weisen, H., Hollenstein, Ch. & Behn, R. 1988 Turbulent density fluctuations in the TCA Tokamak. *Plasma Physics and Controlled Fusion* **30**, 293-309.

Compressible Magnetoconvection in Three Dimensions

P.C. MATTHEWS

Department of Applied Mathematics and Theoretical Physics
University of Cambridge, Silver St., Cambridge, CB3 9EW UK

Convection in a compressible fluid with an imposed vertical magnetic field is studied numerically in a three-dimensional Cartesian
geometry, restricting attention to the weakly nonlinear regime.
Steady convection occurs in the form of two-dimensional rolls when
the field is weak but three-dimensional squares when the magnetic
field is sufficiently strong. In the regime where convection is oscillatory, the preferred planform for moderate fields is found to be
'alternating rolls' – standing waves in both horizontal directions
which are out of phase. For stronger fields, oscillatory convection
takes the form of a two-dimensional travelling wave.

1 INTRODUCTION

This paper is concerned with the effect of an imposed magnetic field on thermal convection. Although this is not directly relevant to dynamo theory, the
question of the interaction of convection with magnetic fields is important
for a full understanding of a convectively driven dynamo. The motivation for
this work is to understand some aspects of convection in the Sun, particularly
in regions of high magnetic field, such as sunspots, where the strong, predominantly vertical field inhibits thermal convection, causing the spot to appear
dark. This work is part of an ongoing collaboration with Michael Proctor and
Nigel Weiss. This brief report summarises some of the main results of this
work; further details will be published in a future paper (Matthews, Proctor
& Weiss 1993).

Linear theory for the onset of magnetoconvection in an incompressible fluid
was discussed extensively by Chandrasekhar (1961). Convection can be oscillatory if the magnetic diffusivity is lower than the thermal diffusivity and the

M.R.E. Proctor, P.C. Matthews & A.M. Rucklidge (eds.)
Theory of Solar and Planetary Dynamos, 211–218
©1993 Cambridge University Press.

field is sufficiently strong. As the field strength is increased, more and more thermal forcing is required to achieve convection and the resulting convection cells become increasingly narrow.

More recent numerical work on the fully nonlinear problem has included compressibility but restricted attention to the two-dimensional case. Steady compressible magnetoconvection was studied by Hurlburt & Toomre (1988). The oscillatory case was investigated by (Hurlburt *et al.* 1989), who showed that at onset, oscillatory convection takes the form of a standing wave for a moderate magnetic field. When the field becomes stronger, the preferred mode is a travelling wave and the magnetic pressure leads to large density fluctuations.

The present work extends these results to three dimensions. An important question is whether or not the convective flow is three-dimensional at onset. For convection in an incompressible fluid with no magnetic field and fixed-temperature boundaries, it is well known that the flow is two-dimensional at onset (Schlüter, Lortz & Busse 1965).

2 EQUATIONS FOR COMPRESSIBLE MAGNETOCONVECTION

The mathematical model consists of a layer of compressible fluid permeated by a vertical magnetic field. It is assumed that the heat capacities c_p and c_v, the thermal conductivity K, the viscosity μ, the magnetic diffusivity η, the permeability μ_0 and the acceleration due to gravity g are all constant. The boundary conditions imposed at the upper and lower surfaces of the layer are that the normal velocity and horizontal stress are zero, that the temperature is fixed and that the magnetic field is vertical. All variables are assumed to be periodic in the horizontal directions.

Dimensionless variables are introduced using the depth of the layer d, the fixed temperature T_0 at the upper surface, the density ρ_0 at the upper surface in the absence of convection, and the strength of the initial vertical magnetic field, B_0. The time unit used is the isothermal sound travel time at the top of the layer, $d/\sqrt{R_*T_0}$ where $R_* = c_p - c_v$ is the gas constant.

This non-dimensionalization introduces the following dimensionless parameters to the problem: the Prandtl number, $\sigma = \mu c_p/K$; the magnetic Prandtl number, $\zeta = \eta c_p \rho_0/K$; the polytropic index, $m = gd/R_*\Delta T - 1$ where ΔT is the temperature difference between the top and bottom of the layer; the dimensionless temperature difference $\theta = \Delta T/T_0$; the ratio of specific heats $\gamma = c_p/c_v$; the dimensionless thermal conductivity $k = K/d\rho_0 c_p\sqrt{R_*T_0}$; and the dimensionless field strength $F = B_0^2/R_*T_0\rho_0\mu_0$, which is twice the ratio of the magnetic pressure to the gas pressure at the top of the layer.

The mid-layer Rayleigh number, representing the ratio of the destabilising superadiabatic temperature gradient to the stabilising effects of thermal and viscous diffusion can be written in terms of the above parameters as

$$Ra = \frac{(m+1)\theta^2}{\sigma k^2 \gamma}(m+1-m\gamma)(1+\theta/2)^{2m-1}. \qquad (1)$$

Convection occurs if the Rayleigh number is greater than some critical value Rc.

Cartesian axes (x, y, z) are chosen so that the z-axis points downwards. The lengths of the periodic box in the x and y directions are denoted by x_{max} and y_{max}.

The equations for mass, momentum, magnetic field and heat are

$$\frac{\partial \rho}{\partial t} = -\nabla \cdot (\rho \boldsymbol{u}), \qquad (2)$$

$$\frac{\partial}{\partial t}(\rho u) = -\nabla (P + FB^2/2) + \nabla \cdot (FBB - \rho uu + \sigma k \tau) + \theta(m+1)\rho \hat{z}, \quad (3)$$

$$\frac{\partial B}{\partial t} = \nabla \times (\boldsymbol{u} \times \boldsymbol{B} - \zeta k \nabla \times \boldsymbol{B}), \qquad (4)$$

$$\frac{\partial T}{\partial t} = -\boldsymbol{u} \cdot \nabla T - (\gamma - 1) T \nabla \cdot \boldsymbol{u} + \frac{\gamma k}{\rho} \nabla^2 T + \frac{k(\gamma - 1)}{\rho}(\sigma \tau^2/2 + F\zeta J^2), \quad (5)$$

where τ is the stress tensor,

$$\tau_{ij} = \frac{\partial u_i}{\partial x_j} + \frac{\partial u_j}{\partial x_i} - \frac{2}{3}\delta_{ij}\frac{\partial u_k}{\partial x_k}, \qquad (6)$$

and $\boldsymbol{J} = \nabla \times \boldsymbol{B}$. The pressure P is given by the equation of state $P = \rho T$ and the magnetic field is constrained to be solenoidal, $\nabla \cdot \boldsymbol{B} = 0$.

The equations (2)–(6) have a stationary polytropic solution in which $\boldsymbol{u} = 0$, $\boldsymbol{B} = \hat{z}$, $T = 1 + \theta z$, $\rho = (1 + \theta z)^m$ and $P = (1 + \theta z)^{m+1}$. This state was used as the initial condition, with a small random perturbation added to the temperature field.

3 NUMERICAL METHODS

The equations were solved by a Fourier collocation / finite-difference method, using an extension and modification of the program of Cattaneo *et al.* (1991). Differentiation is carried out spectrally in the periodic x and y directions, using fast Fourier transforms, and by fourth-order finite differences in the vertical direction. The explicit second-order Adams–Bashforth method was used for the time integration. The time-step is restricted by stability constraints relating to the diffusion time and the wave travel time over a mesh interval. These two limits were found to be similar in magnitude.

For the weakly nonlinear convection discussed in this paper, a fairly low resolution is sufficient. Most of the results were obtained using 16 points in the x and y directions, and 25 points in the z direction.

4 RESULTS

In order to focus attention on the magnetic field, and for the purposes of comparison with earlier work, the parameters governing the polytropic atmosphere are fixed at $\gamma = 5/3$, $m = 1/4$, $\theta = 6$, $\sigma = 0.1$. This means that the temperature increases by a factor of 7 across the layer, the density contrast is only 1.63 and the pressure contrast is 11.4. These parameters are the same as those used by Hurlburt et al. (1989) in their two-dimensional calculations. For the purposes of this short paper, the geometrical parameters x_{max} and y_{max} are fixed at 1.0, so that the computational box is a cube. Different values of the geometrical parameters will be considered in a future paper (Matthews, Proctor & Weiss 1993).

The other parameters of the problem, including the Rayleigh number, the magnetic field strength F and the magnetic Prandtl number ζ are allowed to vary in order to illustrate several different regimes of magnetoconvection. The magnetic Prandtl number ζ is particularly important. Linear theory (Chandrasekhar 1961) for the Boussinesq problem shows that when $\zeta > 1$, convection is steady, but for $\zeta < 1$ oscillatory convection occurs if the field is sufficiently strong. These results remain qualitatively correct for compressible magnetoconvection. The parameter ζ is defined at the top of the layer in section 2; the results below are given in terms of $\hat{\zeta} = \zeta(1 + \theta/2)^m$, which is the effective value of ζ at the middle of the layer. The results are divided into two sections: Steady convection for $\hat{\zeta} = 1$ and oscillatory convection for $\hat{\zeta} = 0.1$.

4.1 Steady convection ($\hat{\zeta} = 1$)

For steady convection the planform at onset may be either two-dimensional rolls, in which the flow depends only on one horizontal coordinate, or three-dimensional squares, in which the variables depend on x and y symmetrically.

Figure 1 shows the linear stability curve for the onset of steady convection as a function of the Rayleigh number Ra and the field strength parameter F. Two regimes are clearly apparent here. When the field is weak, convection is resisted by thermal and viscous diffusion and the magnetic field is essentially passive. In the region where the curve rises steeply, the magnetic field is the dominating influence resisting the convective motions.

In the first regime, convection occurs in the form of two-dimensional rolls (indicated by circles in Figure 1). Figure 2(a) shows the density and flow velocity for a typical roll solution. Rolls were found for $F = 0.1$, $F = 0.2$ and $F = 0.3$, with the Rayleigh number set to $1.5Rc$ in each case. For $F = 0.1$, this roll solution was found to be unstable to a shearing instability as described by Rucklidge & Matthews (1993).

In the second regime, in which convection is resisted by the magnetic field, squares were found to be stable (asterisks in Figure 1). The square planform

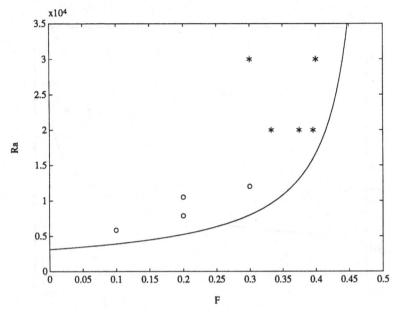

Figure 1. Steady magnetoconvection, $\hat{\zeta} = 1$. Circles indicate a roll solution, and asterisks represent squares. The solid line is the linear stability boundary, below which the static solution is stable.

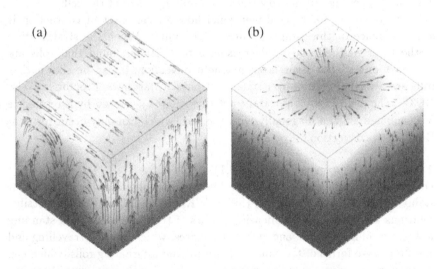

Figure 2. (a) The roll planform ($F = 0.2$, $Ra = 7875$) and (b) the square planform ($F = 0.4$, $Ra = 30000$). Arrows indicate fluid velocity and shading denotes density.

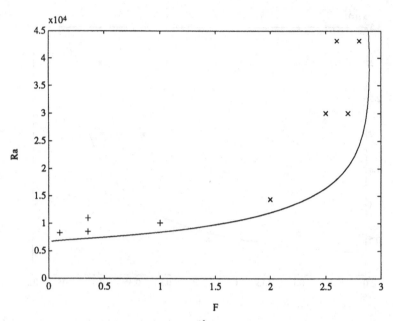

Figure 3. Oscillatory magnetoconvection, $\hat{\zeta} = 0.1$. Plus signs indicate the alternating roll solution, and crosses represent travelling rolls. The solid line is the linear stability boundary.

is illustrated in Figure 2(b). These squares have an up-down asymmetry: the upward-moving plume spreads out rapidly near the upper surface, and the downflow is confined to narrow sheets around the edges of the cell.

This preference for three-dimensional flow at the onset of convection is a consequence of the compressibility of the fluid. When the stratification of the layer was reduced by decreasing θ to 1, it was found that rolls are stable at onset. For Boussinesq magnetoconvection, it can be shown that rolls are always preferred over squares at onset (Matthews, Proctor & Weiss 1993). The asymmetric appearance of the squares is caused by a resonance with squares aligned diagonally with the box, which can only occur for non-Boussinesq convection.

4.2 Oscillatory convection ($\hat{\zeta} = 0.1$)

For oscillatory convection, there are five possible planforms that can occur stably at the onset of convection (Silber & Knobloch 1991). The five possible solutions are the two two-dimensional cases of travelling waves and standing waves; travelling squares and standing squares, which are just travelling and standing wave forms of the square planform; and alternating rolls, which can be thought of as two standing waves, in x and y, with equal amplitude but with a phase difference of $\pi/2$.

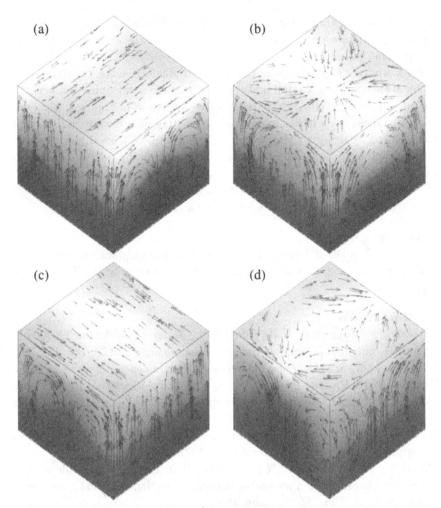

Figure 4. The alternating roll planform at four stages of the cycle, separated by one eighth of the period ($F = 0.35$, $Ra = 7812$). Arrows indicate fluid velocity and shading denotes density.

Figure 3 shows the linear stability curve for the oscillatory case. For $F = 0.1$, $F = 0.35$ and $F = 1.0$, the preferred planform when $Ra = 1.2Rc$ is the alternating roll pattern described by Silber & Knobloch (1991). In this solution, the horizontal dependence of, say, the temperature is proportional to $\cos kx \sin \omega t + \cos ky \cos \omega t$, and the horizontal velocity components u, v are proportional to $\sin kx \sin \omega t$ and $\sin ky \cos \omega t$ respectively. Thus the kinetic energy averaged over the layer is constant in time. The appearance of the convection pattern (see Figure 4) is of rolls aligned with the y-axis at $t = 0$, squares at a time $1/8$ of the period, and rolls aligned with the x-axis at $1/4$

of the period. At a fixed point on the upper surface of the box, the velocity vector appears to rotate (in a clockwise or anticlockwise direction depending on the point chosen).

For a stronger magnetic field, the preferred planform when the initial condition of the system is a small perturbation from the polytropic equilibrium is travelling rolls. Travelling rolls are indicated by crosses in Figure 3. However, in each of these cases it was found that alternating rolls are also stable. This was checked by using an alternating roll solution plus a small perturbation as the initial condition. This result that both alternating rolls and travelling rolls are stable for some parameter values is consistent with the stability analysis of Silber & Knobloch (1991). As the amplitude of convection is increased, these travelling rolls become modulated by a transverse oscillation.

Acknowledgements

I am grateful to Derek Brownjohn for help with the graphics, to Fausto Cattaneo for assistance with the numerical methods and to SERC for financial support.

REFERENCES

Cattaneo, F., Brummell, N.H., Toomre, J., Malagoli, A. & Hurlburt, N.E. 1991 Turbulent compressible convection. *Astrophys. J.* **370**, 282–294.

Chandrasekhar, S. 1961 *Hydrodynamic and Hydromagnetic Stability.* Oxford.

Hurlburt, N.E., Proctor, M.R.E., Weiss, N.O. & Brownjohn, D.P. 1989 Nonlinear compressible magnetoconvection. Part I. Travelling waves and oscillations. *J. Fluid Mech.* **207**, 587–628.

Hurlburt, N.E. & Toomre, J. 1988 Magnetic fields interacting with nonlinear compressible convection. *Astrophys. J.* **327**, 920–932.

Matthews, P.C., Proctor, M.R.E. & Weiss, N.O. 1993 Compressible magnetoconvection in three dimensions: planform selection and weakly nonlinear behaviour. *J. Fluid Mech.* (submitted).

Rucklidge, A.M. & Matthews, P.C. 1993 Shearing instabilities in magnetoconvection. In this volume.

Schlüter, A., Lortz, D. & Busse, F. 1965 On the stability of steady finite amplitude convection. *J. Fluid Mech.* **23**, 129–144.

Silber, M. & Knobloch, E. 1991 Hopf bifurcation on a square lattice. *Nonlinearity* **4**, 1063–1106.

The Excitation of Nonaxisymmetric Magnetic Fields in Galaxies

D. MOSS

Dept. of Mathematics
University of Manchester
Manchester, M13 9PL UK

A. BRANDENBURG

Isaac Newton Institute for Mathematical Sciences
University of Cambridge, 20 Clarkson Rd., Cambridge, CB3 0EH UK

Current address:
HAO/NCAR, P.O. Box 3000
Boulder, CO 80307 USA

The nonaxisymmetric ('bisymmetric spiral') magnetic field observed in the spiral galaxy M81 presents a challenge for mean field dynamo theory. We discuss several relevant mechanisms, and present simple numerical models to illustrate how a dominant $m = 2$ dependence of the turbulent coefficients might produce significant $m = 1$ field structure.

---

1 INTRODUCTION

Very naturally, many of the investigations into astrophysical dynamo theory have been directed to explaining the Solar cycle: after all, this is the system for which the most detailed information, both spatial and temporal, is available. The large scale Solar magnetic field appears to be approximately axisymmetric and so it is appropriate to study strictly axisymmetric dynamos. More recently, evidence has accumulated that magnetic fields with a significant nonaxisymmetric component may be present in late type 'active giant' stars (see, e.g., the discussion in Moss et al. 1991a, and references therein), and also in one or two spiral galaxies, notably M81 (e.g. Krause et al. 1989; Sokoloff et al. 1992). Thus the conditions under which nonaxisymmetric fields can be excited in astrophysical systems are of current interest. Rädler et al. (1990) and Moss et al. (1991a) have recently investigated nonlinear spherical mean field dynamo models in which stable nonaxisymmetric fields may be excited with suitably chosen distributions of alpha effect and differential

M.R.E. Proctor, P.C. Matthews & A.M. Rucklidge (eds.)
Theory of Solar and Planetary Dynamos, 219–224
©1993 Cambridge University Press.

rotation; see also Stix (1971). Rüdiger & Elstner (1992) considered models where the introduction of an anisotropy in the alpha tensor may have a similar effect; see also Rüdiger (1980). However, when constructing galactic dynamo models, there is less freedom, as the differential rotation is then known and cannot be chosen in an arbitrary manner.

In general, differential rotation is known to discriminate against the excitation of nonaxisymmetric dynamo modes compared with axisymmetric, although this effect is very much reduced in thin disc geometry (e.g. Ruzmaikin *et al.* 1988; Moss & Brandenburg 1992). However, galactic mean field dynamo models with axisymmetric distributions of alpha effect and turbulent resistivity have not been found preferentially to excite nonaxisymmetric fields. Of course, it is necessary to show that any given field, once excited, is stable in the nonlinear regime, and so linear theory can give only partial insight into these questions.

There are several mechanisms that might be important when discussing the origin of the bisymmetric spiral (BSS) structure of the fields in spiral galaxies. Firstly, spiral galaxies are distinctly nonaxisymmetric: if the disc turbulence is largely driven by supernovae explosions and young, hot stars in the arms, then the associated mean field dynamo coefficients α and η (turbulent resistivity) will also be nonaxisymmetric. Secondly, growth times for galactic dynamos are not very short compared to galactic lifetimes (in contrast to the situation for stellar dynamos). Further, the observed approximate equipartition between magnetic field and turbulent gas kinetic energies suggests that galactic dynamos are at least mildly nonlinear. Thus the eventually stable configuration may not have yet been attained in 'real' galaxies, and the observed galactic fields may be transients. In this case details of the initial conditions (seed fields) may still be important, and studies restricted to the eventually stable field configurations may be misleading. And thirdly, spiral galaxies sometimes occur in interacting groups. In particular, M81 appears to have undergone a recent close encounter with a companion galaxy. Dynamical models of the encounter suggest the presence of large scale nonaxisymmetric velocities in the disc of M81, which can generate nonaxisymmetric field components from a pre-existing field.

In this paper, attention will be concentrated on the first of these possibilities, although all three are of potential significance. Further discussion of the other two mechanisms is given in Moss *et al.* (1993).

2 THE MODEL

A preliminary investigation of the effects of a nonaxisymmetric perturbation to the α effect was made in Moss *et al.* (1991b). There

$$\alpha = \alpha_0 \cos \theta (1 + \delta_m \cos m\phi), \tag{1}$$

where α_0 is constant and r, θ, ϕ are spherical polar coordinates. Attention was there focussed on the case $m = 1$ and it was found that, in spherical α^2 dynamos, eigenmodes corresponding to $\delta_m > 0.1$ had substantial nonaxisymmetric parts, as measured by the parameter $M = E_{\mathrm{nax}}/E_{\mathrm{tot}}$, where E_{tot} is the total energy and E_{nax} that in the nonaxisymmetric part of the field. E_{nax} was then primarily in the $m = 1$ component. However differential rotation severely reduced M. These models are not directly applicable to galaxies: apart from all other simplifications, spiral structure is not usually of $m = 1$ type, and $m = 2$ is often more appropriate. In contrast, the nonaxisymmetric parts of observed galactic magnetic fields are dominantly of $m = 1$ structure. This appears to present a fundamental difficulty for explaining galactic fields by the influence of spiral arms on the turbulence but, given that no spiral structure is perfectly regular, a representation of α by

$$\alpha = \alpha_0 \cos\theta(1 + \delta_1 \cos\phi + \delta_2 \cos 2\phi), \qquad (2)$$

with $0 < |\delta_1| \ll |\delta_2|$ might be more appropriate. Then when solving the mean field dynamo equation

$$\partial \mathbf{B}/\partial t = \nabla \times (\mathbf{u} \times \mathbf{B} + \alpha\mathbf{B}) - \nabla \times (\eta \times \nabla\mathbf{B}), \qquad (3)$$

the $(\cos\phi, \cos 2\phi)$ interaction in the $\alpha\mathbf{B}$ term gives $\cos\phi$ and $\cos 3\phi$ components, and the latter will decay preferentially. Nonaxisymmetries in the turbulence should also influence the turbulent resistivity, so we also write

$$\eta = \eta_0(1 + \eta_2 \sin\theta \cos\phi). \qquad (4)$$

(The factor $\sin\theta$ is necessary to give regular behaviour of the solutions at the poles. It should also have been included as a coefficient of $\cos m\phi$ in (2), but there its omission does not seem to affect qualitatively the solutions.) We consider either $\eta_2 = \delta_2$ or $\eta_2 = 0$.

3 RESULTS

3.1 Spherical α^2 models

For α^2 dynamos, the single dynamo parameter is $C_\alpha = \alpha_0 R/\eta_0$, where R is the radius of the computational volume. We found eigenmodes of equation (3) by step-by-step integration, using a modification of the code described in Moss et al. (1991a). The dimensionless time is $\tau = t\eta/R^2$. Taking, for example, $C_\alpha = 10$, with $\delta_2 = \eta_2 = 0.5$, $\delta_1 = 0.1$, we found an oscillatory eigensolution (Figure 1, $\tau < 7.5$) with M varying between about 0.25 and 0.95, and E_1/E_2, the ratio of energies in the $m = 1$ and $m = 2$ parts of the field, between about 2 and 9. When η_2 was set to zero, the oscillations disappeared (Figure 1, $\tau > 7.5$). Results are summarised in Table 1.

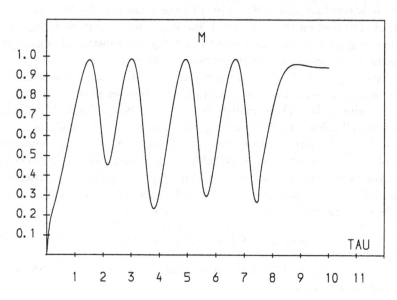

Figure 1. M as a function of time for the model with $C_\alpha = 10$, $\delta_2 = \eta_2 = 0.5$, $\delta_1 = 0.1$.

δ_1	δ_2	η_2	M	E_1/E_2
0.1	0.5	0.5	0.25–0.95	2–9
0.1	0.5	0.0	0.85	12
0.05	0.2	0.2	0.15	6
0.05	1.0	0.0	0.95	33

Table 1. Summary of results for α^2 dynamos in spherical geometry.

We also investigated a simple nonlinear model, with the same parameters as the model of Figure 1, but with an 'α-quenching' nonlinearity, imposed by introducing a factor $(1 + \mathbf{B}^2(r, \theta, \phi, \tau))^{-1}$ into the definition of α, see (2). We then found a steady solution with $M = 0.11$ and $E_1/E_2 \approx 2$.

3.2 Thick disc $\alpha^2\omega$ models

Here we considered a 'thick disc' model (e.g. Brandenburg *et al.* 1992; Moss *et al.* 1993), with α-effect non-zero only within distance $0.25\,R$ of the equatorial plane. Now

$$\alpha = \alpha_0 p^2 (3 - 2p) z (1 + \delta_1 \cos\phi + \delta_2 \cos 2\phi), \qquad (5)$$

where $z = r\cos\theta/R$, $p = (h - |z|)/h$ for $0 < |z| < h$ and $p = 0$ for $|z| > h$, and $h = 0.25$. (Note that now the maximum value of α is much less than α_0, and the critical values of C_α are correspondingly increased.) The rotation law was given by

$$\Omega = \Omega_0[1 + (\varpi/\varpi_0)^2]^{-1/2}, \tag{6}$$

with corresponding dynamo parameter $C_\Omega = \Omega_0 R^2/\eta_0$, $\varpi = r \sin\theta$ and $\varpi_0 = 0.3\,R$. Taking values of the order of $C_\Omega = 400$, $C_\alpha = 400$, $\delta_1 = 0.05$, $\delta_2 = \eta_2 = 0.5$, we found eigensolutions with $M \approx 0.1 - 0.2$ and $E_1/E_2 \approx 10$. The effect of differential rotation in this relatively thick disc appears to reduce the strength of the nonaxisymmetric field.

4 CONCLUSIONS

We have shown that it is possible to produce fields with nonaxisymmetric components of predominantly $m = 1$ structure by a dominant $m = 2$ term in the turbulent coefficients. In spherical models, when the dynamo is modestly supercritical the effect appears more markedly in eigensolutions than in the simple nonlinear solution that we calculated, although we have not made a systematic investigation of the effects of nonlinearities. For example, in linear models with $\delta_1/\delta_2 \approx 0.2$, ratios E_1/E_2 of up to ten can be attained. If the turbulence were largely driven by spiral structure, then values of $|\delta_2|$ of order unity may not be unreasonable. Then relatively small values of the ratio $|\delta_1/\delta_2|$ can give large values of $|E_1/E_2|$. Including non-zero η_2 can change the qualitative behaviour of the α^2 solutions from steady to oscillatory so that the energy ratio then depends on phase, but does not otherwise appear to have important effects.

We have not thoroughly studied 'galactic' models. Nevertheless, our results suggest that such a mechanism, taken in conjunction with long lived but transient structure associated with predominantly nonaxisymmetric seed fields, or with the effects of large scale nonaxisymmetric motions driven by an interaction with a companion galaxy (Moss *et al.* 1993), may introduce a significant $m = 1$ (BSS) field structure in certain spiral galaxies. Obvious modifications might include a spatial structure corresponding to a 'spiral' form for δ_1, δ_2, η_2, and a z-dependence in η_0. Although the structure of the fields of only a few spiral galaxies is known with any confidence, it does appear that coherent BSS structure is quite rare: our computations also suggest that a combination of circumstances may be necessary for nonaxisymmetric fields to be important.

REFERENCES

Brandenburg, A., Donner, K.J., Moss, D., Shukurov, A., Sokoloff, D.D. & Tuominen, I. 1992 Dynamos in discs and halos of galaxies. *Astron. Astrophys.* **259**, 453–461.

Krause, M., Beck, R., & Hummel, E. 1989 The magnetic field structures in two nearby spiral galaxies. II. The bisymmetric spiral field in M81. *Astron. Astrophys.* **217**, 17–30.

Moss, D. & Brandenburg, A. 1992 The influence of boundary conditions on the excitation of disk dynamo modes. *Astron. Astrophys.* **256**, 371–374.

Moss, D., Tuominen, I. & Brandenburg, A. 1991a Nonlinear nonaxisymmetric dynamo models for cool stars. *Astron. Astrophys.* **245**, 129–135.

Moss, D., Brandenburg, A. & Tuominen, I. 1991b Properties of mean field dynamos with non-axisymmetric alpha effect. *Astron. Astrophys.* **247**, 576–579.

Moss, D., Brandenburg, A., Donner, K.J. & Thomasson, M. 1993 Models for the magnetic field of M81. *Astrophys. J.* (submitted).

Rädler, K.-H., Wiedemann, E., Brandenburg, A., Meinel, R. & Tuominen, I. 1990 Nonlinear mean-field dynamo models: Stability and evolution of three-dimensional magnetic field configurations. *Astron. Astrophys.* **239**, 413–423.

Rüdiger, G. 1980 Rapidly rotating α^2-dynamo models. *Astron. Nachr.* **301**, 181–187.

Rüdiger, G. & Elstner, D. 1992 Nonaxisymmetry vs. axisymmetry in dynamo-excited stellar magnetic fields. *Astron. Astrophys.* (submitted).

Ruzmaikin, A.A., Shukurov, A. & Sokoloff, D.D. 1988 *Magnetic Fields of Galaxies*. Kluwer.

Sokoloff, D.D., Shukurov, A. & Krause, M. 1992 Pattern recognition of the regular magnetic field in disks of spiral galaxies. *Astron. Astrophys.* **264**, 396–405.

Stix, M. 1971 A non-axisymmetric α-effect dynamo. *Astron. Astrophys.* **13**, 203–208.

Localized Magnetic Fields in a Perfectly Conducting Fluid

M. NÚÑEZ

Dept. de Análisis Matemático
Universidad de Valladolid
47005 Valladolid, Spain

We study the existence and behaviour of magnetic fields localized within a region of a perfectly conducting fluid whose velocity field u is time-independent. This corresponds mathematically to the part of the spectrum of the induction operator $\mathbf{B} \longrightarrow \nabla \times (\mathbf{u} \times \mathbf{B})$ associated with approximate eigenfunctions with support localized in a given domain of the fluid. Finally we comment on the possible extension of these results to the case of positive resistivity.

1 INTRODUCTION

The induction equation

$$\frac{\partial \mathbf{B}}{\partial t} = \nabla \times (\mathbf{u} \times \mathbf{B})$$

governs the evolution of a magnetic field in an electrically charged fluid with velocity u and zero resistivity. B and u are linked by the remaining magneto-hydrodynamic equations, but the study of the above equation uncoupled from the rest, i.e. assuming a fixed velocity u, is useful to describe fluid motions which may give rise to magnetic enhancement and dynamo action (kinematic dynamo theory: see Ghil & Childress 1987). This must be followed by the study of the inverse problem of how such motions may occur. Magnetic fields which vanish outside a given region of the fluid are common in several phenomena, such as sunspots and resonant magnetospheric waves (Chen & Hasegawa 1974; Kivelson & Southwood 1985). We want to explore the possibility of such fields with a simple harmonic dependence on time and for

225

M.R.E. Proctor, P.C. Matthews & A.M. Rucklidge (eds.)
Theory of Solar and Planetary Dynamos, 225–228
©1993 Cambridge University Press.

a time-independent velocity **u**. This is admittedly an extremely simplified situation, but it can give some qualitative information on the problem. In mathematical terms, we need to study the spectrum of the operator

$$T : \mathbf{B} \longrightarrow \nabla \times (\mathbf{u} \times \mathbf{B}).$$

Since without resistivity the field lines are transported by the fluid as material points, the regions where a magnetic field depending exponentially on time may be localized are necessarily closed by the flow.

In this paper we concentrate on a single closed streamline and see how the spectrum depends on the shape of the flux tubes around it: if they close into themselves the resonance frequencies are real and no dynamo effect occurs. The proofs of the following theorems are lengthy and rely on delicate bounds; hence we omit them in this communication. They will appear elsewhere.

We first define the domain of T. Let U denote the domain filled by the fluid, $\mathrm{H}_0^1(U)$ the Sobolev space of functions vanishing at the boundary whose partial derivatives are square integrable functions. Since we need only $\mathbf{u} \times \mathbf{B}$ to be differentiable, we define

$$D(T) = \{\mathbf{B} \in \mathrm{L}^2(U)^3 \ : \ \mathbf{u} \times \mathbf{B} \in \mathrm{H}_0^1(U)^3, \ \nabla \cdot \mathbf{B} = 0\}.$$

It may be shown that $T : D(T) \longrightarrow \mathrm{L}^2(U)^3$ is a closed operator. It is not self-adjoint and therefore spectral decomposition does not hold; however, the points of the spectrum of T can always be interpreted as resonance frequencies.

2 MAIN RESULTS

From now on, γ will denote a closed streamline of real length L, P any fixed point of it (although all the results are independent of P), s the displacement parameter $((d/ds = \mathbf{u} \cdot \nabla), s \in [0, L])$, ϕ_s the flux function associated with **u**, ϕ_s' its differential, $\varphi = \nabla \cdot \mathbf{u}$, and $\alpha = \int_0^L \varphi \, ds$.

The variational equation through γ

$$\mathbf{w}'(s) = (\nabla \mathbf{u})(\phi_s(P))\mathbf{w}(s)$$

is a linear periodic one. As such, Floquet theory may be used. Let ν be a characteristic exponent of it. (Recall that one exponent is always zero, since $\phi_L'(P)$ takes the vector $\mathbf{u}(P)$ to itself.)

Definition 1. *We will say that ν is admissible if any of the following properties holds:*

    a) ν is not real.

    b) $\nu = \alpha/2$. In this case the three exponents are $(0, \alpha/2, \alpha/2)$.

    c) $\nu = 0$, if $\alpha = 0$.

Theorem 1. *If ν is admissible, all the values $z = \nu - \alpha + 2\pi in/L : n \in Z$ belong to the essential spectrum of T. Moreover, for any $\epsilon > 0$ and any neighbourhood V of γ, there exists $\mathbf{B} \in D(T)$ whose support is contained in V and such that*

$$\|T\mathbf{B} - z\mathbf{B}\|_2 < \epsilon\|\mathbf{B}\|_2\,.$$

When there exists a time-independent density for the fluid, the continuity equation implies $\alpha = 0$. If this does not happen, this part of the spectrum is unstable when $\alpha < 0$, i.e. when the flux tube around γ is contracted by the flow. This is related to the fact that in this case the magnetic field lines are compressed by the fluid.

The above approximate eigenfunctions tend to zero in $L^2(U)$: no function can have its support on γ. Therefore there exists no eigenfunction associated to those spectral points; instead there are eigendistributions. Rather surprisingly, in most instances they turn out to have order one, although in some cases they are measures.

Theorem 2. *Let $\mathbf{h} : [0, L] \longrightarrow U$ be the parametrization of γ, z one of the spectral points described in Theorem 1 such that its characteristic exponent ν is associated to two eigenvectors $\mathbf{B}_1$, $\mathbf{B}_2$ of $\phi'_L(P)$. Let*

$$\mathbf{n}(s) = \phi'_{s,P}(\mathbf{B}_1) \times \phi'_{s,P}(\mathbf{B}_2)\,.$$

Then the eigendistribution of T for z is

$$\mathbf{g} \longrightarrow \int_0^L \exp(-zs)(\nabla \times \mathbf{g}(\mathbf{h}(s))) \cdot \mathbf{n}(s)\, ds\,.$$

Apparently this indicates that resonance with z gives rise to electric fields localized in γ, since for harmonic fields $\nabla \times \mathbf{E}$ is proportional to $\mathbf{B}$. However, when $\alpha = 0$ there exists an eigenmeasure: this is

$$\mathbf{g} \longrightarrow \int_0^L \mathbf{u}(\mathbf{h}(s)) \cdot \mathbf{g}(\mathbf{h}(s))\, ds\,,$$

which means that the limit is a magnetic field localized in γ and proportional there to the fluid velocity.

3 POSITIVE RESISTIVITY

The operator

$$T_\eta : \mathbf{B} \longrightarrow \eta\Delta\mathbf{B} + \nabla \times (\mathbf{u} \times \mathbf{B})$$

is radically different in character from T, no matter how small η. We have not been able to prove a general theorem relating the spectra of T_η and T. However, in the particular cases studied there exists some kind of convergence for the eigenvalues when $\eta \to 0$: this is not true for the eigenfunctions. We will illustrate this point by taking U as the cube $[0, 2\pi]^3$, the velocity field $\mathbf{u}$ as $(u, 0, 0)$ (u constant) and imposing as boundary conditions 2π-periodicity in all the variables. After some algebra, the equation $T\mathbf{B} = \omega\mathbf{B}$ uncouples in the system

$$\begin{cases} \omega\mathbf{B}_2 = -u\dot{\mathbf{B}}_2 \\ \omega\mathbf{B}_3 = -u\dot{\mathbf{B}}_3 \end{cases}$$

where the dot denotes derivative with respect to x, and a scalar equation which for $z \neq 0$ merely defines $\mathbf{B}_1$. If $\mathbf{B}_2$ and $\mathbf{B}_3$ are periodic so is $\mathbf{B}_1$, hence this last equation can be ignored. We see that $\mathbf{B}_2$ and $\mathbf{B}_3$ are periodic when $\omega = ikn$, $k \in Z$, and any function of the form $g(y, z)\exp(-\omega x/u)$, g periodic, is a solution. Obviously these solutions may be concentrated at any streamline merely by shrinking the support of g, which concurs with our previous results.

As for the $\omega\mathbf{B} = T_\eta\mathbf{B}$, as before we are left with the equation

$$\omega\mathbf{B}_2 = \eta\Delta\mathbf{B}_2 - u\dot{\mathbf{B}}_2,$$

and the same for $\mathbf{B}_3$. By using Fourier analysis, we may assume that $\mathbf{B}_2$ depends on x, y, z as

$$\exp(ikx + imy + inz), \quad k, m, n \in Z.$$

Thus

$$\omega = -iuk - \eta(k^2 + m^2 + n^2).$$

Each of these $\omega(\eta)$ tends to $-iuk$ when $\eta \to 0$, thus filling the spectrum of T: however, the aspect of the whole spectrum of T_η is quite different, since it now is distributed in the left half plane and tends to become dense in it. This is natural: T_η has infinite dissipative modes whereas T has none.

As for the eigenfunctions, if ω is fixed, then k and $m^2 + n^2$ are also fixed. Any sum of harmonics $\exp(imy + inz)$ with $m^2 + n^2$ fixed is a finite trigonometric polynomial which cannot be localized around a point (y_0, z_0). Thus the resonant modes for $\eta > 0$ cannot be localized around a streamline with any approximation.

This work was supported in part by grant PB91-0212 of the DGICYT (Spain).

REFERENCES

Ghil, M. & Childress, S. 1987 *Topics in Geophysical Fluid Dynamics: Atmospheric Dynamics, Dynamo Theory and Climate Dynamics.* Springer.

Chen, L. & Hasegawa, A. 1974 A theory of long-period magnetic pulsations 1: Steady state excitation of field line resonance. *J. Geophys. Res.* **79**, 1024–1032.

Kivelson, M.G. & Southwood, D.J. 1985 Resonant ULF waves: a new interpretation. *Geophys. Res. Lett.* **12**, 49–52.

Turbulent Dynamo and the Geomagnetic Secular Variation

O.V. PILIPENKO & B.G. ZINCHENKO

Institute of Physics of the Earth
Moscow 123810, Russia

D.D. SOKOLOFF

Physics Department
Moscow University
Moscow 119899, Russia

Random motions of conducting fluids lead to self-excitation of random magnetic fields if the magnetic Reynolds number exceeds a certain threshold value estimated to be about $10^1 - 10^2$ (the fluctuation dynamo). This dynamo mechanism acts independently of the mean field dynamo up to the first-order approximation. We discuss here applications of the theory of the fluctuation dynamo to the Earth's liquid core. We suggest that intense flux ropes generated by random motions can produce magnetic perturbations at the core-mantle boundary, which are responsible for the geomagnetic secular variations having a time scale of about 100 years.

There exist two ways in which random motions of conducting fluids generate magnetic fields. One of them is the well-known mechanism of the mean field dynamo that requires an α-effect to operate. Another mechanism, the fluctuation dynamo, acts also in the mirror-symmetric flow and produces random, small scale magnetic fields, which are almost independent of the mean magnetic field. Random velocities in the liquid core of the Earth appear to be sufficiently high to sustain the fluctuation dynamo. Indeed, modern estimates of the scale of convective velocities in the outer core are about $v_0 \approx 0.03 - 0.1 \, \mathrm{cm \, s^{-1}}$, given that the space scale of these motions is comparable with the thickness of the outer core, gives $l_0 \approx 1000 - 2000$ km, and the magnetic diffusivity $\nu \approx 2 \cdot 10^4 \, \mathrm{cm^2 \, s^{-1}}$. Then the magnetic Reynolds number in the outer core is of the order of $R_m \approx 10^2 - 10^3$, yielding the threshold value for the fluctuation dynamo. The fluctuation dynamo produces magnetic fields, which are concentrated in ropes having thickness $l_0 R_m^{-1/2} \approx 10^2$ km and diameter about l_0. The equipartition estimate of the magnetic field is

M.R.E. Proctor, P.C. Matthews & A.M. Rucklidge (eds.)
Theory of Solar and Planetary Dynamos, 229–231

about 1 gauss. The characteristic time of the small scale magnetic field is $\tau \sim l_0/v_0 \approx 100$ years. Proceeding from these estimates, Ruzmaikin *et al.* (1989) proposed to relate the small scale magnetic field with the geomagnetic field of the secular variation. At least two problems are connected with this interpretation. First, the amplitude of the secular variation at the Earth's surface is only a few hundredths of a gauss, and its space scale (the size of so called focuses) is of the order of 1000 – 10000 km. It is natural to think that the small scale magnetic field is smoothed while propagating through the thick mantle with its very low conductivity. Due to this process the field magnitude decays and the 100 km space scale disappears. In this paper we give the quantitative analysis of this effect.

Our approach to the origin of the secular variation implies that it should be studied in terms of correlation functions and other averaged quantities. It is convenient to separate from the random geomagnetic field the component which is normal to the core–mantle boundary. Let us introduce

$$w(x) = \langle H_n(r)H_n(r+x)\rangle,$$

the correlation function of the normal component at that boundary. We designate $W(x)$ as the correlation function of the same field component at the Earth's surface. We consider here the simplest model and approximate the Earth's mantle by a plane layer of vacuum.

Suppose we know the distribution of the normal component of the magnetic field at the core–mantle boundary $H_n(r)$. Then the normal component at the Earth's surface can be represented as a Poisson integral:

$$h_n(R) = \frac{1}{2\pi} \int \frac{LH_n(r)dr}{|R-r|^3}, \tag{1}$$

where L is the thickness of the mantle and we integrate over the whole core–mantle boundary. Let us consider now the integral representation of type (1) for two points R_1 and R_2 at the Earth's surface and average their product over a statistical ensemble. As a result we obtain the relation between the correlation functions at the core–mantle boundary and at the Earth's surface:

$$W(|R_1 - R_2|) = \frac{1}{4\pi^2} \int \frac{L^2 w(|r_1 - r_2|)dr_1 dr_2}{|R_1 - r_1|^3 |R_2 - r_2|^3}. \tag{2}$$

Pilipenko & Sokoloff (1991, 1992) evaluated the integral (2) for the correlation function of the small scale magnetic field produced by the fluctuation dynamo. The following results about correlation properties of this field were obtained. The size of a focus of the secular variation (the doubled correlation length) at the Earth's surface is about the thickness of the mantle, $L \approx 3000$ km, which agrees well with the observations. The estimate of the amplitude of the variation is

$$(W(0))^{1/2} \sim -\frac{3}{32}L^{-2} \left(\int_0^\infty x^3 w(x) dx \right)^{1/2} \sim R_m{}^{-5/8}(l/L)^2 (w(0))^{1/2},$$

i.e.

$$H_{var}(R) \approx 2 \cdot 10^{-2}\,\mathrm{G}.$$

A similar approach is also appropriate to examine the tangential component of the fluctuating geomagnetic field. Since the electrical conductivity of the outer core is quite high, it is reasonable to think that for the most part only the normal component of the fluctuating field penetrates from the core into the low conducting mantle. On this assumption Sokoloff & Zinchenko (1992a) estimated that the amplitude of the secular variation for the tangential field at the Earth's surface is about half of that for the normal field. The space scale of the variation appears to be the same for both components, while isolines near the focuses must be oblong for the tangential field, contrary to the round ones for the normal field. All these results seem to agree with the observed data.

Sokoloff & Zinchenko (1992b) demonstrated that the cross-correlation of the normal and tangential magnetic fields vanishes at least in the case when they penetrate through the plane vacuum layer.

REFERENCES

Pilipenko, O.V. & Sokoloff, D.D. 1991 Propagation of the magnetic-field correlation function through the mantle. *Magnetohydrodynamics* **27**, 368–370.

Pilipenko, O.V. & Sokoloff, D.D. 1992 The propagation of fluctuating geomagnetic fields through the mantle. *Geomag. Aeron.* **33**, 61–66.

Ruzmaikin, A.A., Sokoloff, D.D. & Shukurov, A.M. 1989 The nature of secular variations in the primary magnetic field of the Earth. *Geomag. Aeron.* **29**, 1001–1006.

Sokoloff, D.D. & Zinchenko, B.G. 1992a On diffusion of the tangential fluctuations of the geomagnetic field through the mantle. *Astron. Nachr.* **313**, 115–123.

Sokoloff, D.D. & Zinchenko, B.G. 1992b On the correlations between tangential and normal component of the geomagnetic fluctuations. *Magnitnaya Gidrodyn.* **3**, 3–10.

On-Off Intermittency: General Description and Feedback Model

N. PLATT

Code R44, NavSWC
10901 New Hampshire Ave
Silver Spring, MD 20903-5000 USA

There is a large number of physical phenomena exhibiting a pecu-
liar behavior: the system is quiescent for long periods followed by
a burst of activity. This behavior is persistent, and can be characterized by intermittent switching of system variables. A general
model describing intermittent behavior has been found. The simplest version of On-Off intermittency does not involve feedback
of the intermittent signal into the forcing function which makes
it unrealistic in most physical situations. This paper discusses a
method of putting feedback into the system and its applications
to simple dynamical systems.

1 INTRODUCTION

Chaotic dynamical systems can be grouped into two classes according to
the characteristics of their behavior. One class is characterized by aperiodic
modulations of already periodic signals while the other class is characterized
by signals which exhibit apparently random switching between qualitatively
different kinds of behavior. The latter behavior is called intermittency. Examples of intermittency are abundant in nature. They include intermittent
bursts of turbulence in otherwise laminar pipe flow in fluid dynamics, sunspot
activity in astrophysics, and stock market crashes in economics. A model of
intermittency in terms of dynamical systems as well as a partial classification
of some types of intermittency was given by Pomeau & Manneville (1980). In
general, signals produced by this scenario are periodic oscillations interrupted
from time to time by some aperiodic bursts of activity. Another model of intermittency, crisis-induced intermittency, was introduced by Grebogi, Ott,
Romeiras & Yorke (1987). This intermittency involves a collision in phase-

M.R.E. Proctor, P.C. Matthews & A.M. Rucklidge (eds.)
Theory of Solar and Planetary Dynamos, 233–240
©1993 Cambridge University Press.

space of two chaotic attractors as some parameter is varied, and it is again characterized by random switching between different aperiodic oscillations. In both of these models of intermittency, the laminar phase consists of either periodic or aperiodic oscillations, while intermittency itself is manifested by apparently random switching between different kinds of behavior. But, for the case of pipe flow or sunspot activity, the laminar phase is characterized by a flat signal (no cross flow in the pipe during a laminar phase, almost no sunspot activity during a Maunder minimum), while intermittency manifests itself by a short burst of activity. Thus, the laminar phase of such a phenomenon can be approximated by a point in a suitably chosen projection of the phase-space, instead of a limit cycle (Pomeau & Manneville scenario) or chaotic attractor (crisis-induced intermittency). A model of intermittency describing this switching of activity was introduced recently by Platt, Spiegel & Tresser (1992).

2 GENERAL DESCRIPTION OF THE ON-OFF INTERMITTENCY

The mechanism of On-Off intermittency was described in detail elsewhere and will be discussed here only briefly. Consider a simple dynamical system

$$\dot{x} = -\epsilon_1 x^3 + \mu x \tag{1}$$

with $\epsilon_1 > 0$. Then $\mu_c = 0$ is a critical value of the bifurcation parameter μ for the fixed point $x_0 = 0$. For $\mu < 0$ the fixed point $x_0 = 0$ is stable, while at $\mu > 0$ it loses stability and the system bifurcates to either one of the newly created fixed points $x_{1,2} = \pm\sqrt{\mu/\epsilon_1}$. Now replace μ by suitably chosen time-dependent function $\mu(t)$ such that $\mu(t)$ crosses the $\mu = 0$ line irregularly and spends a considerable amount of time below and above $\mu = 0$ threshold line. Then the resulting time series $x(t)$ will display On-Off intermittent behavior.

Let us introduce basic equations used throughout this paper:

$$\begin{aligned}
\dot{x} &= -\epsilon_1 x^3 + (b - y)x, \\
\dot{y} &= q, \\
\dot{q} &= -y^3 + z - \epsilon\nu q, \\
\dot{z} &= -\epsilon(z + a(y^2 - 1)).
\end{aligned} \tag{2}$$

Here, equation (2a) is a dynamical system exhibiting On-Off intermittency, while equations (2b-d) are equivalent to the Lorenz system under a nonlinear change of variables (Marzec & Spiegel 1980). Here the bifurcation parameter μ is replaced by $\mu(t) = b - y(t)$ and b plays a role of a new bifurcation parameter for the onset of the intermittency.

Figure 1 depicts the intermittent signal $x(t)$ and the forcing function $y(t)$ vs. time produced by the system (2).

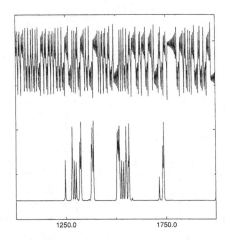

Figure 1. $y(t)$ (upper curve) and $x(t)$ (lower curve) for the system (2).

3 FEEDBACK MODEL

Equations (2) are representative of the simplest version of the On-Off intermittency. This model includes a master system (2b-d) which parametrically modifies the bifurcation parameter of the slave system (2a) and there is no feedback from the slave system into the master equations. In most physical situations this is unrealistic since usually there is coupling, and according to the 'strength' of the burst, it takes a longer time to build up 'energy' for another burst. Here, we would like to model intermittent systems whose dynamical behavior can be broken into the following stages:

I. A long laminar phase when system builds up 'energy' for the burst. In addition, the length of the laminar phase depends on the 'strength' of the previous burst: a stronger burst requires longer time to build up energy for the subsequent burst.

II. After enough energy is stored in the system for the burst, it takes some unpredictable amount of time for some fluctuations (dynamic or stochastic) to trigger the burst.

III. The burst itself is a relatively short phenomenon and the 'strength' of the burst determines the amount of time system spends in stage I.

Real physical situations which can be possibly modeled by such dynamical systems include stress build up in the Earth's core before, during, and after an earthquake or volcanic eruption, or the energy stored in the intermittently bursting pipe flow. Particular examples of equations exhibiting similar dynamical behavior were introduced by Spiegel (1980), and Hughes & Proctor (1990). The method presented here is geometrical in nature and can be generalized to a large variety of dynamical systems.

(a)   (b)

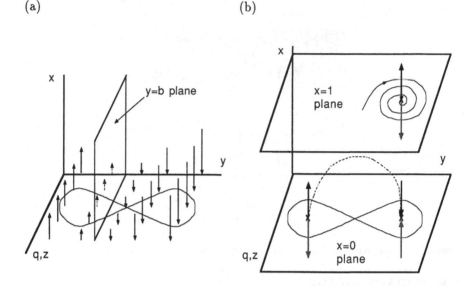

Figure 2. (a) Schematics of the dynamical system (2) flow; (b) schematics of the desired flow with the feedback.

Figure 2(a) shows a schematic of the flow produced by the equations (2). Here, the hyperplane $y = b$ divides the Lorenz system into stable and unstable 'ears' in the transversal x-direction. Now, the flow should be deformed in such a way that after the burst the flow trajectory is deflected towards the fixed point at the center of the stable ear of the Lorenz system (Figure 2(b)). Consider a deformed flow with two invariant subspaces $x = 0$ and $x = 1$. In the $x = 0$ plane the system is a Lorenz system, and the left and right fixed points in the center of the ears are respectively unstable and stable in the x-direction. In the $x = 1$ plane put a fixed point as shown in Figure 2(b). This fixed point is globally attracting in the (y, q, z)-subspace and repelling in the transversal x-direction. Thus any trajectory leaving the unstable (left) ear will be strongly attracted towards this fixed point in the (y, q, z)-subspace and deflected towards the $x = 0$ hyperplane along the x-direction. Moreover, a larger excursion of the trajectory along the x-direction gives the fixed point at $x = 1$ plane more time to affect this trajectory. Thus the reinjection of the system into the $x = 0$ invariant subspace at the beginning of the laminar phase will take place closer to the x-directionally stable fixed point of the Lorenz system.

Now, in the $x = 0$ plane the local behavior should satisfy equations (2) while in the $x = 1$ plane the behavior should be a sink in (y, q, z) subspace and a repeller in x-direction, i.e.

$$
\begin{aligned}
\dot{x} &= \alpha(x - 1), \\
\dot{y} &= \sigma(y - y_0) + \beta(q - q_0), \\
\dot{q} &= -\beta(y - y_0) + \sigma(q - q_0), \\
\dot{z} &= \gamma(z - z_0),
\end{aligned}
\tag{3}
$$

with $\alpha > 0$, $\sigma < 0$, $\gamma < 0$. Therefore smooth connecting functions f and g are needed in a global dynamical system

$$
\dot{X} = F(X)f(x) + G(X)g(x) = H(X),
\tag{4}
$$

with $X = (x, y, q, z)$ and

$$
f(0) = 1, \quad f(1) = 0, \quad g(0) = 0, \quad g(1) = 1.
\tag{5}
$$

Here, F and G describe the local dynamical behavior in the $x = 0$ and $x = 1$ hyperplanes respectively. In addition $H(X)$ should agree with local behaviors F and G in $x = 0$ and $x = 1$ hyperplanes respectively. This requires that the Jacobians and all higher gradients of the global system $\dot{X} = H(X)$ agree with the Jacobians and higher order gradients of the local dynamical systems $\dot{X} = F(X)$ in $x = 0$ plane and $\dot{X} = G(X)$ in $x = 1$ plane. This sets additional restrictions on the connecting functions f and g. For simplicity, here we require that only the Jacobians agree.

Working with the first equation of system (4) we arrive at

$$
\frac{\partial H_1}{\partial x} = \frac{\partial F_1}{\partial x} f(x) + F_1(X)f'(x) + \frac{\partial G_1}{\partial x} g(x) + G_1(X)g'(x).
\tag{6}
$$

Hence,

$$
\left. \frac{\partial H_1}{\partial x} \right|_{x=0} = \left. \frac{\partial F_1}{\partial x} \right|_{x=0} + F_1(0, y, q, z)f'(0) + G_1(0, y, q, z)g'(0).
\tag{7}
$$

Now, $\frac{\partial H_1}{\partial x}\big|_{x=0}$ should agree with $\frac{\partial F_1}{\partial x}\big|_{x=0}$ for all y, q, z. Thus

$$
F_1(0, y, q, z)f'(0) + G_1(0, y, q, z)g'(0) = 0, \quad \forall y, q, z,
\tag{8}
$$

and similarly

$$
F_1(1, y, q, z)f'(1) + G_1(1, y, q, z)g'(1) = 0, \quad \forall y, q, z.
\tag{9}
$$

Thus the additional restrictions on functions f and g are

$$
f'(0) = 0, \quad f'(1) = 0, \quad g'(0) = 0, \quad g'(1) = 0.
\tag{10}
$$

(a)   (b)

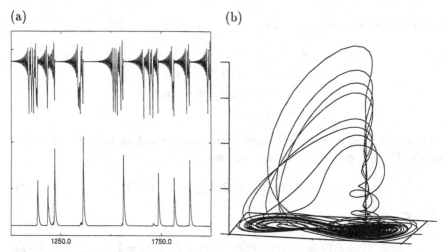

Figure 3. (a) $y(t)$ (upper curve) and $x(t)$ (lower curve) for the system (12); (b) A trajectory of the system (12) in suitably chosen projection of the phase-space.

It is a straightforward computation to show that further partial derivatives of H do not introduce any additional conditions on f and g.

There are infinitely many connecting functions satisfying conditions (5) and (10), for example

$$f(x) = 1 - 3x^2 + 2x^3, \quad \cos^2(\frac{\pi x}{2}); \quad g(x) = 3x^2 - 2x^3, \quad \sin^2(\frac{\pi x}{2}). \tag{11}$$

Thus the following dynamical system with a total of 12 parameters is defined:

$$\begin{aligned}
\dot{x} &= (-\epsilon_1 x^3 + (b - y)x)\cos^2(\frac{\pi x}{2}) + \alpha(x - 1)\sin^2(\frac{\pi x}{2}), \\
\dot{y} &= q\cos^2(\frac{\pi x}{2}) + (\sigma(y - y_0) + \beta(q - q_0))\sin^2(\frac{\pi x}{2}), \\
\dot{q} &= (-y^3 + zy - \epsilon\nu q)\cos^2(\frac{\pi x}{2}) + (-\beta(y - y_0) + \sigma(q - q_0))\sin^2(\frac{\pi x}{2}), \\
\dot{z} &= -\epsilon(z + a(y^2 - 1))\cos^2(\frac{\pi x}{2}) + \gamma(z - z_0)\sin^2(\frac{\pi x}{2}).
\end{aligned} \tag{12}$$

Figure 3(a) shows x and y responses of the system while Figure 3(b) shows the evolution of the trajectory in a suitably chosen 3D projection of the phase-space.

One way to simplify the dynamical system (12) and reduce the number of parameters is to replace the Lorenz forcing by a similar system in 2D. The natural analogy to a Lorenz system in two dimensions is a spiral system

$$\begin{aligned}
\dot{y} &= \frac{A(y^2 - 1)}{2} + Byq, \\
\dot{q} &= \frac{B(1 - y^2)}{2} + Ayq,
\end{aligned} \tag{13}$$

(a)   (b)

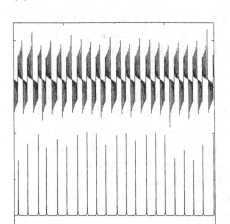
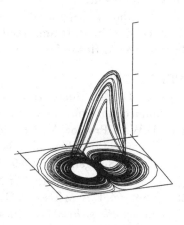

6500   7500

Figure 4. (a) $y(t)$ (upper curve) and $x(t)$ (lower curve) for the system (14); (b) trajectory of the system (14) in the phase-space.

with a pair of fixed points $(1,0)$, $(-1,0)$. Here one fixed point is stable while the other fixed point is unstable. Now, adding a transversal x-direction as before we arrive at the following 3D system with 9 parameters:

$$\dot{x} = (-\epsilon_1 x^3 + (b-y)x)\cos^2(\frac{\pi x}{2}) + \alpha(x-1)\sin^2(\frac{\pi x}{2}),$$

$$\dot{y} = (\frac{A(y^2-1)}{2} + Byq)\cos^2(\frac{\pi x}{2}) + (\sigma(y-y_0) + \beta(q-q_0))\sin^2(\frac{\pi x}{2}), \quad (14)$$

$$\dot{q} = (\frac{B(1-y^2)}{2} + Ayq)\cos^2(\frac{\pi x}{2}) + (-\beta(y-y_0) + \sigma(q-q_0))\sin^2(\frac{\pi x}{2}).$$

Figure 4(a) shows x and y responses of the system while Figure 4(b) shows the evolution of the trajectory in a 3D phase-space.

4 CONCLUSIONS

In this paper a new type of intermittency, called On-Off intermittency, was introduced. In short, a bare-bones model of the intermittency invokes dynamic forcing of a bifurcation parameter of some simple dynamical system in a way that makes the bifurcation parameter spend suitable amounts of time below and above the bifurcation threshold. The result is a time series displaying long periods of inactivity followed by a sudden burst. This simple model does not involve feedback from the intermittent equation into the forcing equations. Thus, the resulting equations are unrealistic in real-life applications since no information about the burst is being transmitted to the driving equations. A geometric model utilizing feedback has been proposed and successfully applied to two- and three-dimensional forcing equations.

Acknowledgements

The author, who was supported through the Office of Naval Technology Post-doctoral Fellowship Program, would like to thank E. Spiegel, C. Tresser, S. Hammel and J. Heagy for a number of fruitful discussions.

REFERENCES

Grebogi, C., Ott, E., Romeiras, F. & Yorke, J.A. 1987 Critical exponents for crisis-induced intermittency. *Phys. Rev. A* **36**, 5365–5380.

Hughes, D.W. & Proctor, M.R.E. 1990 A low-order model of the shear instability of convection: chaos and the effect of noise. *Nonlinearity* **3**, 127–153.

Marzec, C.J. & Spiegel, E.A. 1980 Ordinary differential equations with strange attractors. *SIAM J. Applied Math.* **38**, 403–421.

Platt, N., Spiegel, E.A. & Tresser, C. 1992 On-Off intermittency: a mechanism for bursting. *Phys. Rev. Lett.* (in press).

Pomeau, Y. & Manneville, P. 1980 Intermittent transition to turbulence in dissipative dynamical systems. *Commun. Math. Phys.* **74**, 189–197.

Spiegel, E.A. 1980 A class of ordinary differential equations with strange attractors. *Annals of New York Academy of Sciences* **357**, 305–312.

Dynamo Action in a Nearly Integrable Chaotic Flow

Y. PONTY, A. POUQUET, V. ROM-KEDAR & P.L. SULEM

OCA, CNRS URA 1362
BP 229, 06304 Nice Cedex 4, France

Dynamo action of a time periodic flow with frequency Ω, depending on two space variables, introduced by Galloway & Proctor (1992), is considered when the underlying dynamical system is nearly integrable. Competition between fast and slow dynamos is obtained according to the value of Ω. Fast dynamos produce magnetic sheets located in the chaotic regions near the separatrices of the integrable flow. Slow dynamos lead to magnetic eddies which elongate with increasing magnetic Reynolds number R_m and tend to circumscribe elliptic stagnation points. Sheets and eddies may coexist at moderate R_m. A heuristic argument based on the Melnikov method is used to characterize the frequencies which maximize the efficiency of fast dynamos.

---

1 INTRODUCTION

A simple smooth chaotic flow often used as a candidate for fast dynamos is the 'ABC flow' $\mathbf{u} = (A \sin z + C \cos y, B \sin x + A \cos z, C \sin y + B \cos x)$, where A, B, C are non-zero coefficients. This flow only involves one wavenumber, which considerably reduces the number of operations when the induction equation for the magnetic field

$$\partial_t \mathbf{b} = \nabla \times (\mathbf{u} \times \mathbf{b}) + \eta \Delta \mathbf{b} \tag{1}$$

is solved numerically in Fourier space (Arnold & Korkina 1983; Galloway & Frisch 1986). The dynamo problem is however three-dimensional, which limits the Reynolds number to moderate values.

Examples of flows that seem well-suited to probe the large magnetic Reynolds number limit on present-day computers were recently introduced by Galloway & Proctor (1992), who used the fact that flows depending on only

241

M.R.E. Proctor, P.C. Matthews & A.M. Rucklidge (eds.)
Theory of Solar and Planetary Dynamos, 241–248
©1993 Cambridge University Press.

two space variables can be chaotic if they are time-dependent. We concentrate here on their 'circularly polarized' model (CP)

$$\mathbf{u} = \Big(A\sin(z + \sin\Omega t) + C\cos(y + \cos\Omega t),$$
$$A\cos(z + \sin\Omega t), C\sin(y + \cos\Omega t)\Big). \tag{2}$$

These flows, which display large chaotic regions, can be viewed as a modification of the integrable ABC flow corresponding to $B = 0$, by the introduction of a time periodic phase. For this velocity, magnetic field modes with wavevectors having the same component k_1 in the x-direction evolve independently. Consequently, k_1 can be fixed and the magnetic field computed with a two-dimensional code. For convenience, the (y, z)-periodicity of the magnetic field is taken as that of the flow. There is no periodicity in the x-direction and k_1 can be chosen arbitrarily (but non-zero). For $k_1 = 0.57$, together with $A = C = (3/2)^{1/2}$ and $\Omega = 1$, convincing evidence of fast dynamo was obtained, the magnetic growth rate remaining essentially constant for magnetic Reynolds number $10^2 \le R_m \le 10^4$. We consider here a similar flow, but in a regime where it is nearly integrable. This is obtained by introducing a small parameter ϵ in front of the oscillatory phase of the velocity.

2 THE DYNAMICAL SYSTEM
The fluid trajectories, to be understood mod 2π, obey

$$\dot{y} = A\cos(z + \epsilon\sin\Omega t), \quad \dot{z} = C\sin(y + \epsilon\cos\Omega t) \tag{3}$$

together with $\dot{x} = A\sin(z + \epsilon\sin\Omega t) + C\cos(y + \epsilon\cos\Omega t)$. By dividing (3) by C and rescaling time, it is easily seen that, in addition to the perturbation amplitude ϵ, this dynamical system depends on only two parameters, the reduced frequency $\omega = \Omega/C$ and the ratio $a = A/C$. We concentrate here on the case $a = 1$ for which the unperturbed system has heteroclinic orbits. For convenience, results will be presented in terms of ω.

For $\epsilon = 0$, system (3) admits two elliptic stagnation points $(0, \pi/2)$, $(\pi, 3\pi/2)$, and two hyperbolic ones $(0, 3\pi/2)$, $(\pi, \pi/2)$. For $\epsilon \ne 0$, the points of zero velocity rotate with angular velocity Ω on a circle of radius ϵ, centered at the stagnation points of the unperturbed problem. Useful insight on the system is provided by Poincaré sections in time (stroboscopic views) at $t = n2\pi/\Omega$ with $n \in N$. The size of the chaotic zones is as expected monotonically decreasing with ϵ, yet exhibits a non trivial dependence on ω. Figure 1 shows the section in the (x, y)-plane, for $\epsilon = 0.1$, and various ω. The observable chaotic regions are localized along the heteroclinic connections of the unperturbed system, whereas for $\epsilon = 1$ they cover a large fraction of the domain.

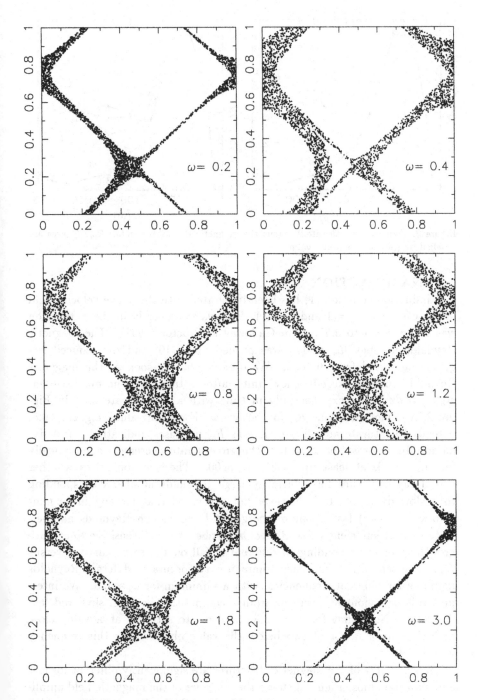

Figure 1. Time Poincaré sections for flow (3), with $a = 1$, $\epsilon = 0.1$.

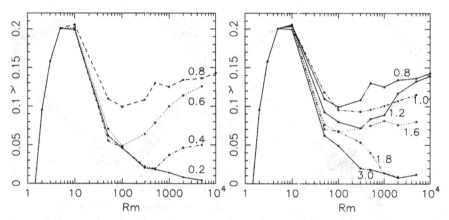

Figure 2. Dynamo growth rate λ versus the magnetic Reynolds number R_m. Curves are labelled by the corresponding value of ω.

3 DYNAMO ACTION

The induction equation (1) has been integrated with the above velocity field in the case $A = C = 1$ and $\epsilon = 0.1$. The wavenumber k_1 in the x-direction was taken equal to 0.57, as in Galloway & Proctor (1992). The magnetic Reynolds number $R_m = 1/\eta$ was pushed up to 10^4 and the reduced frequency ω varied from 0 to 3. We observe that as soon as the magnetic Reynolds number exceeds a few units, after a transient, the magnetic energy $\int \mathbf{b}^2 dx$ grows exponentially in time with a growth rate 2λ. In Figure 2, λ is plotted versus R_m for various ω. For some values, e.g. $\omega = 0.8$, λ tends to saturate at a finite value as R_m increases, indicating a fast dynamo. In contrast for $\omega = 0.2$, the growth rate decreases monotonically for $R_m > 10$, at least up to $R_m = 5000$. The question arises whether this decay continues for arbitrarily large Reynolds number corresponding to a slow dynamo. It is however not precluded that for any finite nonzero ω, a (weak) fast dynamo emerges at large enough Reynolds number. Note that at sufficiently low Reynolds number λ is not sensitive to ω since growth rates corresponding to different ω fall on the same curve λ versus R_m. The separation from this common curve occurs at different Reynolds numbers for different frequencies, with a minimum for $\omega \approx 0.8$. We interpret this behaviour as resulting from a competition between slow and fast dynamos. Note that for $\omega = 0.8$, λ seems to saturate at a value close to 0.15, only a factor of two below the value obtained for this frequency with $\epsilon = 1$.

The geometry of the magnetic structures are significantly different for fast and slow dynamos. Figure 3 shows the contours of the magnetic field amplitude in the (y, z)-plane for various ω at $R_m = 2000$. At $\omega = 0.8$, for which

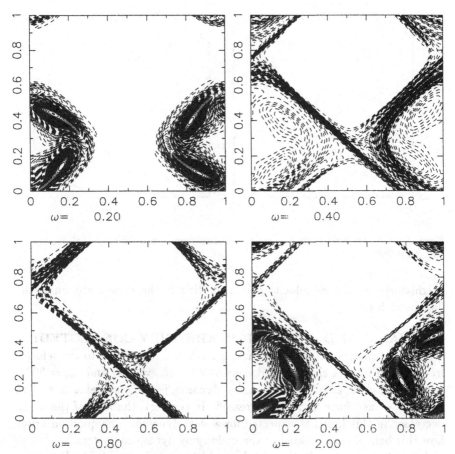

Figure 3. Contours of magnetic field intensity at $R_m = 2000$. Dashed lines refer to levels smaller than half the maximum.

the dynamo appears to be fast, the magnetic structures consist essentially of magnetic layers located in the chaotic regions, near the separatrices of the unperturbed system. As R_m is increased, the thickness of the layers decreases, possibly like $R_m^{-1/2}$ as suggested by a dominant balance argument between stretching and dissipation. Furthermore, the transverse structure of the layer becomes richer with the formation of secondary maxima, and is reminiscent of the structure of the tangled unstable manifold of (3).

For $\omega = 0.2$, for which the dynamo appears to be slow, the magnetic field concentrates in 'magnetic eddies', located in non-chaotic regions, close to the resonance bands of system (3). As R_m is increased, the magnetic eddies, which are rather isotropic at moderate Reynolds numbers, become more elongated and tend to circumscribe the elliptic points. It is noticeable that magnetic eddies, like the resonance bands of (3), appear only in the

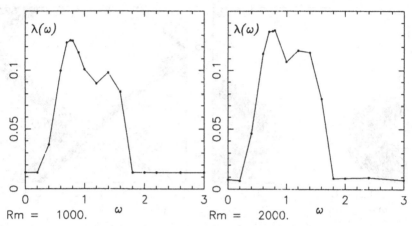

Figure 4. Dynamo growth rate λ versus the reduced frequency ω for $\epsilon = 0.1$, at $R_m = 1000$ and $R_m = 2000$.

neighbourhood of zero velocity points rotating in the same direction as the flow particles.

4 DYNAMO AND CHAOS: HOW ARE THEY CONNECTED?

We already observed that the dynamo growth rate is sensitive to the velocity frequency. Figure 4 shows this dependency for $R_m = 1000$ and $R_m = 2000$. The central peak, associated with a fast dynamo, is maximum at $\omega \simeq 0.8$ and tends to a fixed form as R_m is increased. In contrast, the level of the wings decreases in this limit, as expected for a slow dynamo. The question arises how this behaviour is related to the underlying dynamical system.

A standard characterization of a dynamical system is provided by Lyapunov exponents, because of the analogy between separation of infinitesimally close fluid particles and stretching of the magnetic field at zero magnetic diffusivity. Figure 5(a) shows the largest Lyapunov exponent L versus ω for the trajectories shown in Figure 1. The correlation between this graph and the dynamo growth rate at large magnetic Reynolds number appears to be weak. This confirms that, in the presence of magnetic diffusion, the rate of stretching alone cannot prescribe the efficiency of the dynamo action. Massive cancellation can indeed take place between magnetic field elements stretched in directions which vary strongly from place to place.

One may suspect that the geometry of chaotic zones of the flow and in particular their extent may affect the efficiency of the fast dynamo action. It was suggested by Leonard et al. (1987) and Ottino (1989) that the 'extent of chaos' may be estimated using the Melnikov method. This method is a perturbative calculation of the distance between stable and unstable manifolds resulting from perturbation of homoclinic or heteroclinic trajectories. It

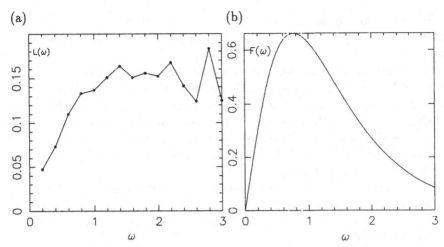

Figure 5. (a) Variation with the reduced frequency ω of the maximum Lyapunov exponent L and (b) of the function F given in (7) for $\epsilon = 0.1$.

is classically used to test the existence of transverse homoclinic orbits which imply the presence of Smale horseshoes and their attendant chaotic dynamics (Guckenheimer & Holmes 1983; Ottino 1989).

By the change of variables $u = z - \pi/2 + \epsilon \sin \omega \tau$, $v = y + \epsilon \cos \omega \tau$ and $\tau = Ct$, and for $a = 1$, (3) becomes

$$\dot{u} = \sin v + \epsilon \omega \cos \omega \tau, \quad \dot{v} = -\sin u - \epsilon \omega \sin \omega \tau. \tag{4}$$

This is the standard form for the implementation of Melnikov method. To leading order in ϵ, the distance $d(\tau_0)$ between the manifolds is proportional to ϵ and to the 'Melnikov function'. For each of the unperturbed heteroclinic solutions (u_0^j, v_0^j), where $j = 1, 2, 3, 4$, this function reads

$$M^j(\tau_0) = -\omega \int_{-\infty}^{\infty} \sin v_0^j \, \sin(\omega(\tau + \tau_0)) \, d\tau + \omega \int_{-\infty}^{\infty} \sin u_0^j \cos(\omega(\tau + \tau_0)) \, d\tau. \tag{5}$$

After some algebra, we get

$$M^j(\tau_0) = F(\omega) \, (S_1^j \sin \omega \tau_0 + S_2^j \cos \omega \tau_0), \tag{6}$$

where $(S_1, S_2) = (-1, 1), (-1, -1), (1, -1), (1, 1)$ for $j = 1, 2, 3, 4$ respectively. Furthermore

$$F(\omega) = \omega \pi \, \text{sech}(\frac{\pi \omega}{2}). \tag{7}$$

Quoting Ottino (1989), 'we expect that an extreme in $F(\omega)$ should maximize the extent of chaos'. This function is plotted in Figure 5(b). We observe that the range of frequencies ω leading to the largest dynamo growth rates

(Figure 4) is located around the maximum of $F(\omega)$. We checked that for sufficiently small ϵ (typically $\epsilon < 0.5$), this behaviour is essentially independent of ϵ.

The Melnikov method has here been used as an heuristic tool to measure the width of the chaotic zones. Further investigations are required to decide whether the location of the maximum dynamo growth rate is indeed correlated with the location of the maximum of the Melnikov function (as found here) or if it is mostly a coincidence. As the next step, we plan to examine the case $a \neq 1$, where heteroclinic connections of the unperturbed problem are replaced by homoclinic orbits.

Computations were performed on the CRAY-YMP of the Institut Mediteranéen de Technologie (Marseille).

REFERENCES

Arnold, V.I. & Korkina, E.I. 1983 The growth of magnetic field in an incompressible flow. *Vestn. Mosk. Univ. Mat. Meckh.* **3**, 43–46 .

Galloway, D.J. & Frisch, U. 1986 Dynamo action in a family of flows with chaotic stream lines. *Geophys. Astrophys. Fluid Dynam.* **36**, 53–83.

Galloway, D.J. & Proctor, M.R.E. 1992 Numerical calculations of fast dynamos for smooth velocity field with realistic diffusion. *Nature* **356** , 691–693.

Guckenheimer, J. & Holmes, P. 1983 *Nonlinear Oscillations, Dynamical Systems and Bifurcations of Vector Fields.* Springer.

Leonard, A., Rom-Kedar, V. & Wiggins, S. 1987 Fluid mixing and dynamical systems, *Nucl. Phys.* **B** *(Proc. Suppl.)* **2**, 179–190.

Ottino, J.M. 1989 *The Kinematics of Mixing: Stretching, Chaos and Transport.* Cambridge University Press.

The Dynamo Mechanism in the Deep Convection Zone of the Sun

T. PRAUTZSCH

Universitäts-Sternwarte Göttingen
Geismarlandstraße 11
D-3400 Göttingen, Germany

The helioseismological results about the Solar law of rotation pose some serious problems for dynamo theory. However, if the magnetic flux is bounded in the lower part of the convection zone and the α-effect is concentrated at the equator, it is possible to obtain correct butterfly diagrams. This model seems to be a natural combination of the new law of rotation, the suggested storage of the magnetic flux at the bottom of the convection zone, the trapping of flux tubes at low latitudes and the induction effect of magnetostrophic waves.

1 THE DYNAMO IN THE CONVECTION ZONE

In 1969, Steenbeck & Krause presented results of the first hydrodynamic dynamo model acting in the turbulent convection zone (CZ) and based on the idea of mean field electrodynamics. They introduced two spherical shells for the induction effects: in the inner, there is the differential rotation ($\Omega \sim r$) and in the outer, one has the turbulent rotating matter ($\alpha \sim \cos \vartheta$). This simple model is in agreement with most of the observed magnetic patterns, such as the butterfly diagram (Figure 1), Hale's polarity rule and the 22 year period of the Solar cycle.

During half a cycle, i.e. eleven years, the activity belts, as a measure of the toroidal field, move from about $\pm 30°$ latitude towards the equator. In the vicinity of the pole, there are no active regions. But observations of torsional oscillations (Howard & LaBonte 1982) and Solar wind (Legrand & Simon 1991) suggest that the toroidal field starts the reversal of its polarity there (Schüssler 1981).

M.R.E. Proctor, P.C. Matthews & A.M. Rucklidge (eds.)
Theory of Solar and Planetary Dynamos, 249–256
©1993 Cambridge University Press.

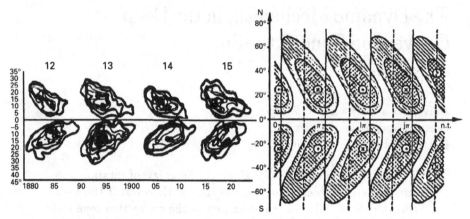

Figure 1. Butterfly diagrams (contour lines as a function of latitude and time). Left: Observed latitude drift of sunspot occurrence (Scheffler & Elsässer 1974). Right: Calculated toroidal field (Steenbeck & Krause 1969).

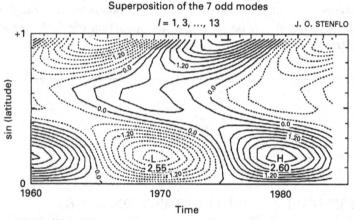

Figure 2. Contours of observed radial magnetic field as a function of latitude and time (Stenflo 1988).

The radial field component at the equator behaves like the toroidal field but with a phase lag of π (Stix 1987). Furthermore, it is possible to find polar branches (Figure 2) which are responsible for the migration towards the pole of the magnetic structures on the surface of the Sun (e.g. Stix 1974).

2 PROBLEMS

Further observations and theoretical considerations about turbulent convection, field structure, poloidal field and the law of rotation did not match the concept of the simple model of Steenbeck & Krause (1969).

Numerical calculations (Köhler 1973) and the mixing length formalism (Stix 1976) showed considerably larger values for the turbulent diffusivity

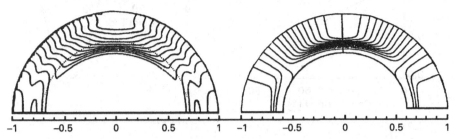

Figure 3. Contours of constant angular velocity in a quadrant section across the Sun. Left: Helioseismological results (Libbrecht 1988). Right: Expansion of Legendre functions applied in the following calculations.

($\eta_t = 10^{12} - 10^{13}$ cm^2 s^{-1}). These yield periods of the Solar cycle from half a year up to five years for a model with distributed induction effects.

Magnetoconvection leads to a concentration of the magnetic flux in small-scaled structures, the flux tubes (Galloway & Weiss 1981). This affects the mathematical requirements for the α-effect and in addition, leads to storage problems for the magnetic fields in the CZ (e.g. Schmitt 1992; Weiss 1993).

Recent results of helioseismology give us a law of rotation (Figure 3). Only in the lowest part of the CZ does the relative velocity due to the differential rotation dominate the turbulent velocity. However, this is necessary for the well ordered generation and for the observed dominance of the toroidal field.

3 THE OVERSHOOT REGION

Because of the problems mentioned above, it was suggested (Galloway & Weiss 1981) that the mean part of the magnetic flux is concentrated in a thin subadiabatic layer below the CZ, called the overshoot region (OR). There, the turbulent convection and the turbulent diffusivity should drastically decrease. Numerical calculations of Moreno-Insertis *et al.* (1992) confirm the existence of realistic values for the magnetic field ($\simeq 10^5$ G), the depth ($\simeq 10^4$ m) and the subadiabaticity ($\nabla - \nabla_{ad} \leq -10^{-5}$) of this layer.

Also, the new law of rotation with a strong radial gradient at the bottom of the CZ and a rigidly rotating core as well as the observed quasi-rigid rotation of the recurrence of the magnetic patterns on the surface (Stenflo 1990) support this idea of a magnetic source layer in this region.

4 MODEL

The following assumptions will be made: incompressibility and an isotropic α-effect. Furthermore, the kinematic approach will be applied and solutions with dipole symmetry will be chosen. For α, two dependencies on the colatitude ϑ are applied (Figure 4, top). The models in the CZ (Figure 4, middle) and the OR (Figure 4, bottom) differ in the radial distribution of α and η_t.

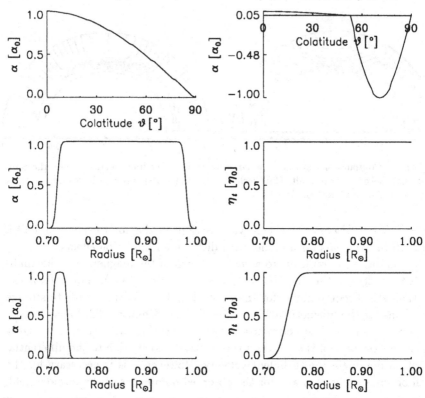

Figure 4. Top: The two applied α-dependencies on ϑ: $\alpha \sim \cos\vartheta$ (left) and α concentrated at the equator (right). Middle: The radial distribution of α (left) and η_t (right) in the CZ. Bottom: The radial distribution of α (left) and η_t (right) in the OR.

The law of rotation is modelled by expansion of Legendre functions (Figure 3, right side).

5 RESULTS

With $\eta_t = $ const., there are only oscillating solutions if α is positive in the northern hemisphere. In this case, dynamo waves move outwards along the isorotation lines (e.g. Stix 1976) which point mostly in the radial direction in the CZ. Therefore, the equatorwards migration of the toroidal field must be evoked by diffusion and different dynamo wave speeds (Figure 5, top). If $\alpha \sim \cos\vartheta$, we find the strongest dynamo action with the fastest dynamo wave and therefore the maximum of the toroidal field in the butterfly diagram at $\pm 50°$ (Figure 5). As a result, the equatorial branches are too weak and the polar ones too strong. However, we get big polar branches for the radial field component (Figure 6).

With η_t reduced in the lower part of the CZ, the magnetic field is essentially concentrated in this region (Figure 7, bottom). Therefore, the dynamo waves

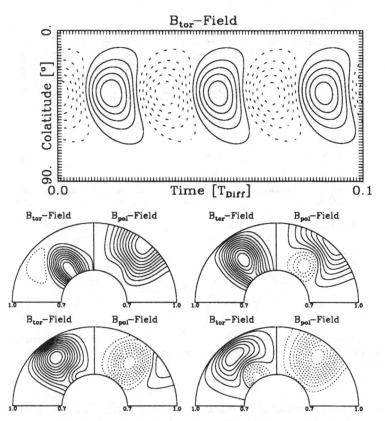

Figure 5. Dynamo model in the CZ with $\alpha \sim \cos \vartheta$. Top: Butterfly diagram of the toroidal field. Bottom: Contours of constant toroidal field strength (left) and poloidal lines of force (right). The four pictures with time differences of $T/8$ cover half a cycle.

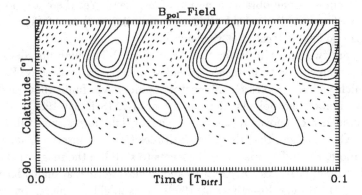

Figure 6. Butterfly diagram of the radial field for a dynamo model in the CZ with $\alpha \sim \cos \vartheta$.

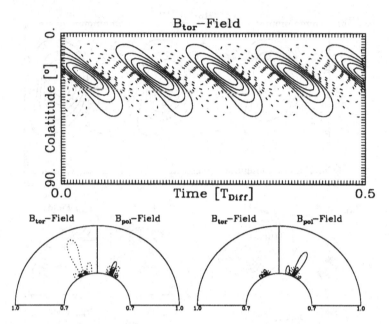

Figure 7.   Dynamo model in the OR with $\alpha \sim \cos\vartheta$. Top: Butterfly diagram of the toroidal field. Bottom: Meridional field pictures as in Figure 5. The two pictures with a time differences of $T/4$ cover half a cycle.

migrate in the latitudinal direction (Figure 7, top). Due to the reduced turbulent diffusivity, the period is about three to five times larger.

In order to obtain the observed equatorwards migration of dynamo waves at low latitudes, α has to be negative north of the equator. In that case, we obtain periodic solutions if α is concentrated at the equator (Figure 8). The resulting butterfly diagram is in good agreement with observations. With $\eta_t = 10^{12}$ cm^2 s^{-1}, one obtains a period of 27 years. Yet there are no polar branches. The meridional field pictures are similar to those of Figure 7 but the magnetic field is located at the equator.

There are two possibilities for that kind of α: i) The high flux density in the OR requires a dynamic *ansatz* instead of the kinematic one (e.g. Parker's loop which yields $\alpha \sim \cos\vartheta$). Magnetostrophic waves fulfill that condition and may induce the poloidal field as we need it (Schmitt 1987). ii) Due to the influence of buoyancy, curvature and Coriolis forces, flux tubes with $B \leq 10^5$ G oscillate in the OR at low latitudes (Moreno-Insertis *et al.* 1992). Furthermore, the differential rotation seems to stabilize the magnetic field in the OR at lower latitudes ($\partial\Omega/\partial r > 0$) and to destabilize it at higher latitudes ($\partial\Omega/\partial r < 0$) (Moreno-Insertis *et al.* 1992). This might be the reason for the concentration of the dynamo action around the equator.

B_{tor}–Field

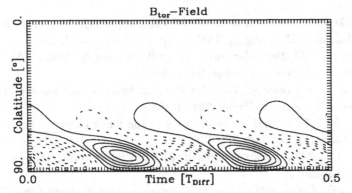

Figure 8. Butterfly diagram of the toroidal field for a dynamo model in the OR with α concentrated at the equator.

6 CONCLUSION

Small-scale structure of magnetic field, strong turbulence and the law of rotation yielded by helioseismology suggest that most of the dynamo mechanism takes place at the bottom of the convection zone. Assuming this, we only get realistic butterfly diagrams if the α-effect is concentrated at the equator and negative in the northern hemisphere. Magnetostrophic waves or the trapping of flux tubes at lower latitudes could lead to that kind of induction effect for the poloidal field. The polar branches of that field perhaps can originate in the convection zone by a turbulent α-effect.

REFERENCES

Galloway, D.J. & Weiss, N.O. 1981 Convection and magnetic fields in stars. *Astrophys. J.* **243**, 945–953.

Howard, R. & LaBonte, B.J. 1982 Torsional waves on the Sun and the activity cycle. *Solar Physics* **75**, 161–178.

Köhler, H. 1973 The Solar dynamo and estimates of the magnetic diffusivity and the α-effect. *Astron. Astrophys.* **25**, 467–476.

Legrand, J.P. & Simon, P.A. 1991 A two-component Solar cycle. *Solar Physics* **131**, 187–209.

Libbrecht, K.G. 1988 Solar p-mode frequency splitting. In *Seismology of the Sun and Sun-Like Stars* (ed. E.J. Rolfe), ESA SP-286, pp. 131–136. European Space Agency.

Moreno-Insertis, F., Schüssler, M. & Ferriz Mas, A. 1992 Storage of magnetic flux tubes in a convective overshoot region. *Astron. Astrophys.* (in press).

Scheffler, H. & Elsässer, H. 1974 *Physik der Sterne und der Sonne*, p. 207. Bibliographisches Institut.

Schmitt, D. 1987 An $\alpha\omega$-dynamo with an α-effect due to magnetostrophic waves. *Astron. Astrophys.* **174**, 281–287.

Schmitt, D. 1992 The Solar dynamo. In *The Cosmic Dynamo* (ed. F. Krause, K.H. Rädler & G. Rüdiger), IAU-Symp. 157 (in press). Reidel.

Schüssler, M. 1981 The Solar torsional oscillation and dynamo models of the Solar cycle. *Astron. Astrophys.* **94**, L17–L18.

Steenbeck, M. & Krause, F. 1969 Zur Dynamotheorie stellarer und planetarer Magnetfelder. *Astron. Nachr.* **291**, 49–84.

Stenflo, J.O. 1988 Global waves patterns in the Sun's magnetic field. *Astrophys. Space Science* **144**, 321–336.

Stenflo, J.O. 1990 Time invariances of the Sun's rotation rate. *Astron. Astrophys.* **233**, 220–228.

Stix, M. 1974 Comments on the Solar dynamo. *Astron. Astrophys.* **37**, 121–133.

Stix, M. 1976 Differential rotation and the Solar dynamo. *Astron. Astrophys.* **47**, 243–254.

Stix, M. 1987 On the origin of stellar magnetism. In *Solar and Stellar Physics* (ed. E.H. Schröter & M. Schüssler), pp. 15–33. Springer.

Weiss, N.O. 1993 Solar and stellar dynamos. In *Lectures on Solar and Planetary Dynamos* (ed. M.R.E. Proctor & A.D. Gilbert), Cambridge University Press.

Shearing Instabilities in Magnetoconvection

A.M. RUCKLIDGE & P.C. MATTHEWS

Department of Applied Mathematics and Theoretical Physics
University of Cambridge, Silver St., Cambridge, CB3 9EW UK

Recent numerical simulations of two-dimensional convection (compressible and Boussinesq) in the presence of a vertical magnetic field reveal that in some circumstances, narrow rolls are unstable to horizontal shear: tilted rolls are observed, as well as oscillating shearing motion. During the oscillation, the rolls tilt over and are replaced by a vigorous horizontal streaming motion, which decays, and the rolls are reformed, only to tilt over again, either in the same or in the opposite direction. A low-order model of this problem is constructed by truncating the PDEs for Boussinesq magnetoconvection. In the model, oscillatory shearing motion is created either in a Hopf bifurcation from untilted rolls, in which case the rolls tilt first one way and then the other, or in a Hopf bifurcation from tilted rolls, in which case the rolls always tilt in the same direction. Oscillations of the second type are converted into oscillations of the first type in a gluing bifurcation. This scenario is interpreted in terms of a Takens–Bogdanov bifurcation.

1 MOTIVATION

The interaction between convection and magnetic fields plays a central role in the theory of stellar dynamos. In order to investigate this interaction in detail, we consider a simplified problem: two-dimensional convection in a vertical magnetic field. To represent the astrophysical situation, in which there are no sidewalls, we consider a box with periodic boundary conditions in the horizontal direction, allowing horizontal flows. It is found that convection can be unstable to a horizontal shearing motion. Figure 1 (after Proctor *et al.* 1993) shows a periodic oscillation in which steady convection in a stratified layer is unstable to shear: the rolls tip over to the left, ending up in a state of vigorous shearing motion with all rolls turning over in one direction. This motion is opposed by the Lorentz forces of the magnetic field, which has been stretched out by the shearing flow, so the fluid comes to an almost complete halt before redeveloping rolls and tipping over in the opposite direction. At each stage in the oscillation, the pattern drifts, but the oscillation has an overall symmetry:

M.R.E. Proctor, P.C. Matthews & A.M. Rucklidge (eds.)
Theory of Solar and Planetary Dynamos, 257–264
©1993 Cambridge University Press.

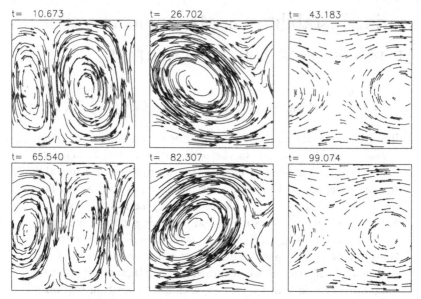

Figure 1. Two-dimensional compressible convection in a vertical magnetic field (after Proctor *et al.* 1993). The frames are (approximately) equally spaced in time over one oscillation (reading from left to right) and each frame shows the instantaneous velocity vectors for one complete horizontal period. This calculation was done in a square box.

the motion is invariant under a shift of half a period in time followed by a reflection in a vertical plane. This type of oscillation is termed a direction-reversing travelling wave by Landsberg & Knobloch (1991), or a pulsating wave by Proctor & Weiss (1993). Note that the oscillations described here occur for magnetic fields for which the initial bifurcation is to steady convection.

The onset of the shearing instability can be understood by the following simple feedback mechanism. Convection rolls that are tilted to the left carry the left-moving fluid to the top of the layer (and the right-moving fluid to the bottom), leading to a net shear flow through the layer, which enhances the leftwards tilt of the rolls.

2 THE SHEARING INSTABILITY IN BOUSSINESQ MAGNETOCONVECTION

Similar behaviour is observed in two-dimensional Boussinesq convection in a vertical magnetic field, and since the partial differential equations (PDEs) for Boussinesq convection are much simpler than those for compressible convection, enabling more detailed numerical experiments to be carried out and facilitating the construction of truncated model equations, we shall concentrate on this case. It should be noted, however, that the equations for Boussinesq convection are symmetric under reflections in the horizontal mid-plane (unlike compressible convection), which implies that unless this up–down symmetry

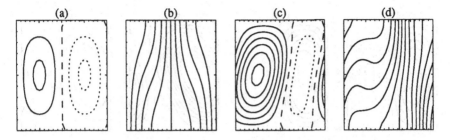

Figure 2. Two-dimensional Boussinesq convection in a vertical magnetic field. (a) $R = 1.15R_0$, streamlines and (b) magnetic field lines for steady untilted convection; (c) $R = 1.30R_0$, streamlines and (d) magnetic field lines for steady tilted convection. Solid streamlines indicate clockwise rolls, dotted streamlines indicate counter-clockwise rolls, and $\Psi = 0$ on the dashed streamline.

is broken, the pulsating waves described above will not travel and will have point symmetry about the centre of the roll.

The PDEs for Boussinesq convection in a vertical magnetic field are:

$$\frac{\partial \omega}{\partial t} + \mathrm{J}\left(\Psi, \omega\right) = \sigma\nabla^2\omega - \sigma R\frac{\partial \theta}{\partial x} - \sigma\zeta Q\left(\frac{\partial \nabla^2 A}{\partial z} + \mathrm{J}\left(A, \nabla^2 A\right)\right), \quad (1)$$

$$\frac{\partial \theta}{\partial t} + \mathrm{J}\left(\Psi, \theta\right) = \nabla^2\theta + \frac{\partial \Psi}{\partial x}, \quad (2)$$

$$\frac{\partial A}{\partial t} + \mathrm{J}\left(\Psi, A\right) = \zeta\nabla^2 A + \frac{\partial \Psi}{\partial z}, \quad (3)$$

where $\omega = -\nabla^2\Psi$ is the vorticity, Ψ is the streamfunction, θ is the deviation from the conducting temperature profile, A is the deviation of the flux function from a uniform vertical magnetic field, and x, z and t are the horizontal, vertical and time coordinates respectively (Knobloch, Weiss & Da Costa 1981). The nonlinearities in the equations are in the Jacobian operator $\mathrm{J}(f, g) = (\partial f/\partial x)(\partial g/\partial z) - (\partial g/\partial x)(\partial f/\partial z)$. The physical parameters are the Prandtl number σ and magnetic Prandtl number ζ, the Rayleigh number R (proportional to the temperature difference across the layer) and the Chandrasekhar number Q (proportional to the square of the imposed magnetic field). The boundary conditions are chosen for mathematical convenience: $\Psi = \omega = \theta = \partial A/\partial z = 0$ on the top and bottom walls ($z = 0, 1$). We impose periodic horizontal boundary conditions with spatial period $\lambda = 2\pi/k$.

Illustrative solutions of the PDEs for Boussinesq magnetoconvection are given in Figures 2–4, which show the behaviour of the system as the Rayleigh number R is increased with the other parameters fixed at $\sigma = 0.5$, $\zeta = 0.2$, $\lambda = 0.756$ and $Q = 63.1$. For these values, $R_0 = 7125$, where R_0 is the critical Rayleigh number for Bénard convection in rolls of this width – with Q equal to the value given, steady convection sets in at $R = 1.1R_0$. These solutions were computed using a spectral code with eight Fourier modes in each of the

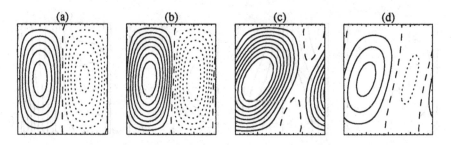

Figure 3. One complete vacillation at $R = 1.45R_0$ (streamlines). Note how the direction of tilt and the sense of the rolls does not change. The contour levels are the same throughout.

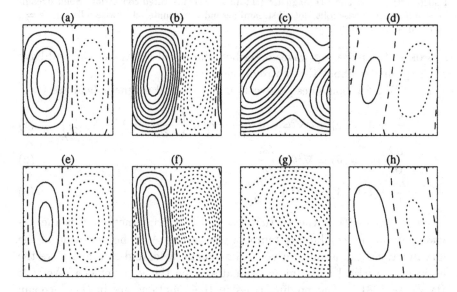

Figure 4. Oscillatory behaviour at $R = 1.50R_0$ (streamlines): a pulsating wave. The direction of tilt changes through the cycle while the sense of the rolls does not change, although in (c) and (g), one of the two rolls has been swamped by the vigorous shearing motion.

two directions. Although the resolution is low, it is sufficient for these mildly supercritical calculations.

At $R = 1.15R_0$ (Figure 2a and b), steady convection is observed, with vertical roll boundaries; at $R = 1.30R_0$ (Figure 2c and d), the steady rolls have tilted boundaries. The steady untilted solution is invariant under reflection in the vertical mid-plane; at the higher Rayleigh number, this symmetry is broken. The two possible tilted solutions are related by the broken reflection symmetry. In the Boussinesq calculation, these steady tilted rolls are stationary, but in the compressible case, the absence of the up–down symmetry implies that tilted rolls always drift (cf. Proctor & Weiss 1993).

As R is increased ($R = 1.45R_0$, Figure 3), the system undergoes a Hopf bifurcation and vacillates about the (unstable) steady tilted solution. This periodic solution is always tilted in the same direction. At still higher R ($R = 1.50R_0$, Figure 4), the symmetry is restored in a gluing bifurcation: the system now spends an equal amount of time tilted in each direction, and is invariant under a reflection in the vertical mid-plane followed by a shift in time of half a period. At the gluing bifurcation, there is a pair of homoclinic connections from the (unstable) untilted solution back to itself: if the system started near the untilted solution, it would tip over (to the left or to the right) and then return to a neighbourhood of the untilted solution.

3 MODEL EQUATIONS

Ordinary untilted convection is represented by terms like $\Psi \propto \sin kx \sin \pi z$. A term proportional to $\sin \pi z$ generates a horizontal shear, and nonlinear interactions between these two modes generate a $\cos kx \sin 2\pi z$ term, which gives the rolls a tilted appearance. We therefore pose the eleven-mode truncation:

$$\Psi = \Psi_{11} \sin kx \sin \pi z + \Psi_{01} \sin \pi z + \Psi_{12} \cos kx \sin 2\pi z, \qquad (4)$$

$$\theta = \theta_{11} \cos kx \sin \pi z + \theta_{02} \sin 2\pi z + \theta_{12} \sin kx \sin 2\pi z, \qquad (5)$$

$$A = A_{11} \sin kx \cos \pi z + A_{20} \sin 2kx$$
$$+ A_{01} \cos \pi z + A_{12} \cos kx \cos 2\pi z + A_{10} \cos kx, \qquad (6)$$

where the mode amplitudes are functions only of time. This truncation yields an eleventh-order set of ordinary differential equations (ODEs) that includes as invariant subsystems the Lorenz (1963) equations and the equations of Howard & Krishnamurti (1986) for sheared Bénard convection (no magnetic field), and the fifth-order truncated model of magnetoconvection without shear of Knobloch, Weiss & Da Costa (1981). Lantz (1993) has studied the analogous truncated model for sheared convection in a horizontal field.

In general, the model ODEs have three types of fixed point: the trivial solution (with all modes equal to zero), the two steady untilted solutions ($\Psi_{01} = \Psi_{12} = \theta_{12} = A_{01} = A_{12} = A_{10} = 0$) and the four tilted steady solutions (all modes nonzero). The multiplicity of the solutions is due to the symmetry of the problem: the rolls may go clockwise or counter-clockwise, and they may tilt to the left or to the right. As R is increased, the pitchfork bifurcation from the trivial solution to steady convection is followed by a second pitchfork to tilted steady convection if Q is not too large. This pitchfork bifurcation occurs at small values of Ψ_{11} (that is, soon after the initial pitchfork bifurcation) if the rolls are narrow. The number of free parameters in the problem is reduced by taking the limit of very narrow rolls, using the approach of Hughes & Proctor (1990) and of Rucklidge (1992). While difficult to justify physically, taking this limit has the advantage of reducing the model from eleventh to fifth order while retaining the essential dynamics. The resulting set of ODEs is:

$$\dot{\Psi}_{11} = \mu\Psi_{11} + \Psi_{11}\theta_{02} - \Psi_{01}\Psi_{12},$$

$$\dot{\theta}_{02} = -\theta_{02} - \Psi_{11}^2,$$

$$\dot{\Psi}_{12} = (\mu - \frac{9\sigma}{4(1+\sigma)})\Psi_{12} + \Psi_{11}\Psi_{01},$$

$$\dot{\Psi}_{01} = -\frac{\sigma}{4}\Psi_{01} - \frac{\sigma Q}{4\pi^2}A_{01} + \frac{3(1+\sigma)}{4\sigma}\Psi_{11}\Psi_{12},$$

$$\dot{A}_{01} = \frac{\zeta}{4}\Psi_{01} - \frac{\zeta}{4}A_{01},$$

(7)

where μ is proportional to the amount by which the Rayleigh number exceeds critical and the mode amplitudes have been scaled by the small width of the rolls. This model clarifies the nonlinear interactions that are occurring in the oscillations. Suppose Ψ_{11}, Ψ_{01} and Ψ_{12} are all positive at some stage (for example, in Figure 3b or Figure 4a). The nonlinear terms in the $\dot{\Psi}_{01}$ and $\dot{\Psi}_{12}$ equations both act as to increase the shear and the tilt (Figure 3c or Figure 4b–c), so Ψ_{01} and Ψ_{12} will grow. The nonlinear interaction between these modes in the $\dot{\Psi}_{11}$ equation will damp the convective motion (Figure 3d or Figure 4d), so Ψ_{11} will decrease, followed by Ψ_{01} and Ψ_{12} as they are damped by linear viscous terms. At this point, there are four possibilities: the three amplitudes may grow again without changing sign (as in Figure 3), Ψ_{01} and Ψ_{12} may change sign (as in Figure 4e–h), Ψ_{11} and Ψ_{12} may change sign (which occurs if $Q = 0$ – see Howard & Krishnamurti 1986), or Ψ_{11} and Ψ_{01} may change sign – such an oscillation can arise after a global bifurcation.

A partial unfolding diagram for (7) with $\sigma = 0.5$ and $\zeta = 0.2$ is shown in Figure 5. The initial bifurcation to steady convection occurs at $\mu = 0$. If Q is not too large ($0 < Q \leq 5$), then the next bifurcation as μ increases is a pitchfork bifurcation to steady tilted rolls, followed by a Hopf bifurcation and a gluing bifurcation to form pulsating waves in exactly the same sequence as the Boussinesq PDEs described in section 2. This is followed by a complicated sequence of global bifurcations that will be discussed in detail elsewhere (Rucklidge & Matthews 1993). The Hopf and the gluing bifurcations originate at the Takens–Bogdanov point C where they intersect the line of pitchfork bifurcations to tilted convection. Also emerging from this point is a line of Hopf bifurcations directly to pulsating waves – this bifurcation has yet to be observed in the PDEs. Near the point C, the Hopf and gluing bifurcations have a different character from that when Q is small: at C, the Hopf bifurcation is subcritical and the gluing bifurcation involves unstable pulsating waves. These orbits, created to the right of the gluing bifurcation near the point C, are destroyed in a saddle-node bifurcation (not shown in the figure) when they collide with the stable pulsating waves created in the Hopf bifurcation

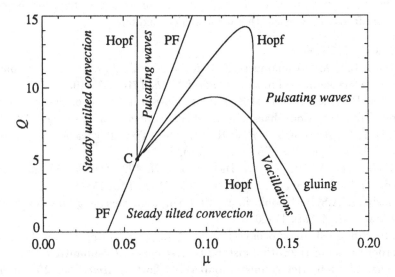

Figure 5. Partial unfolding diagram for (7) with $\sigma = 0.5$ and $\zeta = 0.2$ – the attracting solutions in certain regions of the unfolding plane are indicated. Only the first few bifurcations are shown.

from steady untilted convection. Thus the transition between the two routes to pulsating waves is organised by the Takens–Bogdanov bifurcation point.

4 CONCLUDING REMARKS

We have shown that with periodic horizontal boundary conditions, one may observe steady and oscillatory sheared convection – this type of motion seems to be favoured just above the initial onset of convection if the rolls are narrow and if Q is not too large. This behaviour is also observed in three-dimensional convection – preliminary three-dimensional calculations (Matthews, Proctor & Weiss 1993) show two-dimensional shearing motion as described in this paper. Further interesting behaviour is observed: two-dimensional rolls along one axis are destroyed by vigorous shearing motion to be replaced by two-dimensional rolls along a perpendicular axis. Another issue is how the sheared rolls interact with travelling convection, which is also possible with periodic boundary conditions. A low-order truncated model including sheared and travelling rolls would require 21 modes: complex versions of most of the modes included here, as well as $\Psi_{02} \sin 2\pi z$ and $A_{02} \cos 2\pi z$. Julien, Brummell & Hart (1993) have considered sheared travelling waves in Bénard convection. In the limit of narrow rolls, they obtained model ODEs analogous to (7), but with $Q = 0$ and the $\dot{A}_{01}$ equation removed.

This work represents part of an ongoing study of magnetoconvection in collaboration with N.O. Weiss and M.R.E. Proctor. We thank D.P. Brownjohn for kindly providing the data for Figure 1, and we are grateful for financial support from SERC while this work was carried out.

REFERENCES

Howard, L.N. & Krishnamurti, R. 1986 Large-scale flow in turbulent convection: a mathematical model. *J. Fluid Mech.* **170**, 385–410.

Hughes, D.W. & Proctor, M.R.E. 1990 A low-order model of the shear instability of convection: chaos and the effects of noise. *Nonlinearity* **3**, 127–153.

Julien, K.A., Brummell, N.H. & Hart, J. 1993 Travelling waves in convection with large-scale flows (in preparation).

Knobloch, E., Weiss, N.O. & Da Costa, L.N. 1981 Oscillatory and steady convection in a magnetic field. *J. Fluid Mech.* **113**, 153–186.

Landsberg, A.S. & Knobloch, E. 1991 Direction-reversing traveling waves. *Phys. Lett. A* **159**, 17–20.

Lantz, S.R. 1993 Mgnetoconvection dynamics in a stratified layer. II. A low-order model of the tilting instability. *Astrophys. J.* (submitted).

Lorenz, E.N. 1963 Deterministic nonperiodic flow. *J. Atmos. Sci.* **20**, 130–141.

Matthews, P.C., Proctor, M.R.E. & Weiss, N.O. 1993 Compressible magnetoconvection in three dimensions: planform selection and weakly nonlinear behaviour. *J. Fluid Mech.* (submitted).

Proctor, M.R.E. & Weiss, N.O. 1993 Symmetries of time-dependent magnetoconvection. *Geophys. Astrophys. Fluid Dynam.* (in press).

Proctor, M.R.E., Weiss, N.O., Brownjohn, D.P. & Hurlburt, N.E. 1993 Nonlinear compressible magnetoconvection. Part 2. Instabilities of steady convection. *J. Fluid Mech.* (submitted).

Rucklidge, A.M. 1992 Chaos in models of double convection. *J. Fluid Mech.* **237**, 209–229.

Rucklidge, A.M. & Matthews, P.C. 1993 Analysis of the shearing instability in nonlinear magnetoconvection (in preparation).

On the Role of Rotation of the Internal Core Relative to the Mantle

ALEXANDER RUZMAIKIN

Isaac Newton Institute for Mathematical Sciences
University of Cambridge, 20 Clarkson Rd., Cambridge, CB3 0EH UK

The outer fluid core of the Earth can be considered as a fluid between two hard spheres (the internal core and the rock mantle)
rotating with different but close angular velocities. In the incompressible, nonconducting almost inviscid limit a singular cylindrical surface having the radius of the internal sphere appears (the
Proudman solution). A shear layer forming around this surface in
the non-ideal fluid may have important implications for the geodynamo.

1 INTRODUCTION

The aim of this short paper is to attract attention to one feature in the Earth's
fluid core. The feature is an internal shear layer induced by a relative rotation
of the inner core. Large gradients of the velocity around this layer may be
important for the geodynamo. Note, in particular, that in the geodynamo
model-Z without an account of the inner core rotation one of the basic sources
(the α-effect) is assumed to be concentrated near the core-mantle boundary
(Braginsky 1993).

The inner core of the Earth can be considered as a hard iron ball of radius
approximately $0.2R$, where R is the Earth's radius. The rest of the planet is
occupied by the outer liquid core and the rock mantle in the form of spherical
shells of almost equal width, $0.4R$. The other iron-rock planets (Mercury,
Mars), except probably Venus, also have inner cores (Stevenson 1983). As
the source of compositional convection (Loper & Roberts 1983) the inner core
is apparently a necessary part of the planetary dynamo.

M.R.E. Proctor, P.C. Matthews & A.M. Rucklidge (eds.)
Theory of Solar and Planetary Dynamos, 265–270
©1993 Cambridge University Press.

A new problem appears when we take into account the rotation of the inner core. There is no reason to expect that the inner core rotates with the same angular velocity as the mantle. The difference in the rotations may arise in the process of core formation or may be due to a difference in the Lunar and Solar tidal action on the core and mantle if we take into account a small difference in ellipticities of the real core and mantle. However the actual difference between the angular velocities of the inner core and mantle is determined not by these slow evolutional and tidal reasons but by the viscous and magnetic coupling through the outer liquid core. Thus, theoretically speaking, to find this difference one has to solve the complicated MHD problem in the spherical shell with the appropriate boundary conditions at the conducting inner core and nonconducting mantle. Practically, we can expect that the difference is small, of the order of the westward drift of the nondipole modes of the geomagnetic field (Bullard *et al.* 1950). This assumption essentially simplifies the problem.

Consider first an extremely simplified situation of an incompressible, non-conducting, and almost inviscid fluid (the low Ekman number limit, $E = \nu/\Omega R^2 \ll 1$) between two hard concentric spheres rotating with slightly different angular velocities, Ω and $\Omega(1+\varepsilon)$. This pure fluid mechanical problem was considered by Ian Proudman in the first issue of *Journal of Fluid Mechanics* (Proudman 1956). At first glance the solution should have the form of geostrophic balance almost everywhere in the fluid volume except near the spherical boundaries with the inner core and mantle:

$$v_s = 0, \quad v_z = v_z(s), \quad p = p(s),$$

with the angular velocity constant on the cylinders parallel to the axis of rotation, $v_\varphi = v_\varphi(s)$. Here cylindrical coordinates s, φ, z are used. However Proudman discovered that there is no smooth axisymmetric solution consistent with a natural condition for the symmetry across the equatorial plane and non-slip boundary conditions. A self-consistent solution must include a singular cylindrical surface that touches the inner sphere and is parallel to the axis of rotation (Figure 1).

On this singular surface velocity gradients are very large. Outside the cylinder the fluid rotates as a rigid body with the same angular velocity as the outer sphere. Inside the cylinder the fluid rotates with an angular velocity intermediate to the angular velocities of the spheres, and there is also a meridional circulation, $v_z = v_z(s)$. Later Stewartson (1966) presented a detailed investigation of the structure of the shear layer near the cylindrical surface. The possible geophysical importance of these solutions was noted much later (Ruzmaikin 1989). A situation with non-parallel axes of rotation was studied by Nikitina (1990) who showed that in a stationary state the angle between the axes is small, of the order of ε. Recent numerical simulations of

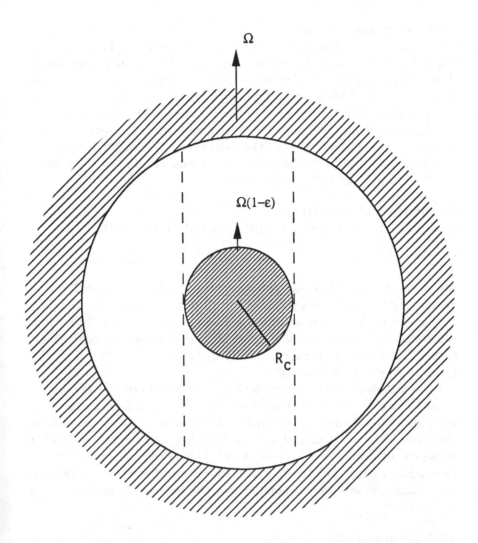

Figure 1. The singular cylinder appears when one tries to organise a transition from the mantle to the rotating inner core.

the Proudman–Stewartson problem also demonstrate the appearance of the singular cylinder (Hollerbach 1992; Brandenburg unpublished).

Will this singularity survive when we turn from the ideal situation to the real planetary core with a density stratification, convection and magnetic fields? The estimates and speculations below show how these factors could modify the singular cylinder.

2 DISCUSSION

2.1 Density stratification
The relative density change over the outer fluid core of the Earth, $\Delta\rho/\rho \approx$ 0.1, is due to the gravitational field of the internal core. It is immediately clear that the density stratification is unessential within thin boundary and shear layers. A direct study of the hydrodynamical equations beyond the layers shows (Nikitina & Ruzmaikin 1990) that outside the singular cylinder, $s > R_c$, where R_c is the radius of the inner core, the fluid rotates as before with the mantle angular velocity. The stratification only slightly changes the distribution of differential rotation and meridional circulation in the region $s < R_c$, i.e., over and under the inner core.

2.2 Influence of convection
Small-scale turbulent convection produces a turbulent viscosity,

$$\nu_T \approx \frac{1}{3}lv,$$

where $v \approx 10^{-4}$ m s^{-1} is a characteristic turbulent velocity, and $l \approx 10^6$m is a mixing length. This value is very large compared to the molecular viscosity in the liquid core. The fast rotation which tends to make the convective motion two-dimensional must essentially reduce this value. However even the above (upper limit) value gives only a quantitative change. The Ekman number corresponding to ν_T is about 10^{-7}.

Large-scale laminar convection in the outer Earth's core has the form of banana rolls elongated parallel to the axis of rotation (see, for example, Busse & Cuong 1979). Thus it does not violate the global symmetry of the problem under consideration. A small deviation of the cells' axes from the z-direction induces a differential rotation over the whole fluid core, as was shown in a preliminary study by Nikitina & Ruzmaikin (1992). However, this effect may be considered as superimposed on the basic flow between the rotating mantle and inner core.

2.3 Magnetic effects
Actually, the fluid in planetary cores is a good conductor and there is a magnetic field so the MHD approach is needed. The solution of this MHD problem includes an ideal conducting region and several boundary and shear layers. The Ekman–Hartmann boundary layers near a spherical surface were studied by Loper (1970). The axisymmetric MHD flow with all boundary and shear layers between two concentric spheres rotating with different angular velocities is studied in a recent paper by Kleeorin *et al.* (1992). (For a non-axisymmetric numerical study see Hollerbach & Proctor 1993; for the axisymmetric case see also Hollerbach & Jones 1992.) The various types of

interactions and dynamics of rotation of conductive fluid and the solid parts of system with variable conductivity are considered. In particular, magnetic fields inside the Ekman–Hartmann layers cause an electromagnetic relaxation of differential rotation of the fluid between the two spheres while the magnetic fields generated outside the boundary layers excite the differential rotation of the fluid. It is shown that the properties of the internal cylindrical shear layer are essentially modified by the magnetic field generated inside this layer. For example, in the nonconductive fluid the maximal thickness of the cylindrical shear layer is $\simeq R_c \cdot E^{1/4}$ (Stewartson 1966). In MHD the maximal thickness of the layer is reduced to $\simeq R_c \cdot E^{1/2} q^{-1/2}$, where $q = H_s^2 (4\pi \rho \Omega R \nu_m)^{-1}$.

REFERENCES

Braginsky, S.I. 1993 The nonlinear dynamo and model-Z. In *Lectures on Solar and Planetary Dynamos* (ed. M.R.E. Proctor & A.D. Gilbert), Cambridge University Press.

Bullard, E.C., Freedman, C., Gellman, H. & Nixon, J. 1950 The westward drift of the Earth's magnetic field. *Phil. Trans. R. Soc. Lond.* A **243**, 67–92.

Busse, F.H. & Cuong, P.G. 1979 Convection in rapidly rotating spherical fluid shells. *Geophys. Astrophys. Fluid Dynam.* **8**, 17–41.

Hollerbach, R. 1992 A direct spectral solution of the Ekman and Stewartson layers in a rotating spherical shell. *Geophys. Astrophys. Fluid Dynam.* (submitted).

Hollerbach, R. & Jones, C.A. 1992 A geodynamo model incorporating a finitely conducting inner core. *Phys. Earth Planet. Inter.* (in press).

Hollerbach, R. & Proctor, M.R.E. 1993 Non-axisymmetric shear layers in a rotating spherical shell. In this volume.

Kleeorin, N.I., Rogachevskii, I.V. & Ruzmaikin, A.A. 1992 MHD flow between two rotating spheres. *J. Fluid Mech.* (submitted).

Loper, D.E. 1970 General solution for the linearised Ekman–Hartmann layer on a spherical boundary. *Phys. Fluids* **13**, 2995–2998.

Loper, D.E. & Roberts, P.H. 1983. Compositional convection and the gravitationally powered dynamo. In *Stellar and Planetary Magnetism* (ed. A.M. Soward), pp. 297–327. Gordon and Breach, New York.

Nikitina, L.V. 1990 On the angle between axes of rotation of the mantle and inner core. *Geomag. Aeron.* **30**, 832–836.

Nikitina, L.V. & Ruzmaikin, A.A. 1990 A flow inside the Earth created by a relative rotation of the mantle and internal solid core. *Geomag. Aeron.* **30**, 127–131.

Nikitina, L.V. & Ruzmaikin, A.A. 1992 Differential rotation of the fluid Earth's core. *Geomag. Aeron.* **32**, 140–144.

Proudman, I. 1956 The almost-rigid rotation of viscous fluid between concentric spheres. *J. Fluid Mech.* **1**, 505–516.

Ruzmaikin, A.A. 1989 A large-scale flow in the Earth's core. *Geomag. Aeron.* **29**, 299–303.

Stevenson, D.J. 1983 Planetary magnetic fields. *Rep. Prog. Phys.* **46**, 555–620.

Stewartson, K. 1966 On almost rigid rotation. *J. Fluid Mech.* **26**, 131–144.

Evolution of Magnetic Fields in a Swirling Jet

A.M. SHUKUROV

Computing Center
Moscow University
Moscow 119899, Russia

D.D. SOKOLOFF

Physics Department
Moscow University
Moscow 119899, Russia

Using the asymptotic forms of the eigenfunctions, we solve, for $R_m \gg 1$ and $t \to \infty$ (with R_m the magnetic Reynolds number), the Cauchy problem for the kinematic screw dynamo. It is demonstrated that for a spatially localized seed magnetic field the field grows at different rates within the region of localization and outside it.

The screw dynamo is one of the simplest examples of a conducting fluid flow in which magnetic field can be self-excited provided the magnetic Reynolds number is sufficiently large (see, e.g., Roberts 1993). Such a flow can be encountered in some astrophysical objects and also in such technological devices as breeder reactors. For example, jet outflows in active galaxies and near young stars can be swirling. A flow of this type is used for modelling the dynamo effects in laboratory conditions (Gailitis 1993). The generation of magnetic fields by a laminar flow with helical streamlines was discussed by Lortz (1968), Ponomarenko (1973), Gailitis & Freiberg (1976), Gilbert (1988), Ruzmaikin et al. (1988) and other authors as an eigenvalue problem. Below we use the results of the asymptotic analysis of this problem for large R_m by Ruzmaikin et al.

We introduce an axisymmetric velocity field whose cylindrical polar components are $(0, r\omega(r), v_z(r))$, with (r, ϕ, z) the cylindrical coordinates. We

271

M.R.E. Proctor, P.C. Matthews & A.M. Rucklidge (eds.)
Theory of Solar and Planetary Dynamos, 271–274
©1993 Cambridge University Press.

consider smooth functions $v_z(r)$ and $\omega(r)$ vanishing as $r \to \infty$. Both $v_z(0)$ and $\omega(0)$ are assumed to be of order unity.

For $R_m \gg 1$, an eigenmode of the screw dynamo represents a dynamo wave concentrated in a cylindrical shell of thickness $\simeq R_m^{-1/4}$ at a certain radius r_0. The radius r_0 is related to the longitudinal wave number of the eigenmode, k:

$$k = -\frac{m\omega'(r_0)}{v'(r_0)}, \qquad (1)$$

where m is the azimuthal wave number and the prime denotes the derivative with respect to r. The axial component of the phase velocity of the dynamo wave is given by

$$V_{\mathrm{ph}} = -\frac{\mathrm{Im}(\gamma)}{k},$$

where $\mathrm{Im}(\gamma) \simeq R_m^0$ is the oscillation frequency. For a smooth velocity profile the growth rate, $\mathrm{Re}(\gamma)$, is of the order of $R_m^{-1/2}$. It is important that the eigenvalue spectrum is continuous with respect to the longitudinal wave number k because the flow is considered to be unbounded along the z-axis.

These results apply to the eigensolutions with moderate azimuthal wave number m. For large m ($\simeq R_m^\sigma$, where $\sigma > 1/4$) the eigensolutions behave in a different way (Gilbert 1988).

Under certain conditions, the solution of the eigenvalue problem is insufficient to describe the behaviour of the dynamo. In particular, this happens when the seed field is concentrated in space, so that the propagation of magnetic field to $|z| \to \infty$ can take a long time; this represents the physically interesting problem. In the context of the screw dynamo, we consider the seed field represented by a magnetized region within the flow which is narrow along the z-axis (the *magnetic blob*). Since the screw dynamo is a slow one, i.e., $\mathrm{Re}(\gamma) \propto R_m^{-1/2}$, the advection time is $R_m^{1/2}$ times shorter than the dynamo growth time, so that the intermediate regime of the spreading of the magnetic field along the jet from the blob is physically interesting. Furthermore, there arises the question of whether or not a rapidly advected magnetic blob can leave magnetized the fluid behind it.

A formal solution of the Cauchy problem for the kinematic screw dynamo can be represented in terms of the eigenfunctions as follows:

$$\mathbf{H}(\mathbf{r}, t) = \int_0^\infty c(r_0) \mathbf{h}(r, r_0) \exp[\lambda(r_0)t] \exp[i\nu(r_0)t + ik(r_0)z + im\phi] r_0 dr_0, \qquad (2)$$

where $\mathbf{H}$ is the magnetic field, $\mathbf{h}(r, r_0)$ is the radial eigenfunction which concentrates at $r = r_0$, $c(r_0)$ represents the weights of different eigenfunctions in the initial magnetic field distribution, $\nu = \mathrm{Im}(\gamma)$ and $\lambda = \mathrm{Re}(\gamma)$. An appropriate asymptotic form of the radial eigenfunctions is discussed in detail by Ruzmaikin *et al.* (1988). We consider an intermediate asymptotic solution of the Cauchy problem for $t \gg 1$ but before the solution reduces to a single leading eigenfunction. Therefore, the integrand in (2) represents a rapidly oscillating function and the solution can be obtained using the stationary phase method.

Two physically distinct cases should be considered separately when evaluating the integral (2). Firstly, consider a point at a fixed position within the jet, $z = \text{const.} \ll \nu t$. Then the dominant contribution to the integral (2) comes from those modes which concentrate near the radius $r = R_j$ governed by

$$\left.\frac{d\nu(r_0)}{dr_0}\right|_{r=R_j} = 0. \tag{3}$$

Secondly, consider the magnetic blob itself. Suppose that it moves along the z-axis at the velocity V. Since the dynamo is a slow one, magnetic diffusion plays a crucial role and we cannot suppose that $V = v_z$ beforehand but we should determine the velocity of the magnetic blob by solving the Cauchy problem. Within the blob, $z = Vt \gg 1$ and the dominant contribution to (2) comes from the modes which concentrate near the radius $r = R_b$ governed by

$$v_z(R_b) = V \tag{4}$$

(see Reshetnyak *et al.* 1991 for details). The leading eigenmode is among those solving equation (4), so that the magnetic field within the moving blob is dominated by the leading eigenmode and the blob is advected at the flow velocity at the position where the leading eigenfunction concentrates, or $R_b = r_{0,\text{max}}$, $V = v_z(r_{0,\text{max}})$, where $\lambda(r_{0,\text{max}}) \equiv \lambda_{\text{max}} \equiv \max(\lambda)$. In a certain sense, the magnetic blob is frozen into the flow!

Consider (3) in more detail. The implication is that the magnetic field behind the blob is dominated by the eigenmode which has the smallest oscillation frequency. Obviously, in the general case this mode differs from that having the largest growth rate, so that the growth rate of the magnetic field in the jet behind the blob is smaller than that within it. Moreover, the jet can remain unmagnetized after the passage of the blob if the growth rate at $r = R_j$ is negative. This is due to interference effects: rapidly oscillating modes cancel each other even though they grow rapidly, except those which have the lowest possible frequency of oscillations.

These arguments discussed in detail by Reshetnyak *et al.* (1991) can be formulated in terms of the group velocity of the magnetic wave packet, or the magnetic blob. The group velocity is given by $V_g = d\nu/dk$. The magnetic field outside the jet, at $z \ll \nu t$, grows provided magnetic field does not leave this region, or $V_g = 0$; this reduces to (3) with the help of (1). In a moving frame within the blob we have $\nu_{\text{eff}} = \nu + kV$, so that the condition $V_g = 0$ reduces now to (4).

We note that this successful application of the concept of the group velocity is not trivial because, unlike the wave equation, the induction equation governing the dynamo waves is a parabolic one. In particular, this means that a concept similar to that of the speed of light as a limiting velocity cannot be introduced for dynamo waves.

Our analysis of the evolution of a magnetic blob in a swirling jet can be regarded also as an asymptotic estimation of the Green's function for the screw dynamo. The Green's function $G(t, z, t_0, z_0)$ is the solution for the initial condition at $t = t_0$ concentrated at the position $z = z_0$. As follows from our discussion above, the Green's function is not vanishingly small in two regions. First, this occurs for $z - z_0 \simeq v_z(r_{0,max})(t - t_0)$. In this region G is of the order of $\exp[\lambda_{max}(t - t_0)]$, where $\lambda_{max} = \max(\lambda)$. As follows from (3), the second region is $0 < z - z_0 \ll v_z(t - t_0)$, where the Green's function is of the order of $\exp[\lambda(R_j)(t - t_0)]$.

REFERENCES

Gailitis, A. 1993 Experimental aspects of laboratory scale liquid sodium dynamo model. In this volume.

Gailitis, A. & Freiberg, J. 1976 On the theory of helical MHD dynamo. *Magnitnaya Gidrodinamika* **10**, 15–19.

Gilbert, A. 1988 Fast dynamo action in the Ponomarenko dynamo. *Geophys. Astrophys. Fluid Dynam.* **44**, 241–258.

Lortz, D. 1968 Exact solutions of the hydromagnetic dynamo problem. *Plasma Phys.* **10**, 967–972.

Ponomarenko, Yu.B. 1973 On the theory of hydromagnetic dynamo. *Zh. Prikl. Mekh. Tekhn. Fiz. (USSR)* **6**, 47–51.

Reshetnyak, M., Sokoloff, D. & Shukurov, A. 1991 Evolution of a magnetic blob in a helical flow. *Astron. Nachr.* **312**, 33–39.

Roberts, P.H. 1993 Fundamentals of dynamo theory. In *Lectures on Solar and Planetary Dynamos* (ed. M.R.E. Proctor & A.D. Gilbert), Cambridge University Press.

Ruzmaikin, A., Sokoloff, D. & Shukurov, A. 1988 Hydromagnetic screw dynamo. *J. Fluid Mech.* **197**, 39–56.

Analytic Fast Dynamo Solution for a Two-dimensional Pulsed Flow

ANDREW M. SOWARD

Dept. of Mathematics and Statistics
University of Newcastle upon Tyne
Newcastle upon Tyne, NE1 7RU UK

Recent results concerning the amplification of magnetic field frozen to a two-dimensional spatially periodic flow consisting of two distinct pulsed Beltrami waves are summarised. The period α of each pulse is long ($\alpha \gg 1$) so that fluid particles make excursions large compared to the periodicity length. The action of the flow is reduced to a map $\mathbf{T}$ of a complex vector field $\mathbf{Z}$ measuring the magnetic field at the end of each pulse. Attention is focused on the mean field $\langle \mathbf{Z} \rangle$ produced. Under the assumption, $\langle \mathbf{T}^{K+2}\mathbf{Z} \rangle - |\lambda_\infty|^2 \langle \mathbf{T}^K \mathbf{Z} \rangle \to 0$ as $K \to \infty$, an asymptotic representation of the complex constant λ_∞ is obtained, which determines the growth rate $\alpha^{-1}\ln(\alpha|\lambda_\infty|)$. The main result is the construction of a family of smooth vector fields $\mathbf{Z}_N$ and complex constants λ_N with the properties (for even N), $\langle \mathbf{T}^{K+2}\mathbf{Z}_N \rangle - |\lambda_N|^2 \langle \mathbf{T}^K \mathbf{Z}_N \rangle = O(\epsilon^{1+(N/2)})$ and $\lambda_N - \lambda_\infty = O(\epsilon^{1+(N/2)})$ for all integers $K(> 0)$, where $\epsilon = \alpha^{-3/2}$. The relation of $\mathbf{Z}_N$ and λ_N to the modes of the corresponding dissipative problem with the fastest growth rates is discussed.

M.R.E. Proctor, P.C. Matthews & A.M. Rucklidge (eds.)
Theory of Solar and Planetary Dynamos, 275–286
©1993 Cambridge University Press.

1 INTRODUCTION

The key characteristic of a fluid motion necessary for fast dynamo action is the existence of a positive Liapunov exponent. Childress (1992) calls a motion with this property a stretching flow and it is generally manifest by chaotic particle paths. For steady flows the regions of exponential stretching are often small, as they are, for example, in the case of the spatially periodic flows discussed by Dombre *et al.* (1986). The numerical demonstration of fast dynamos in such flows has proved difficult and Galloway & Frisch's (1984, 1986) results were inconclusive even at the largest values of the magnetic Reynolds number reached. The more recent frozen field results of Gilbert (1992), however, appear to confirm that fast dynamo action occurs.

To establish the existence of fast dynamos simply, a more promising line is the investigation of time periodic rather than steady flows. Then even two-dimensional spatially periodic flows,

$$\mathbf{u}(x,y,t) = \mathbf{u}(x+2\pi,y,t) = \mathbf{u}(x,y+2\pi,t) = \mathbf{u}(x,y,t+2\alpha), \qquad (1)$$

can exhibit highly chaotic particle paths. Here our dimensionless units are based on the periodicity length $2\pi L$, the magnitude U of the velocity, the time L/U, while $2\alpha L/U$ is the period. An immediate advantage of this simplification is that the magnetic induction equation admits separable solutions

$$\mathbf{B}(x,y,z,t) = \Re\{\widehat{\mathbf{B}}(x,y,t)\exp(-ikz - \eta k^2 t)\} \qquad (k>0) \qquad (2)$$

for some positive vertical wavenumber k. Here η is the inverse magnetic Reynolds number or equivalently the magnetic diffusivity measured in units LU. Following Otani (1988), the numerical results of Galloway & Proctor (1992) obtained by time stepping clearly demonstrate the existence of fast dynamos of this type. In this note we summarise an asymptotic solution (Soward 1993a) of the fast dynamo problem for the case of the pulsed Beltrami waves

$$\mathbf{u}(x,y,t) = \begin{cases} (0, \ \sin x, \ \cos x) & (0 \leq t < \alpha), \\ \\ (\sin y, \ 0, \ -\cos y) & (\alpha \leq t < 2\alpha), \end{cases} \qquad (3)$$

each with positive helicity unity. A similar pulsed flow was investigated previously by Bayly & Childress (1988) using numerical methods.

During the first pulse, the magnetic induction equation is satisfied when the horizontal xy-components, denoted by the subscript H, of the complex vector $\widehat{\mathbf{B}}$ in (2), namely $\widehat{\mathbf{B}}_H[= (\widehat{B}_x, \ \widehat{B}_y)]$, satisfy

$$\mathcal{L}\widehat{B}_x = 0, \qquad \mathcal{L}\widehat{B}_y = i(\partial\mathcal{L}/\partial k)\widehat{B}_x \qquad (0 \leq t < \alpha), \qquad (4)$$

where $\mathcal{L}$ is the advection-diffusion operator

$$\mathcal{L} \equiv (\partial/\partial t) + (\sin x)(\partial/\partial y) - ik(\cos x) - \eta\nabla_H^2 \qquad (5)$$

for the flow (3). The term on the right of (4) corresponds to the stretching of magnetic field and is the usual source of dynamo activity. Over the complete 2α-period of the two pulses the following $S^{\pm}(\chi)$ symmetries are preserved:

$$\pm \begin{pmatrix} \widehat{B}_x \\ \widehat{B}_y \end{pmatrix} (-x, -y, t) = \begin{pmatrix} \widehat{B}_x \\ \widehat{B}_y \end{pmatrix} (x, y, t) = e^{2i\chi} \begin{pmatrix} \widehat{B}_x^* \\ -\widehat{B}_y^* \end{pmatrix} (\pi - x, \pi + y, t), \quad (6)$$

where the star denotes complex conjugate, χ is a real constant, and $\widehat{\mathbf{B}}_H$ has the same 2π spatial periodicity as the flow. The symmetries are such that only the $S^+(\chi)$-modes have non-zero horizontal mean which is given by

$$\langle \widehat{\mathbf{B}}_H \rangle = B_0 e^{i\chi}(1, \, -iH), \quad (7)$$

where B_0 and H are real functions of time. The corresponding magnetic field is

$$\overline{\mathbf{B}}_H = B_0[\cos k(z - z_0), \, - H \sin k(z - z_0)] \quad (kz_0 = \chi), \quad (8)$$

where the compact bar notation, like the angle brackets, denotes mean value. The constant H measures the magnetic helicity

$$B_0^2 k H = \overline{\mathbf{B}}_H \cdot \nabla \times \overline{\mathbf{B}}_H. \quad (9)$$

Consider the initial field

$$\widehat{\mathbf{B}}_H(x, y, 0) = \mathbf{Z}(x, y). \quad (10)$$

The field at the end of the first pulse can be used to define the map

$$\mathbf{TZ}(x, y) = \alpha^{-1}[\widehat{B}_y^*, \, \widehat{B}_x^*](-y, -x, \alpha), \quad (11)$$

where we have anticipated that a further repetition of the map gives the field

$$\mathbf{T}^2\mathbf{Z}(x, y) = \alpha^{-2}\widehat{\mathbf{B}}_H(x, y, 2\alpha) \quad (12)$$

at the end of the complete period 2α. The result follows because during the second interval, $\alpha \leq t < 2\alpha$, the equations governing $\widehat{\mathbf{B}}_H$ are obtained from (4) by replacing $\widehat{B}_x$, $\widehat{B}_y$ by $\widehat{B}_y^*$, $\widehat{B}_x^*$ provided that x, y in (5) are replaced by $-y, -x$. Accordingly the $S^+(\chi)$-normal modes are the solutions of the eigenvalue problem

$$\mathbf{TZ} = \lambda\mathbf{Z}, \qquad i\lambda e^{2i\chi} = \Lambda, \quad (13)$$

where the symmetries (6) imply that the parameter Λ is real. With (2) it defines the magnetic field growth rate

$$p = \alpha^{-1}\ln(\alpha|\lambda|) - \eta k^2 \quad (14)$$

determined by the spectrum of the eigenvalues $\lambda(\alpha, \eta, k)$ of (13). The dynamo is fast if, for the fastest growing mode,

$$\alpha|\lambda| \to \alpha|\lambda_\infty| > 1 \qquad \text{as} \qquad \eta \downarrow 0. \qquad (15)$$

Now, whereas there is a corresponding positive limit p_∞ for the growth rate, there is no well defined eigenfunction $\mathbf{Z}_\infty(x, y)$ to which $\mathbf{Z}$ converges. Note that implicit in (15) is the assumption that ηk^2 tends to zero with η. There are, of course, modes for which ηk^2 remains finite in that limit, but they are of no concern to us here since they do not have the largest growth rate.

The role of α is interesting. As $\alpha \downarrow 0$ the particle paths converge to the streamlines of the steady flow defined by the average of the two flows (3). That integrable flow has exponential stretching only on the streamlines connecting the stagnation points. Soward (1987) showed that dynamo action, restricted to boundary layers of width $\eta^{1/2}$ containing those streamlines, is not quite fast. For small non-zero α, the heteroclinic connections are broken and a chaotic web emerges. Childress (1993) argues that the small region of stretching flow in the web can support a fast dynamo. The web thickens with α so that in the large α limit,

$$\alpha \gg 1, \qquad (16)$$

the chaos is considerable. Though chaotic the long unidirectional motion in each pulse causes the magnetic field at the end of each pulse to lie largely in planes. In this note we take advantage of this property in our study of the large α limit. For simplicity we only analyse the frozen field limit $\eta = 0$ but use our results to determine the growth rate and a weak (or generalised) eigenfunction for the finite diffusivity limit $\eta \downarrow 0$. Further background can be found in the reviews of Childress (1992), Roberts & Soward (1992), Soward (1993b).

2 THE CAUCHY SOLUTION
For perfectly conducting fluids $\eta = 0$, the Cauchy solution of (4) gives

$$\mathbf{TZ}(x, y) = \mathbf{M}(y)\mathbf{Z}^*(-y, -x + \alpha \sin y) \exp(-i\zeta \cos y), \qquad (17)$$

in which

$$\mathbf{M}(y) = \begin{pmatrix} \cos y & \alpha^{-1} \\ \alpha^{-1} & 0 \end{pmatrix}, \qquad \zeta = k\alpha. \qquad (18)$$

Repeated application of the map yields, after N iterates, the result

$$\mathbf{Z}^{(N)}(x, y) \equiv \mathbf{T}^N \mathbf{Z} = \mathcal{M}^{(N)}\mathbf{Z}^{(0)(*N)}(-y^{(N-1)}, y^{(N)}) \exp(-i\zeta \psi^{(N)}). \qquad (19)$$

Here $(*N)$ denotes the operation of complex conjugation N times and

$$\mathcal{M}^{(K+1)} = \mathcal{M}^{(K)}\mathbf{M}(y^{(K)}), \qquad \psi^{(K+1)} = \psi^{(K)} + (-1)^K \cos(y^{(K)}),$$

$$y^{(K+1)} = y^{(K-1)} + \alpha \sin(y^{(K)}) \tag{20}$$

for $K \geq 0$, where

$$\mathcal{M}^{(0)} = \mathbf{I}, \qquad \psi^{(0)} = 0, \qquad y^{(-1)} = -x, \qquad y^{(0)} = y. \tag{21}$$

With $y^{(K)}$ regarded as a function of x and y, it follows that

$$\partial y^{(K)}/\partial y = \mathrm{O}(\alpha^K) \qquad (K \geq 0). \tag{22}$$

This means that each individual matrix $\mathbf{M}(y^{(K)})$ composing $\mathcal{M}^{(N)}$ responds to variations of y on the length scale

$$\delta(K+1) = \mathrm{O}(\alpha^{-K}) \qquad (K \geq 0). \tag{23}$$

Accordingly (19) to (23) highlight the scale separation, which results from each application of the map and is caused by the large order α excursions of fluid particles during each pulse.

Consider the action of the first pulse on an initially uniform field $\overline{\mathbf{Z}}^{(0)}$ [see (28) below] largely in the x-direction. The horizontal y-motion stretches (S) that field into a strong y-directed field, which is folded (F) so alternating its sign as x increases by π. This field has zero mean. Remember, however, that the original uniform magnetic field $\Re[\overline{\mathbf{Z}}^{(0)} \exp(-ikz)]$ alternates its sign as z increases by π/k. Consequently the up and down shearing (S) motion can reorganise the alternating horizontal field so that a non-zero horizontal mean field results. To optimise the reconstruction of mean magnetic field we anticipate that the maximum separation 2α caused by the up and down motions should be comparable with π/k. That gives

$$\zeta = k\alpha = \mathrm{O}(1) \tag{24}$$

roughly equal to $\frac{1}{2}\pi$, which is comparable with the quantitative result (29) below. This process is the basis of the SFS-map introduced by Bayly & Childress (1988) and constitutes a very efficient fast dynamo mechanism. To quantify its effectiveness we evaluate

$$\overline{\mathbf{Z}}^{(1)} = \langle \mathbf{T}\overline{\mathbf{Z}}^{(0)} \rangle = [\mathbf{A}_0 \overline{\mathbf{Z}}^{(0)}]^*, \tag{25}$$

where from (17) and (18) $\mathbf{A}_0$ is the matrix

$$\mathbf{A}_0 = \mathcal{D}[J_0(\zeta)], \qquad \mathcal{D} = \begin{pmatrix} -i\partial/\partial\zeta & \alpha^{-1} \\ \alpha^{-1} & 0 \end{pmatrix} \tag{26}$$

and $J_N(\zeta)$ is the Bessel function of the first kind of order N. If $\overline{\mathbf{Z}}^{(0)} \in \mathrm{S}^+(0)$, the horizontal average of (13) yields

$$\mathbf{A}_0 \overline{\mathbf{Z}}^{(0)} = i\Lambda \overline{\mathbf{Z}}^{(0)*} \tag{27}$$

with a solution of the form (7), namely

$$\overline{\mathbf{Z}}^{(0)} = (1, -iH), \qquad H = -J_0(\zeta)/(\alpha\Lambda), \qquad \Lambda(\zeta,\alpha) = J_1(\zeta)/(1-H^2). \quad (28)$$

Since α is large, H is small of order α^{-1}. The growth of mean field is maximised for $\zeta = \zeta_{max}$, the first zero of $J_1'(\zeta)$ [remember $J_1(\zeta) = -J_0'(\zeta)$]. There we have

$$\Lambda = \Lambda_{max} = J_1(\zeta_{max}) = 0.5819\ldots, \qquad \zeta_{max} = 1.8412\ldots. \quad (29)$$

The key issue is whether the amplification determined by (28) continues under subsequent iterations of the map. Surprisingly after the second pulse it holds exactly because

$$\overline{\mathbf{Z}}^{(2)} = \mathbf{A}_0^* \mathbf{A}_0 \overline{\mathbf{Z}}^{(0)}. \quad (30)$$

For higher order iterates, this result no longer holds. Nevertheless, if we average $\mathbf{T}\mathbf{Z}^{(N-1)}(= \mathbf{Z}^{(N)})$ defined by (19) with respect to the shortest length scale variable $y^{(N-1)}$, the result, when $\overline{\mathbf{Z}}^{(0)}$ is the solution of (27), is

$$\mathbf{T}^{(N-1)}[\mathbf{A}\overline{\mathbf{Z}}^{(0)}]^* + O(\alpha^{-1}) = -i\Lambda\mathbf{Z}^{(N-1)} + O(\alpha^{-1}). \quad (31)$$

This powerful result, for fixed N, suggests that the value λ, which determines the growth rate, is given correct to order α^{-1} by (28). In the following section we tighten up this result to achieve accuracy to any required order in inverse powers of α.

3 REDUCTION TO A MEAN FIELD PROBLEM

Suppose we take an arbitrary initial field and iterate it indefinitely. In the case of finite diffusivity, $\eta \neq 0$, the fastest growing mode eventually dominates providing us with the eigensolution of (13). For our frozen field ($\eta = 0$) solution (19), we still anticipate that this procedure will identify the amplification factor Λ. On the other hand, since no eigenfunction $\mathbf{Z}$ can emerge, we only demand convergence of some measure of it, namely the mean magnetic field $\overline{\mathbf{Z}}$.

Our method relies on the Fourier representation

$$\mathbf{Z}(x,y) = \sum_{m=-\infty}^{\infty} \sum_{n=-\infty}^{\infty} \mathbf{Z}_n^m \exp[i(mx + ny)], \quad (32)$$

in which $\mathbf{Z}_n^m$ are complex constant vectors. Application of the map (17) to $\mathbf{Z}^{(0)}$ yields $\mathbf{Z}^{(1)}$, whose Fourier coefficients are

$$\mathbf{Z}^{(1)}{}_N^M = \sum_{l=-\infty}^{\infty} \{\mathcal{D}[\Pi_{N-l}^M]\mathbf{Z}^{(0)}{}_M^l\}^*, \quad (33)$$

where $\mathcal{D}$ is defined by (26) and

$$\Pi_N^M = \langle \exp i(Ny + \zeta \cos y + \alpha M \sin y) \rangle = i^N \exp(iN\phi_M)J_N(R_M),$$

$$\zeta + iM\alpha = R_M \exp(i\phi_M). \tag{34}$$

Bayly & Childress (1988) derived a similar map and iterated it numerically. The key to our asymptotic solution for large α lies in the fact that, whereas the $M = 0$ matrices

$$\mathbf{A}_N = \mathcal{D}[\Pi_N^0], \qquad \Pi_N^0 = i^N J_N(\zeta), \tag{35}$$

are of order unity, the remaining $M \neq 0$ matrices

$$\epsilon \mathbf{B}_N^M = \mathcal{D}[\Pi_N^M], \qquad \epsilon = \alpha^{-3/2}, \tag{36}$$

$$(-1)^N \Pi_N^M = J_N(M\alpha) + (M\alpha)^{-1}\{-iN\zeta J_N(M\alpha) + \tfrac{1}{2}\zeta^2 J_N'(M\alpha)\} + O(\alpha^{-2}),$$

are small. Specifically, for $M(\neq 0)$, N of order unity, we have $J_N(M\alpha) = O(\alpha^{-1/2})$ giving $\epsilon \mathbf{B}_N^M = O(\alpha^{-3/2})$. To emphasise this fact, we have introduced the small scale factor ϵ in (36). A word of caution: this estimate is not uniformly valid for all M, N and leads to delicate questions of convergence for sums involving large M, N.

When the initial field has only a mean part $\mathbf{Z}^{(0)} = \overline{\mathbf{Z}}^{(0)}$, the above apparatus yields a relation of the form

$$\overline{\mathbf{Z}}^{(L)} = \sum_{K=1}^{L} \epsilon^{[K]}[\mathcal{C}^{(K)}\overline{\mathbf{Z}}^{(L-K)}]^{(*K)} \qquad (L \geq 1), \tag{37}$$

where $[K] = \tfrac{1}{2}K$ (K even) and $\tfrac{1}{2}(K-1)$ (K odd). General formulas for the matrices $\mathcal{C}^{(K)}$ are given by Soward (1993a). The first five are

$$\mathcal{C}^{(1)} = \mathbf{A}_0, \qquad\qquad\qquad \mathcal{C}^{(2)} = 0,$$

$$\mathcal{C}^{(3)} = \mathbf{A}_{-p}\mathbf{B}_0^{p*}\mathbf{A}_p, \qquad\qquad \mathcal{C}^{(4)} = \mathbf{A}_{-p}^*\mathbf{B}_{-q}^p\mathbf{B}_p^{q*}\mathbf{A}_q,$$

$$\mathcal{C}^{(5)} = \mathbf{A}_{-p}\{\mathbf{B}_0^{p*}\mathbf{A}_{p-q}\mathbf{B}_0^{q*} + \epsilon\mathbf{B}_{-r}^{p*}\mathbf{B}_{p-q}^r\mathbf{B}_r^{q*}\}\mathbf{A}_q, \tag{38}$$

where every triply repeated sub-super script implies summation over all non-zero integer values both positive and negative. Given $\overline{\mathbf{Z}}^{(0)}$, the mean values $\overline{\mathbf{Z}}^{(L)}(L = 1, 2, \ldots)$ may be determined in sequence from (37). The values of the remaining coefficients $\mathbf{Z}^{(L)}{}_N^M$ can then be calculated from the formulas used to construct (37) itself.

The interesting feature of (37) is its 'poor memory'. By that we mean that the $(L-K)$-th iterate only provides a small order $\epsilon^{[K]}$ contribution to the L-th iterate, which follows K iterates later. This suggests that asymptotically $\overline{\mathbf{Z}}^{(N)}$ develops a self-similar structure with

$$\overline{\mathbf{Z}}^{(N+1)} - \lambda^{(*N)}\overline{\mathbf{Z}}^{(N)} \to 0 \qquad \text{as} \qquad N \to \infty, \tag{39}$$

under some suitable normalisation of $\overline{\mathbf{Z}}^{(N)}$. Within this framework we assume that (39) holds exactly. So for given L, we set

$$|\lambda|^L \overline{\mathbf{Z}}^{(N)} = \lambda^{[N+1]} \lambda^{[N]*} \overline{\mathbf{Z}}, \tag{40}$$

which upon substitution into (37) yields $\mathbf{W}_L(\lambda; \overline{\mathbf{Z}}) = 0$ for $L \geq 1$, where

$$
\mathbf{W}_L(\lambda; \overline{\mathbf{Z}}) = \lambda \Big[\mathbf{I} - \sum_{K=1}^{[L]} \epsilon^K |\lambda|^{-2K} \mathbf{C}^{(2K)} \Big] \overline{\mathbf{Z}}
$$
$$
- \Big[\sum_{K=0}^{[L-1]} \epsilon^K |\lambda|^{-2K} \mathbf{C}^{(2K+1)} \overline{\mathbf{Z}} \Big]^* . \tag{41}
$$

Since the self-similar behaviour only holds as $N \to \infty$, we focus attention on the limiting solutions $\lambda = \lambda_\infty$, $\overline{\mathbf{Z}} = \overline{\mathbf{Z}}_\infty$ of

$$\lim_{N \to \infty} \mathbf{W}_N(\lambda; \overline{\mathbf{Z}}) = 0. \tag{42}$$

We argue that the solution of this equation determines the dynamo growth rate $p_\infty = \alpha^{-1} \ln(\alpha |\lambda_\infty|)$ [see (14)]. A similar reduction to a mean field problem has recently been employed by Gilbert (1993). In the following section we show that the solutions of $\mathbf{W}_N(\lambda; \overline{\mathbf{Z}}) = 0$ for finite N also have important implications [see (52) below].

4 THE GENERALISED EIGENFUNCTION

Our similarity assumption has yielded a precise growth rate p_∞ and for that matter a mean field $\overline{\mathbf{Z}}_\infty$. Moreover, the formulas which lead to the mean fields also define the harmonics. Accordingly all the harmonics have limiting values $\mathbf{Z}_{\infty n}{}^m$. On the other hand, since repeated application of the map reduces the length scale indefinitely, there is no eigenfunction and the Fourier sum (32) cannot converge. At best these Fourier coefficients define a generalised function $\mathbf{Z}_\infty$. In this section, we define a family of vector fields $\mathbf{Z}_N$ with readily identifiable properties, whose limiting value as $N \to \infty$ is $\mathbf{Z}_\infty$. Specifically, we construct $\mathbf{Z}_N$ such that

$$\overline{\mathbf{Z}}_N^{(1)} \equiv \langle \mathbf{T}\mathbf{Z}_N \rangle = \lambda_N \langle \mathbf{Z}_N \rangle, \tag{43}$$

and

$$\overline{\mathbf{Z}}_N^{(K+1)} = \lambda_N^{(*K)} \overline{\mathbf{Z}}_N^{(K)} + O(\epsilon^{[N+2]}) \qquad (K \geq 1) \tag{44}$$

under suitable normalisation. This generalises the earlier result (31) and sharpens the error estimate order α^{-1} to $\epsilon^{[N+2]}$, which can be made as tight as we like by an appropriate choice of N.

To appreciate the construction of $\mathbf{Z}_N$, it is helpful to imagine pre-iterates that begin at $t = -N\alpha$, prior to the initial instant $t = 0$, with mean field

$$\mathbf{Z}_N^{(-N)} = \overline{\mathbf{Z}}_{N0}. \tag{45}$$

Subsequently, at time $t = (L - N)\alpha$, we define new fields

$$\mathbf{Z}_N^{(L-N)} = \mathbf{T}\mathbf{Z}_N^{(L-N-1)} + \epsilon^{[L+1]}\overline{\mathbf{Z}}_{NL} \qquad (N \geq L \geq 1). \qquad (46)$$

The idea is that not only is $\mathbf{Z}_N^{(K)}$ $(-1 \geq K \geq -N)$ mapped in the usual way but we add to it some mean field so as to ensure that the relation

$$\overline{\mathbf{Z}}_N^{(K+1)} = \lambda_N^{(*K)}\overline{\mathbf{Z}}_N^{(K)} \qquad (-1 \geq K \geq -N) \qquad (47)$$

holds exactly. We emphasise that this is a conceptual device to picture the nature of the actual initial ($t = 0$) magnetic field

$$\hat{\mathbf{B}}_H(x, y, 0) = \mathbf{Z}_N^{(0)}(x, y) \equiv \mathbf{Z}_N = \sum_{L=0}^{N} \epsilon^{[L+1]}\mathbf{T}^{N-L}\overline{\mathbf{Z}}_{NL}, \qquad (48)$$

which is constructed from (46). The pre-iterates ($t < 0$) are not solutions of the magnetic induction equation.

To determine $\overline{\mathbf{Z}}_{NL}$ $(N \geq L \geq 0)$ in terms of $\overline{\mathbf{Z}}_N^{(K)}$ $(0 \geq K \geq -N)$, we average (46) and solve the N equations (45), (46). The result is

$$\epsilon^{[L+1]}\overline{\mathbf{Z}}_{NL} = \overline{\mathbf{Z}}_N^{(L-N)} - \sum_{K=1}^{L} \epsilon^{[K]}\left[\mathbf{C}^{(K)}\overline{\mathbf{Z}}_N^{(L-K-N)}\right]^{(*K)} \qquad (N \geq L \geq 1). \qquad (49)$$

In addition $\overline{\mathbf{Z}}_{N0}$ is given by (45), while repeated application of (47) links it to the mean value of the initial field (48),

$$\lambda^{[N]}\lambda^{[N+1]*}\overline{\mathbf{Z}}_{N0} = \overline{\mathbf{Z}}_N. \qquad (50)$$

Further substitution of (47) into (49) yields

$$\lambda^{[N-L+2]}\lambda^{[N-L+1]*}\overline{\mathbf{Z}}_{NL} = \epsilon^{-[L+1]}\mathbf{W}_L(\lambda_N; \overline{\mathbf{Z}}_N) \qquad (N \geq L \geq 1), \qquad (51)$$

where $\mathbf{W}_L$ is defined by (41). In this way the initial field (48) is determined in terms of its own mean value. To determine $\overline{\mathbf{Z}}_N$ itself, we appeal to the identity (43), which is (47) extended to the case $K = 0$. Since the first iterate of the magnetic field is simply (46) for the case $L = N + 1$ with $\overline{\mathbf{Z}}_{NL} = 0$, our result (51) leads to the eigenvalue problem

$$\mathbf{W}_{N+1}(\lambda_N; \overline{\mathbf{Z}}_N) = 0 \qquad (N \geq 0). \qquad (52)$$

For $0 \leq L \leq N$, it is easy to show using (52) that $\mathbf{W}_L(\lambda_N; \overline{\mathbf{Z}}_N)$ is of order $\epsilon^{[L+1]}$. As a consequence, (50), (51) imply that $\overline{\mathbf{Z}}_{NL}$ $(N \geq L \geq 0)$ are each of order unity which, in turn, motivates the choice of small parameter in the series (48). For $N + 2 \leq L$, it can also be shown that $\mathbf{W}_L(\lambda_N; \overline{\mathbf{Z}}_N)$ is of order $\epsilon^{[N+2]}$ and that fixes the error estimate in (44) for the second and subsequent iterates. Furthermore, they show convergence to the limiting solution of (42), specifically

$$\lambda_N - \lambda_\infty = O(\epsilon^{[N+2]}), \qquad \overline{\mathbf{Z}}_N - \overline{\mathbf{Z}}_\infty = O(\epsilon^{[N+2]}). \tag{53}$$

We discuss briefly low-order cases. For $N = 0$, (52) coincides with the eigenvalue problem (27). For $N > 0$, (52) continues to possess the same symmetries and $S^+(0)$-eigensolutions exist of the form

$$\mathbf{Z}_N = (1, -iH_N), \qquad i\lambda_N = \Lambda_N, \tag{54}$$

with H_N, Λ_N real as before. Since $\mathcal{C}^{(2)}$ vanishes [see (38)], the $N = 1$ eigenvalue problem is identical to the $N = 0$ case consistent with the result (30). For $N = 2$, use of (38) shows that (52) has the form

$$[\mathbf{A}_0 + \epsilon\Lambda_2^{-2}\mathbf{A}_{-p}\mathbf{B}_0^{p*}\mathbf{A}_p]\overline{\mathbf{Z}}_2 = i\Lambda_2\overline{\mathbf{Z}}_2^*. \tag{55}$$

Use of the definitions (35), (36) of $\mathbf{A}_p$ and $\mathbf{B}_0^p$ yields the result

$$\left\{1 - \alpha^{-2}\frac{|J_0(\zeta)|^2}{|J_1(\zeta)|^2}\right\}\Lambda_2 = J_1(\zeta) +$$

$$\frac{\alpha^{-3/2}}{|J_1(\zeta)|^2}\sqrt{\frac{8}{\pi}}\sum_{n=1}^{\infty}\frac{(-1)^n\zeta|J_n'(\zeta)|^2}{n^{3/2}}\sin(n\alpha - \pi/4) + O(\alpha^{-5/2}), \tag{56}$$

where we have utilised the asymptotic expansions for Bessel functions of large argument. The oscillatory dependence of the coefficients on α indicates the sensitivity of the growth rate to the displacement distance α modulo the periodicity length 2π. In other words, it relates to the location of the folds in the SFS-map.

Though (53) gives tight bounds on the sizes of λ_N and $\overline{\mathbf{Z}}_N$, the size of the Fourier coefficients, which define $\mathbf{Z}_N$, is not so clear. Still, arguments suggest that

$$\mathbf{Z}_{N_n}{}^m - \mathbf{Z}_{\infty_n}{}^m = \begin{cases} o(1) & (m, n \ll [\delta(N)]^{-1}), \\ O(1) & \text{(otherwise)}. \end{cases} \tag{57}$$

By that we mean that convergence is assured on the length scales large compared to the shortest length scale $\delta(N)$ of $\mathbf{Z}_N$, which is dominated by $\mathbf{T}^N\overline{\mathbf{Z}}_{N0}$, but not on shorter length scales. The necessity of the order unity errors for sufficiently large m and n is obvious because of the scale reduction by subsequent iterates.

When the magnetic diffusivity is small but non-zero ($\eta \neq 0$), we may define a diffusive length scale

$$\delta = (\alpha\eta)^{1/2} \tag{58}$$

based on the pulse time α. Only length scales longer than δ are resolvable. We may speculate that the best approximation to the diffusive eigenvalue problem (13) is determined by the solution of (52), where N is chosen such that

$$\delta(N+1) \ll \delta \ll \delta(N). \tag{59}$$

As $\eta \downarrow 0$, this gives $N \sim -\frac{1}{2}(\ln\eta)/(\ln\alpha)$ so that the error estimates in (53) become order $(\alpha\eta)^{3/8}$. Since this is large compared to η, it is at least consistent with the order η error in our use of the frozen field map (17). Of course, pointwise convergence to the eigenfunctions of (13) is impossible but some convergence of the Fourier modes in the sense of (57) is to be expected.

The research summarised here was initiated during a visit to the Courant Institute of Mathematical Sciences of New York University between 24 March and 17 April 1991 with a continuation between 15 and 27 September 1991 under the support of NSF grant, DMS-8922676. I am grateful to Steve Childress, who focussed my attention on the interesting possibilities of the analytic treatment of fast dynamos in pulsed flows. I have benefited from many helpful discussions with Andrew Gilbert.

REFERENCES

Bayly, B. & Childress, S. 1988 Construction of fast dynamos using unsteady flows and maps in three dimensions. *Geophys. Astrophys. Fluid Dynam.* **44**, 211–240.

Childress, S. 1992 Fast dynamo theory. In *Topological Aspects of the Dynamics of Fluids and Plasmas* (ed. H.K. Moffatt, G.M. Zaslavsky, P. Comte & M. Tabor), vol. 218, pp. 111–147. NATO ASI Series E: Applied Sciences, Kluwer.

Childress, S. 1993 On the geometry of fast dynamo action in unsteady flows near the onset of chaos. In preparation.

Dombre, T., Frisch, U., Greene, J.M., Hénon, M., Mehr, A. & Soward, A.M. 1986 Chaotic streamlines in ABC flows. *J. Fluid Mech.* **167**, 353–391.

Galloway, D.J. & Frisch, U. 1984 A numerical investigation of magnetic field generation in a flow with chaotic streamlines. *Geophys. Astrophys. Fluid Dynam.* **29**, 13–18.

Galloway, D.J. & Frisch, U. 1986 Dynamo action in a family of flows with chaotic streamlines. *Geophys. Astrophys. Fluid Dynam.* **36**, 53–83.

Galloway, D.J. & Proctor, M.R.E. 1992 Numerical calculations of fast dynamos in smooth velocity fields with realistic diffusion. *Nature* **356**, 691–693.

Gilbert, A.D. 1992 Magnetic field evolution in steady chaotic flows. *Phil. Trans. R. Soc. Lond. A* **339**, 627–656.

Gilbert, A.D. 1993 Towards a realistic fast dynamo: models based on cat maps and pseudo-Anosov maps. *Proc R. Soc. Lond. A* (in press).

Otani, N.J. 1988 Computer simulation of fast kinematic dynamos. *Trans. Am. Geophys. Union* **69**, No 44, Abstract No SH51-15, 1366.

Roberts, P.H. & Soward, A.M. 1992 Dynamo theory. *Ann. Rev. Fluid Mech.* **24**, 459–512.

Soward, A.M. 1987 Fast dynamo action in a steady flow. *J. Fluid Mech.* **180**, 267–295.

Soward, A.M. 1993a An asymptotic solution of the fast dynamo in a two-dimensional pulsed flow. *Geophys. Astrophys. Fluid Dynam.* (submitted).

Soward, A.M. 1993b Fast dynamos. In *Lectures on Solar and Planetary Dynamos* (ed. M.R.E. Proctor & A.D. Gilbert), Cambridge University Press.

On Magnetic Dynamos in Thin Accretion Disks Around Compact and Young Stars

T.F. STEPINSKI

Lunar and Planetary Institute
3600 Bay Area Blvd.
Houston, Texas 77058 USA

A variety of geometrically thin accretion disks commonly associated with such astronomical objects as X-ray binaries, cataclysmic variables, and protostars are likely to be seats of MHD dynamo actions. Thin disk geometry and the particular physical environment make accretion disk dynamos different from stellar, planetary, or even galactic dynamos. We discuss those particular features of disk dynamos with emphasis on the difference between protoplanetary disk dynamos and those associated with compact stars. We then describe normal mode solutions for thin disk dynamos and discuss implications for the dynamical behavior of dynamo-magnetized accretion disks.

1 INTRODUCTION

It is widely appreciated that magnetic fields can play an important role in accretion disk dynamics. Shakura & Sunyaev (1973), in their well known paper, pointed to magnetic fields as the source of a viscous couple necessary for the accretion to take place. Disk magnetic fields have also been invoked to explain spectra of compact X-ray sources, as a source of coronal heating, and as a source of wind production. In the context of the Solar nebula, which is widely assumed to represent a typical protoplanetary disk, the existence of a magnetic field is inferred from the residual magnetization of primitive meteorites, which are assumed to owe their magnetization to nebular magnetic fields. However, in a typical accretion disk, the timescale for ohmic dissipation is much smaller than the typical radial infall time, thus it is difficult to see how any magnetic field contained in the gas that falls onto the disk can persist long enough to be dynamically or otherwise important, unless it is regenerated by a dynamo cycle. The particular dynamo mechanism

M.R.E. Proctor, P.C. Matthews & A.M. Rucklidge (eds.)
Theory of Solar and Planetary Dynamos, 287–294
©1993 Cambridge University Press.

	Accretion Disks Around Compact Stars	Protoplanetary Accretion Disks
radius	10^9 cm	10^{15} cm
thickness	10^6 cm	10^{12}–10^{13} cm
temperature	10^6–10^7 K	10^2–10^3 K
ionization	thermal	cosmic rays
R_m	10^9–10^{11}	10–10^3
t_{diff}	1–10^3 years	10–10^3 years

Table 1. Typical parameters of disks around compact stars (taken here to be a black hole of $10 M_\odot$) and protoplanetary disks (taken here to be a disk around a T Tauri star of $1 M_\odot$). R_m denotes the magnetic Reynolds number, t_{diff} denotes the characteristic time of ohmic diffusion, and $M_\odot$ denotes the Solar mass.

that is relevant here is the $\alpha\omega$ dynamo, which relies on a combination of nonuniform, Keplerian rotation and helical turbulence to regenerate a magnetic field.

Accretion disks are not all alike, even if we limit ourselves to geometrically thin, optically thick disks. Disks around compact stars such as white dwarfs, neutron stars, and black holes, are relatively small, hot, and well ionized, whereas disks around very young, low-mass stars (stars that are assumed to become Sun-like stars) are relatively large, cool, and, for the most part, weakly ionized. Table 1 shows a comparison of typical parameters of both classes of accretion disks.

For a typical accretion disk around a compact star the value of the magnetic Reynolds number R_m is higher than values of R_m for the Earth's liquid core, the Solar convection zone, or even the galactic gaseous disk. Such a high value of R_m is achieved despite a very small characteristic length scale (taken here to be equal to the disk thickness), and is due to the fast rotational velocity of highly ionized gas. The magnitude of R_m in protoplanetary disks is, *on average*, about as high as in the Earth's core, despite very low ionization levels. This is due to the large characteristic length scale (of the order of 1 A.U.) and fast, Keplerian motion. Thus, the magnetic field and the gas are closely coupled *everywhere* in an accretion disk around a compact star, and at least *somewhere* in a protoplanetary disk, pointing to the possibility that a disk's magnetic field can *in principle* be maintained by an MHD dynamo for as long as there is adequate turbulent flow in the disk.

2 THE ISSUE OF TURBULENCE AND ANGULAR MOMENTUM TRANSPORT IN ACCRETION DISKS

One of the main shortcomings of accretion disk theory is the uncertainty as to the nature of the angular momentum transport responsible for accretion of the mass onto the central object. It is usually assumed that accretion disks are turbulent. This assumption is based primarily on the very large value of their Reynolds numbers, which are as high as 10^{14} for disks around compact stars. The question, however, remains: what is the source of the turbulence? The most obvious candidate is the differential rotation. It has been largely disregarded as a possible source of turbulence because the Keplerian rotation shear is stable with respect to *linear*, infinitesimal perturbation. However, it may be unstable with respect to *nonlinear*, finite amplitude perturbations. Another candidate is convection driven by a superadiabatic temperature gradient across the disk. Again, it is uncertain whether such a gradient can be maintained throughout the significant portion of the disk. Other, more exotic sources of disk turbulence are also highly controversial.

These uncertainties led recently to attempts to construct a disk magnetic dynamo model that does not depend on the existence of turbulent flow in a disk (Vishniac *et al.* 1990; Tout & Pringle 1992). The rationale for such work is that if the successful, *nonturbulent* dynamo can operate in a disk, then a generated magnetic field will provide the necessary angular momentum transport, thus making the issue of turbulence mostly irrelevant to accretion disk theory. In their most recent paper, Tout & Pringle (1992) have proposed a disk magnetic dynamo, in which they invoke the Balbus–Hawley (BH) and Parker instabilities instead of turbulence to close the dynamo loop. This approach capitalizes on the attention that has recently been drawn (Balbus & Hawley 1991) to an instability present in any cylindrical shear flow with an arbitrarily small vertical component of the magnetic field and for which $\partial \omega / \partial r < 0$, where ω is the angular velocity and r the radial coordinate. If present in astrophysical disks (for the argument that BH instability may in fact be absent whenever an azimuthal magnetic field is present see Knobloch 1992), this instability would provide a powerful mechanism of magnetic field amplification, thus making the studies of turbulent disk dynamos mostly academic. Assuming that the BH instability exists, in what accretion disks can it can operate? The BH instability will be damped by a sufficiently high resistive diffusivity η. The condition that damping is unimportant is $V_A^2 \gg 3\omega\eta$, where V_A is the Alfvén speed. On the other hand, there exists a critical wavelength in the BH instability, below which the instability is suppressed. Clearly, this critical wavelength must be smaller than the disk thickness, leading to the condition $V_A < \sqrt{6}c_s/\pi$, where c_s is the speed of sound. These two conditions must be met *simultaneously* in order for the BH instability to exist. In addition, the resulting values of the magnetic field must be such that magnetic pressure is smaller than gas pressure, otherwise the assumption of a thin disk is violated. For an accretion disk around a typical compact star those conditions are met, and the BH instability may in fact provide the most important mechanism of magnetic field generation. In a protoplanetary disk, however, those conditions are irreconcilable except very close to the star, leaving a turbulent dynamo as the leading mechanism of magnetic field regeneration. We consider turbulent dynamos for both types of disk, keeping in mind that in the case of a highly ionized disk it may not be a dominant mechanism of regeneration.

Following Shakura & Sunyaev (1973) we will encapsulate our ignorance about the nature of turbulence into a single parameter $\alpha_\nu = (l_0/h)(v_0/c_s)$, where h is disk half-thickness, l_0 and v_0 are turbulent mixing length and turbulent velocity, respectively. Shakura *et al.* (1978) suggested that irrespective of the source of turbulence, $v_0 \approx \omega l_0$, so $l_0/h \approx v_0/c_s = M_t$, where M_t is the turbulent Mach number. Thus, in our considerations we will take $\alpha_\nu = M_t^2$. The magnetic diffusivity η_t is taken to be identical to the general turbulent diffusion coefficient for a scalar field and equal to $l_0 v_0$, or $\eta_t = M_t^2 h^2 \omega$. The helicity of turbulence, α, is taken on the basis of qualitative order of magnitude arguments given by Ruzmaikin *et al.* (1988) to be equal to $M_t^2 z \omega$, where z is a vertical coordinate. These two statistical parameters η_t and α, together with α_ν, approximate the largely unknown physics of turbulence in an accretion disk. They do not reflect any specific origin of turbulence, nor the influence of rotation on it. In obtaining those parameters, the thin disk approximation $c_s \approx h\omega$ has been explicitly used.

3 COMPUTATIONAL METHOD

Solving the dynamo equation for disk models presents us with many cumbersome mathematical difficulties. First, difficulties arise in handling the boundary condition at infinity in the cylindrical disk geometry. Close to the disk faces, the potential that describes the magnetic field outside the disk interior decays exponentially with distance from the disk surface. On the other hand, far away from a disk, the magnetic field generated by currents in the disk should resemble a multipole and decay as $r^{-(n+1)}$, where $n \geq 2$. We would consider only the approximate boundary conditions: toroidal field and vertical derivative of poloidal vector potential vanish at the disk surfaces. Using such an approximation has only a marginal effect on the character of the generated field inside the disk, and thus on the magnetic angular momentum transport – the major incentive for studying disk dynamos. Even with such simplified boundary conditions, a straightforward numerical solution to the dynamo equation is technically unrealistic due to the tremendous amount of computation required. The practical way of solving the dynamo equation for accretion disks is to take advantage of the great difference between vertical and horizontal dimensions in such disks (see Table 1) and to adopt asymptotic methods. Ruzmaikin *et al.* (1985) pointed out that the so-called adiabatic approximation can be applied to solve the disk dynamo. Stepinski & Levy (1991) formulated a more general and rigorous algorithm based on an idea of adiabatic approximation, which can be directly applied to solving the dynamo equation in the context of an accretion disk. The basic idea is that in the zeroth approximation, when magnetic field diffusion along the disk is considered to be negligibly small as compared with diffusion across the disk, the radial derivatives of the magnetic field can be neglected. The problem thus becomes 'local', with radial coordinate r entering the generation equation only parametrically through radially varying coefficient $D_{eff} = \alpha(r\partial\omega/\partial r)h^3/(\eta + \eta_t)^2$, where η is the resistive magnetic diffusivity. Solving the dynamo equation in this zeroth approximation yields γ, the local rate of magnetic field exponential growth. The zeroth approximation solutions describe the vertical distribution of the magnetic field for fixed or infinitely slow (adiabatic) changes of radial coordinate. The radial structure and the global growth rate Γ of the magnetic field are then determined in the first approximation by the solution of the radial equation with the local growth rate γ used as a 'potential' function subject to the boundary conditions at the inner and outer disk edges. Adiabatic approximation thus provides a way to obtain critical dynamo numbers and radial, as well as vertical, distributions of magnetic field normal modes for accretion disk dynamo described by appropriate α, ω, h, η and η_t.

4 DISKS AROUND COMPACT STARS

Let's consider, as an example, an accretion disk around a black hole with mass $M = 10M_{\odot}$ (the mass inferred for the black hole candidate Cyg X-1). The inner radius of such a disk is at $3R_g \approx 10^7$ cm, and the outer radius is at about $10^3 R_g \approx 10^{10}$ cm (here R_g is the Schwarzschild radius). Taking $M_t \approx 0.1$ and assuming that the mean opacity is well approximated by Kramers' law, the disk temperature is given by $T \approx 6.2 \times 10^4(r/10^{10}\text{cm})^{-3/4}$, and it changes from about 10^7 K at the inner edge to 6×10^4 K at the outer edge. For those temperatures the magnetic resistive diffusivity changes from 10^2 to 10^5 in cgs units. On the other hand, the turbulent diffusivity η_t changes from 10^{14} to 10^{16} in cgs units. Clearly, the magnetic diffusion in such a disk is totally dominated by turbulent diffusion. The radially varying local dynamo number D_{eff} becomes radially constant and equal to $-1.5M_t^{-2} = -1.5/\alpha_\nu \approx -150$. Since the local critical dynamo number is about -12 (Stepinski & Levy 1991), a magnetic field can be maintained by the turbulent dynamo everywhere in a disk around a compact star.

What is the radial structure of the magnetic field in such a disk? Despite the fact that in those disks the local dynamo number remains constant along the disk, the local growth rate, $\gamma(r)$, does change with the radius, and is proportional to ω. Stepinski & Levy (1991) have shown that in a disk where $\gamma(r) \sim \omega \sim r^{-3/2}$ normal modes are localized in the inner parts of the disk. In a disk with $D_{eff} = -150$, a very large number of growing modes will be excited. The fastest-growing mode is confined to the innermost parts of the disk; other progressively less overcritical modes are confined between the inner edge of the disk and some cutoff radius, beyond which the mode is evanescent. The least overcritical mode spans the entire disk. Although those conclusions are obtained in the kinematic limit, the implications for dynamical behavior are clear – the magnetic field is best described as a set of many quasilocalized states each evolving separately from the others. Does magnetic stress dominate viscous stress and thus drive an accretion? The dominant component of the magnetic stress tensor is $t_{\phi r}^m = B_\phi B_r / 4\pi$. Assuming that the field intensity is given by the balance of Coriolis and Lorentz forces, $t_{\phi r}^m$ is of the same order as the viscous stress. However, at magnetic Reynolds numbers about 10^{10}, the magnitude of small-scale, random fields can exceed the magnitude of the mean field, and magnetic stress due to the small-scale fields can completely dominate other stresses and control the underlying disk structure.

5 DISKS AROUND YOUNG STARS

The typical steady-state accretion disk around a $1 M_\odot$ T Tauri star extends approximately from the star's surface to about 100 A.U. and is parameterized by $\alpha_\nu \approx 0.01$ and an accretion rate of about $10^{-6} M_\odot$ per year. At disk locations where the temperature is above about 200 K, the opacity is dominated by grains such as silicate and Fe metal grains, whereas water ice provides the dominant opacity at locations with lower temperatures. In general, the temperature of a disk is above 1000 K from the inner edge of the disk up to a radius of about 1 A.U., but, since the temperature decreases as $r^{-3/4}$, the extended parts of the disk are indeed too cool to ionize the disk's gas thermally. Stepinski (1992) calculated the ionization state of a protoplanetary disk on the basis that cosmic rays and radioactive nuclei are the dominant sources of ionization. According to those calculations the disk's resistive diffusivity η is about 10^9 (cgs units) in the innermost parts of a disk, but increases to about 10^{17} at a radius of about 1 A.U. where it achieves the maximum. From 1 A.U. outward η slowly decreases to a value of about 10^{16} in the outermost disk. On the other hand, the turbulent diffusivity η_t changes from about 10^{14} in the inner disk to about 10^{15}–10^{16} in the outer disk. Thus, with the exception of the disk's innermost part, magnetic diffusion is dominated by resistive losses. The local dynamo number D_{eff} is not radially constant, as was the case in the disk where magnetic diffusion was dominated by turbulence. Instead it varies radially, falling below its critical value in the intermediate part of a disk, which typically extends from about 2 to 5 A.U. This suggests that a turbulent dynamo operating in an accretion disk around the young star can maintain a magnetic field in the inner and outer regions, but not in the intermediate region of such a disk. The radial structure of the generated field is again likely to consist of a large number of quasilocalized states, because in both inner and outer parts of a disk, where the magnetic field is generated, its growth rate changes steeply with radius. With a moderately large magnetic Reynolds number, the magnitude of small-scale, random magnetic fields and large-scale, average fields is about the same. The magnetic stress is the major mechanism of angular momentum transport, but not so overwhelmingly as was the case in disks around compact stars. In significant portions of the outer disk the magnetic field intensity is limited by an ambipolar diffusion on a lower level than would be inferred from the balance of Coriolis and Lorentz forces.

6 CONCLUSIONS

If accretion disks are indeed turbulent, the standard MHD dynamo can, in general, maintain magnetic fields strong enough to be important to the structure and evolution of those disks. In hot disks around black holes, neutron stars, or white dwarfs, a dynamo can operate successfully throughout the entire radial extent of a disk, but its role in magnetic field regeneration may be secondary if the Balbus–Hawley instability is present. Protoplanetary disks are too poor conductors for the BH instability to work, thus a turbulent dynamo seems, at present, the only mechanism capable of maintaining magnetic field there. The poor conductivity of those disks makes it difficult to establish how effective the dynamo process is there and any conclusions depends strongly on estimates of nonthermal ionization levels. The basic structure of all disk dynamo modes is their spatial localization. This suggests that, contrary to stellar or planetary magnetic fields, disk fields may be composed of many quasilocally generated states, which are unlikely to evolve quietly near equilibrium. It is more likely that disk magnetic fields are characterized by highly variable and episodic dynamical behaviors.

This work was done while the author was a Staff Scientist at the Lunar and Planetary Institute, which is operated by USRA under Contract No. NASW-4574 with NASA. This is Lunar and Planetary Institute Contribution No. 792.

REFERENCES

Balbus, S.A. & Hawley, J.F. 1991 A powerful local shear instability in weakly magnetized disks. I. Linear analysis. *Astrophys. J.* **376**, 214–222.

Knobloch, E. 1992 On the stability of magnetized accretion discs. *Mon. Not. R. Astron. Soc.* **255**, 25p–28p.

Ruzmaikin, A.A., Sokoloff, D.D. & Shukurov, A.M. 1985 Magnetic field distribution in spiral galaxies. *Astron. Astrophys.* **148**, 335–343.

Ruzmaikin, A.A., Shukurov, A.M. & Sokoloff, D.D. 1988 *Magnetic Fields of Galaxies.* Kluwer Academic Publishers, Dordrecht.

Shakura, N.J. & Sunyaev, R.A. 1973 Black holes in binary systems. Observational appearance. *Astron. Astrophys.* **24**, 337–355.

Shakura, N.J., Sunyaev, R.A. & Zilitinkevich, S.S. 1978 On the turbulent energy transport in accretion disk. *Astron. Astrophys.* **62**, 179–187.

Stepinski, T.F. & Levy, E.H. 1991 Dynamo magnetic field modes in thin astrophysical disks: an adiabatic computational approximation. *Astrophys. J.* **379**, 343–355.

Stepinski, T.F. 1992 Generation of dynamo magnetic fields in the primordial Solar nebula. *Icarus* **97**, 130–141.

Tout, C.A. & Pringle, J.E. 1992 Accretion disc viscosity: a simple model for a magnetic dynamo. *Space Telescope Science Institute Preprint Series*, **No. 656**.

Vishniac, E.T., Jin, L. & Diamond, P. 1990 Dynamo action by internal waves in accretion disks. *Astrophys. J.* **365**, 648–659.

The Strong Field Branch of the Childress–Soward Dynamo

M.G. ST. PIERRE

Dept. of Mathematics
University of California, Los Angeles
Los Angeles, CA 90024 USA

Current address:
Dept. of Earth and Planetary Science
Harvard University
Cambridge, MA 02138 USA

The Childress–Soward dynamo, which uses rotating Bénard convection to maintain a magnetic field against Ohmic decay, is investigated numerically. A converged three-dimensional solution of
the strong field branch is presented for very small Ekman number.
For strong rotation, the system is able sustain convection and act
as a dynamo even for a Rayleigh number substantially less than
critical. It is found that the dominant forces tend to cancel, and
that the magnitudes of the curls of the Lorentz and Coriolis forces
remain virtually identical.

---

1 INTRODUCTION

Numerical computations comprise an increasingly important tool in the understanding of the Earth's dynamo, and, with the increased accessibility of
supercomputers, direct, realistic simulations of the geodynamo are not far
off. Any such simulation must solve the equations governing a three dimensional, rapidly rotating, dynamically consistent dynamo with Lorentz force
$\mathbf{J} \times \mathbf{B}$ present in the dominant balance of forces. The simplest dynamo with
these characteristics, first proposed by Childress & Soward (1972), uses the
convective motions of rapidly rotating Bénard convection to drive a dynamically consistent MHD dynamo. Computationally, the Childress–Soward dynamo has the advantage of permitting the expansion of the unknown fields in
Fourier series in all directions, allowing three dimensional fast Fourier transforms (FFT's) to be used in calculating the nonlinear terms. Since no fast
Legendre transform exists at the moment, the resulting programs will be

295

M.R.E. Proctor, P.C. Matthews & A.M. Rucklidge (eds.)
Theory of Solar and Planetary Dynamos, 295–302
©1993 Cambridge University Press.

faster than more realistic spherical dynamo simulations, while at the same time reflecting the important features of these models.

In this paper, the strong field branch of the Childress–Soward dynamo is investigated using direct numerical simulations. In section 2, the model is described mathematically, the governing equations are given, and previous results discussed. The numerical methods used are outlined in section 3, and the results presented in section 4. In particular, we will show that the model is indeed capable of sustaining a strong magnetic field, even when the Rayleigh number is less than the critical value obtained from linear stability theory. In this case, the thermal convection is not possible in the absence of a magnetic field, which has obvious important consequences for the theory of planetary cooling.

2 THE CHILDRESS–SOWARD DYNAMO

Childress & Soward (1972) first proposed using rapidly rotating Bénard convection to drive a dynamically consistent MHD dynamo. The system they envisaged is that of a uniform Boussinesq fluid of density ρ, magnetic diffusivity η, kinematic viscosity ν, thermal diffusivity κ and coefficient of thermal expansion α confined between two parallel, infinite, stationary planes separated by a distance πL and heated from below. The boundaries, located at $z = 0$ and $z = \pi L$, are stress free, and maintained at constant temperatures $T_0 + \Delta T$ and T_0 respectively. The exterior of the dynamo is perfectly conducting, both electrically and thermally. The fluid undergoes a constant rotation $\mathbf{\Omega} = \Omega \mathbf{1}_z$, and is subject to a constant downward gravity field $\mathbf{g} = -g\mathbf{1}_z$, where $\mathbf{1}_k$ is the unit normal in the k direction.

2.1 Governing equations

Scaling distance by L, and time by the magnetic decay time scale L^2/η, the system is governed by the nondimensional equations

$$E\frac{d}{dt}\mathbf{v} = -\nabla P + \sigma_\eta \left[\mathbf{v} \times \mathbf{1}_z + \mathcal{R}\theta\mathbf{1}_z + E\mathcal{M}^2(\nabla \times \mathbf{B}) \times \mathbf{B}\right] + \sigma_\eta E\nabla^2\mathbf{v},$$

$$\frac{d}{dt}\theta = u_z + p\nabla^2\theta, \tag{1}$$

$$\frac{\partial}{\partial t}\mathbf{B} = \nabla \times (\mathbf{v} \times \mathbf{B}) + \nabla^2\mathbf{B},$$

$$\nabla \cdot \mathbf{v} = 0, \qquad \nabla \cdot \mathbf{B} = 0,$$

where $d/dt = \partial/\partial t + \mathbf{v} \cdot \nabla$, $\mathbf{v}$ and $\mathbf{B}$ are the nondimensional velocity and magnetic fields, P is the nondimensional pressure, θ is the temperature departure from linear conductive profile, $E = \nu/(2\Omega L^2)$ is the Ekman number, $\sigma_\eta = \nu/\eta$ is the magnetic Prandtl number and $p = \kappa/\eta$ is the Roberts number. Of particular interest in the momentum equation are the parameters

$\mathcal{R} = g\alpha(\Delta T)L/2\Omega\eta$, a modified Rayleigh number, and the Hartmann number $\mathcal{M} = \mathcal{B}L/\sqrt{\mu_0\rho\eta\nu}$ where $\mathcal{B}$ is the characteristic magnetic field strength.

2.2 Previous results

The behavior of system (1) depends critically on the dynamical importance of the Lorentz force, as measured by the Hartmann number $\mathcal{M}$. It is a well known result of linear stability theory that for the nonmagnetic system, the first unstable modes have critical horizontal wave number k_c of order $E^{-1/3}$, and critical modified Rayleigh number $\mathcal{R}_c$ of the same order. This behavior follows from the Proudman–Taylor theorem, which states that in the limit of rapid rotation, the flow is independent of the z coordinate. This theorem clearly must fail for convection to be possible. One manner of achieving this is to make the viscous forces, neglected in the proof of the theorem, important, and hence the need for small scale structures. The large critical Rayleigh number then results from the need to drive the fluid against this strong viscous damping.

That this behavior persists in the presence of weak magnetic fields was demonstrated by Soward (1974), who studied the system (1) in the asymptotic range $\mathcal{M} = \mathcal{O}(1)$, known as the weak field régime, and found a variety of stable solutions evolving on the slow time scale of magnetic diffusion. Fautrelle & Childress (1982), studying the effect of larger magnetic fields, found that fields of magnitude $\mathcal{M} = \mathcal{O}(E^{-1/6})$ are sufficient to destabilize the system, leading to rapid field growth.

2.3 The strong field branch

The unstable nature of the intermediate field branch of Fautrelle & Childress is simply a reflection of the fact that the asymptotic expansions used are no longer valid, and the reason for this is straightforward: as the magnetic field grows, it eventually reaches a point where the Lorentz force $\mathbf{J} \times \mathbf{B}$ is capable of violating the Proudman–Taylor theorem by itself. The system is thus liberated from the small convection cell constraint, and these cells grow. The resulting decrease in the viscous force allows the motions and thus the magnetic field to grow rapidly. This growth proceeds until a new dominant balance between forces is struck, this time between the Coriolis, pressure, buoyancy and Lorentz forces. This balance is called the 'strong field régime'.

This strong field branch, characterized by the range $\mathcal{M}^2 = \mathcal{O}(E^{-1})$, presents the intriguing possibility that convection is now possible for a Rayleigh number $\mathcal{R}$ *less* than the linear critical Rayleigh number $\mathcal{R}_c$ described above. This is hinted at by system (1). Assuming that the buoyancy force is comparable to both the Lorentz and Coriolis forces, and that both $\mathbf{v}$ and $\mathbf{B}$ remain bounded, this balance implies $\mathcal{R} = \mathcal{O}(1)$ as $E \to 0$, much smaller than the critical value $\mathcal{R}_c = \mathcal{O}(E^{-1/3})$. In section 4 below, we present the results of a computer run for which a dynamo is obtained for $\mathcal{R} < \mathcal{R}_c$.

3 NUMERICAL METHODS

To approximate solutions of (1) numerically, we first expand the velocity $\mathbf{v}$ and the magnetic field $\mathbf{B}$ in their Cartesian poloidal and toroidal scalar fields:

$$
\begin{aligned}
\mathbf{v} = \mathbf{v}_P + \mathbf{v}_T + \mathbf{U}_H = \nabla \times \nabla \times \varsigma \mathbf{1}_z + \nabla \times \tau \mathbf{1}_z + \mathbf{U}_H(z), \\
\mathbf{B} = \mathbf{B}_P + \mathbf{B}_T + \mathbf{B}_H = \nabla \times \nabla \times \mathcal{S} \mathbf{1}_z + \nabla \times \mathcal{T} \mathbf{1}_z + \mathbf{B}_H(z),
\end{aligned}
\tag{2}
$$

where inclusion of the horizontally constant fields $\mathbf{U}_H(z) = (U_{Hx}(z), U_{Hy}(z), 0)$ and $\mathbf{B}_H(z)$ is necessary due to the incompleteness of the Cartesian poloidal-toroidal separation.

The dynamo region is of infinite horizontal extent, and the structure of the fields $\mathbf{v}$, $\mathbf{B}$ and θ in this direction is not known; for this reason, we *assume* horizontal periodicity, which permits the expansion of scalar fields $\mathcal{S}$, $\mathcal{T}$, ς, τ and θ in Fourier series of the form

$$
\gamma(x, y, z, t) = \sum_{l,m,n} \hat{\gamma}_{lmn}(t) e^{i(2klx + 2kmy + nz)},
$$

where k is the fundamental horizontal wave number. Since we are now dealing with a strongly nonlinear system, the thin cells characteristic of the nonmagnetic and weak field régimes are no longer necessary, and we may further assume $k = \mathcal{O}(1)$. Truncation of the series in Fourier space yields a system of ordinary differential equations of the form

$$
\frac{d}{dt}\mathbf{Y} = \mathcal{L}\mathbf{Y} + \mathcal{N}(\mathbf{Y}),
$$

where the vector $\mathbf{Y}$ consists of subvectors $\mathbf{Y}_{lmn} = (\hat{\tau}_{lmn}, \hat{\varsigma}_{lmn}, \hat{\theta}_{lmn}, \hat{\mathcal{T}}_{lmn}, \hat{\mathcal{S}}_{lmn})^T$, and the linear operator $\mathcal{L}$ is block diagonal with 5 by 5 blocks, each block representing a given Fourier mode (l, m, n). The boundary conditions imply that the series for fields $\mathcal{S}$, ς, and θ contain only terms in $\sin nz$, while the series for $\mathcal{T}$, τ, $\mathbf{B}_H$ and $\mathbf{U}_H$ contain only cosines in z. The resulting equations are integrated using a third-order Adams–Bashforth/Adams–Moulton $P(EC)^1E$ predictor/corrector method, with the linear operator $\mathcal{L}$ being treated analytically. The time step changes throughout each run, in such a way as to conform with the Courant–Friedrichs–Lewy stability condition for this problem, as derived by St. Pierre (1993).

4 RESULTS

In this section, we present the results of the most successful simulation to date, the parameters for which are given in Table 1. It is to be noted that the Ekman number E for this simulation is roughly two orders of magnitude smaller than for recent studies by Brandenburg *et al.* (1990) and Nordlund

E	5×10^{-6}
$\mathcal{R}$	100
$\mathcal{R}_c$	110.56
σ_η, p	2
k	3

Table 1. Governing parameters for the dynamo simulation.

et al. (1992), which investigated very strongly forced compressible turbulence similar to the overshoot region of the Solar dynamo. A further important aspect of this calculation is that the modified Rayleigh number $\mathcal{R}$ is roughly 0.9 of the critical value, indicating the vital role that the magnetic field may play in the theory of nonlinear thermal convection. The computer run was repeated for the nonmagnetic case with identical initial conditions in **v** and θ, and it was found that the motions quickly die out, and the system decays towards the purely conductive state. The system was solved for 51^3 Fourier modes for each scalar variable, and the nonlinear terms were computed in real space at 63^3 grid points.

The simulation was started with random initial conditions $\mathbf{v}(\mathbf{x}, 0)$, $\mathbf{B}(\mathbf{x}, 0)$ and $\theta(\mathbf{x}, 0)$. Since the dynamo is operating in the subcritical range, it was not possible to let the nonmagnetic convection evolve to a near steady state before adding a small seed magnetic field, as proposed by Nordlund *et al.* (1992). After a short initialization, the system settled down into a relatively well behaved convective state. Rotational constraints are still important, as is reflected by the Fourier space distributions of the kinetic energy and temperature perturbation, both of which are confined to vertical wave number $n \leq 5$, and thus exhibit no small scale structures in the z direction. Small scales are present in the horizontal direction, however, and to help convergence of these modes, a relatively large fundamental horizontal wave number k of 3 was chosen. This anisotropy is not present for the magnetic field, however, which tended to be equally well converged in the horizontal and vertical directions, with roughly 90% of the magnetic energy being confined to regions $\sqrt{l^2 + m^2 + n^2} \leq 18$ in Fourier space.

In Figure 1, we plot the dimensional magnetic and kinetic energies with respect to time for the run. It may be seen that the system is well past the equipartition of energy characteristic of Solar dynamo calculations, with the magnetic energy dominating the kinetic energy, as is to be expected of rapidly rotating, strong-field dynamos.

We now confront the important question of whether we are in fact dealing with a dynamo. The time plotted in Figure 1 is with respect to the free decay time of the $l = 1, m = 0, n = 0$ mode. This is *not* the slowest decay mode for the

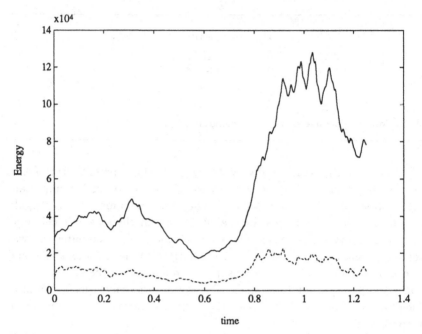

Figure 1. Plot of the magnetic energy E_M (solid line), and kinetic energy E_K (dashed line) *vs* time. Note that $E_M \gg E_K$ throughout the run.

system, which corresponds to the $l = 0$, $m = 0$, $n = 1$ mode. During the run, however, it was found that very little energy is present in these horizontally constant fields. The velocity field $\mathbf{U}_H$ remains negligible throughout the run, while the energy contained in $\mathbf{B}_H$ rapidly levels off to a value of approximately 0.4% of the total magnetic energy. For this reason, the horizontally constant fields $\mathbf{U}_H$ and $\mathbf{B}_H$ were judged to be relatively unimportant to the behavior of the system. This implies that the decay time associated with the $l = 1$, $m = 0$, $n = 0$ mode yields a good indication of the fundamental decay time scale of the system as a whole. Further, following the arguments put forward by Cattaneo, Hughes & Weiss (1991), we looked at the characteristic time of Joule dissipation τ_M, defined by

$$\tau_M = \frac{\langle \mathbf{B} \rangle^2}{\eta \langle \nabla \times \mathbf{B} \rangle^2}.$$

The present run corresponds to a time of the order of $70\tau_M$, which again is a strong indication that the system is indeed acting as a dynamo.

In Figure 2, we plot the L_2 norms of the curls of the dominant forces as well as the curl of their sum. It may be seen that the magnitudes of the curled Lorentz and Coriolis forces are virtually identical, remaining within 3% of each other throughout the run. The curl of the buoyancy force is consistently

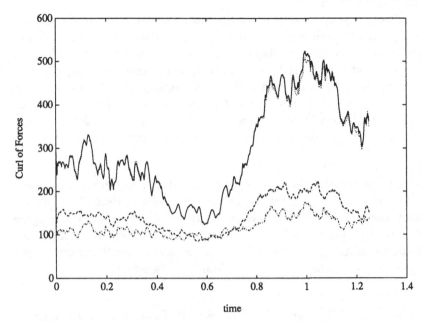

Figure 2. Plot of the curl of the dominant forces *vs* time. The Lorentz (solid curve) and Coriolis (dotted curve) are the largest and are of virtually identical magnitudes. The curl of the sum of the dominant forces (dotted-dashed curve) is smaller than either of the Lorentz, Coriolis or buoyancy (dashed curve) forces.

a factor of two or so less than the above two fields. The fact that the total of the three fields is generally less than either of the individual fields indicates a tendency for the dominant forces to cancel, and may imply that the system is tending towards a magnetostrophic state.

A study of the the system in real space shows that the magnetic field is concentrated near the boundaries, roughly in the upper and lower quarter of the dynamo region, with the middle half of the layer being a relatively quiet zone with respect to **B**. This is thought to be due both to the fact that the field is 'trapped' near the boundary, being unable to penetrate outside the layer due to the boundary conditions, and to the generation of the field being concentrated near the boundaries where the velocity field gradients are greatest. A video animation of the dynamo was produced, and it was found that the dominant features are short-lived retrograde columns extending throughout the layer. The thickness of the columns is typically a quarter of the width of a convection cell, and they seem to evolve on time scales comparable with the time scale of viscous diffusion associated with features of this size. Strong shear is produced when two columns collide, resulting in intense generation and/or destruction of **B** at these locations.

5 CONCLUSIONS

As detailed in the above section, we have found strong evidence that rapidly rotating Bénard convection is capable of driving a strong field, dynamically consistent dynamo, even for Rayleigh numbers less than critical. This is the first example of a converged three dimensional dynamo operating in such a subcritical régime, and the first with an Ekman number E small enough to allow a good separation between the strong and weak field branches. Further investigation of the Childress–Soward dynamo should lend valuable insight into the dynamo process in general, and the geodynamo in particular.

Acknowledgements

This work was performed as part of my doctoral thesis research at the University of California, Los Angeles, under the supervision of Prof. P.H. Roberts, without whose guidance and encyclopedic knowledge of the subject, progress would not have been feasible. I would also like to thank the Los Alamos National Laboratories and the Institute of Geophysics and Planetary Physics for their support under LANL/IGPP Grant UC-89-4-E-78.

REFERENCES

Brandenburg, A., Nordlund, A., Pulkkinen, P., Stein, R.F. & Tuominen, I. 1990 3-D Simulation of turbulent cyclonic magneto-convection. *Astron. Astrophys.* **232**, 277–291.

Cattaneo, F., Hughes, D.W. & Weiss, N.O. 1991 What is a stellar dynamo? *Mon. Not. R. Astron. Soc.* **253**, 479–484.

Childress, S. & Soward, A.M. 1972 Convection driven hydromagnetic dynamo. *Phys. Rev. Lett.* **29**, 837–839.

Fautrelle, Y. & Childress, S. 1982 Convective dynamos with intermediate and strong fields. *Geophys. Astrophys. Fluid Dynam.* **22**, 235–279.

Nordlund, A., Brandenburg, A., Jennings, R., Rieutord, M., Ruokolainen, J., Stein, R.F. & Tuominen, I. 1992 Dynamo action in stratified convection with overshoot. *Astrophys. J.* **392**, 647–652.

St. Pierre, M.G. 1993 The stability of the magnetostrophic approximation I: Taylor state solutions. *Geophys. Astrophys. Fluid Dynam.* **67**, 99–115.

Soward, A.M. 1974 A convection driven dynamo I. The weak field case. *Phil. Trans. R. Soc. Lond.* A **275**, 611–651.

Evidence for the Suppression of the Alpha-effect by Weak Magnetic Fields

L. TAO

Department of Physics
University of Chicago
Chicago, IL 60637 USA

F. CATTANEO & S.I. VAINSHTEIN

Department of Astronomy and Astrophysics
University of Chicago
Chicago, IL 60637 USA

We present results from fully self-consistent numerical simulations of the equations of magnetohydrodynamics at moderate Reynolds numbers. The kinematic calculation show that there is a nonzero turbulent α-effect. However, dynamical calculations including the Lorentz force term give evidence that even weak fields can severely suppress this turbulent α-effect.

---

1 INTRODUCTION

The nature of turbulent magnetic diffusion and the α-effect has been a puzzle for several decades. Until recently, virtually all the work in this subject has been based on analytical theory, but the advent of ready access to supercomputers now allows us to address the question of turbulent magnetic diffusion and the turbulent α-effect from the perspective of numerical experiments. In this paper, we shall describe numerical simulations of an idealized model of mean field dynamos.

In order to understand how fields are generated, the mean field theoretical approach is widely used (see Moffatt 1978). This two-scaled approach conveniently parametrizes the effects of small scale turbulence on large scale fields into two coefficients, α and β. The central problem of mean field electrodynamics is to calculate these transport coefficients from the statistical properties of the flow and the magnetic diffusivity, η. Explicit in these calculations is that the fluid flow is not affected by the presence of magnetic fields.

303

M.R.E. Proctor, P.C. Matthews & A.M. Rucklidge (eds.)
Theory of Solar and Planetary Dynamos, 303–310
©1993 Cambridge University Press.

In typical magnetofluid circumstances, this is assumed to be the case, unless the magnetic energy of the large scale component is comparable to the energy in the flow. Recent two-dimensional simulations suggest that this is not the case and that turbulent diffusivity can be severely suppressed even when the mean field is $R_m^{1/2}$ less than the equipartition field value (see Cattaneo & Vainshtein 1991).

Within the framework of mean field electrodynamics, there are two ways one can compute α analytically. One instance is the short correlation time limit. The existence of a small parameter allows one to do the asymptotic calculation of the small scale magnetic field fluctuations and, therefore, to compute α.

In low magnetic Reynolds number situations one also can compute α analytically using the first order smoothing approximation (see Moffatt 1978). In this limit, terms in the evolution equation for the fluctuations of the magnetic field can be neglected and the fluctuations computed. To date, it has not been clear whether or not there is a nonzero α-effect in moderate to high magnetic Reynolds number cases. In this note, we demonstrate numerically the existence of the turbulent α-effect at moderate Reynolds numbers and we investigate how the α coefficient is altered when we include the effects of the magnetic back reaction on the fluid flow.

2 PHENOMENOLOGICAL CONSIDERATIONS

Recently, Vainshtein & Cattaneo (1992) have argued on general grounds that a large scale magnetic field can suppress turbulent transport even when the energy of the large scale field is small compared to the kinetic energy (equipartition). It is important to realize that in this framework the vigour of the turbulent motions are not appreciably reduced, only the transport properties are suppressed. Since the present calculations are motivated by these ideas we provide here a brief outline of the arguments.

The effect of a turbulent velocity on a *passively* advected quantity (scalar or vector) is rapidly to generate small scale fluctuations. The cascade to ever decreasing scales is eventually halted by diffusion at a scale of order $\delta = \ell(U\ell/\eta)^{-1/2}$, where U and ℓ are the velocity amplitude and correlation length respectively and η is the (molecular) diffusivity of the advected quantity (Batchelor 1959). The scale δ is obtained by assuming a balance, on average, between the rates of advection and diffusion.

In the case of a magnetic field the generation of small scales is accompanied by the stretching, and therefore by the amplification, of the field lines. We can thus imagine another possibility where the cascade to small scales is eventually halted *dynamically* by the Lorentz force. On general grounds we expect a balance to be reached when the magnetic energy becomes comparable to the kinetic energy. Since this balance is dynamical rather than diffusive, we have the possibility that the smallest scales thus generated, δ_{eff} say, may be larger than δ and therefore that diffusive processes, including reconnection, may be greatly reduced.

For two-dimensional fields, Cattaneo & Vainshtein (1991) estimate that (large scale) fields stronger than $U/R_m^{1/2}$, where R_m is the magnetic Reynolds number, reach a dynamical balance *before* a diffusive one and that therefore, for such fields the (large scale) evolution is mostly non-diffusive, i.e. the turbulent diffusion becomes suppressed. In three dimensions the stretching process can be more effective than in two and the estimate above should be regarded as an upper bound. Since the α-effect, like turbulent diffusion, relies on efficient line reconnection, Vainshtein & Cattaneo (1992) have argued that when the one becomes suppressed so does the other. They have further proposed an analogy between the rate of suppression of turbulent diffusion and the α-effect, leading to the estimate

$$\alpha_{eff} = \alpha_T \left(\frac{R_m B_0^2}{U^2}\right)^{-1}. \tag{1}$$

Here, B_0 is the strength of the large scale field (measured in terms of the Alfvén velocity) and α_T is the value of α in the kinematic regime – assuming it exists. It is clear from equation (1) that the suppression of the α-effect is very effective even for very weak large scale fields. The above results are based on general considerations and assume only that the magnetic and kinetic Reynolds numbers are comparable and large and that the velocity in the kinematic regime is turbulent.

3 MODEL DESCRIPTION

The evolution equations for the velocity **u** and the magnetic field **B** in an incompressible fluid of unit density can be written as

$$\frac{\partial \mathbf{u}}{\partial t} + (\mathbf{u} \cdot \nabla)\mathbf{u} = -\nabla p + \frac{1}{M^2}(\nabla \times \mathbf{B}) \times \mathbf{B} + \nu \nabla^2 \mathbf{u} + \mathbf{F}, \tag{2}$$

$$\frac{\partial \mathbf{B}}{\partial t} = \nabla \times (\mathbf{u} \times \mathbf{B}) + \eta \nabla^2 \mathbf{B}, \quad \nabla \cdot \mathbf{u} = 0, \quad \nabla \cdot \mathbf{B} = 0, \tag{3}$$

where p is the pressure, ν and η are the viscosity and magnetic diffusivity, respectively (assumed constant in space), **F** is a forcing term, and M is the Alfvénic Mach number.

We solve equations (2) and (3) numerically on a periodic cube of side length 2π. Initially, the magnetic field is uniform and along the x direction. Because of Stokes's theorem, this uniform component does not change even though **B** does due to the growth of fluctuations. We choose the strength of the uniform component as the unit of magnetic field intensity so that M^2 in equation (2) gives the ratio between the kinetic energy and the magnetic energy of the imposed uniform field. The forcing function $\mathbf{F}$ is chosen to have support in phase space on a band of wavenumbers with $2 \leq k \leq 4$ (seventeen wavenumbers in all). The amplitude of each forcing component is a stationary random process with unit correlation time and satisfying global constraints ensuring that the forcing helicity $\mathbf{F} \cdot (\nabla \times \mathbf{F})$ is positive at all times, and that the resulting (random) velocity had an amplitude of order unity. With these choices the parameters ν and η in equations (2) and (3) are of the same order as the inverse of the kinetic and magnetic Reynolds numbers.

Equations (2) and (3) are solved by standard pseudo-spectral techniques where nonlinear terms are evaluated in configuration space while derivatives are evaluated in phase space (see, for instance, Canuto *et al.* 1988). Different time-advance methods are used: explicit for nonlinear terms and implicit for linear terms. The pressure term in equation (2) is not explicitly computed; rather we ensure solenoidality of both the velocity and magnetic fields by adopting a description in phase space based on the Craya decomposition (see Lesieur 1987). Adequate numerical resolution is achieved by 64^3 collocation points.

It is important to realize exactly what quantities are being measured. Our objective is to calculate α and to see how its value varies as the strength of the uniform component of the magnetic field, as measured by M, increases. It is well known that α is a meaningful quantity only if a separation of scales exists between a large scale field and its fluctuating component. In the present experiment this separation is achieved by identifying the small scale (defining the fluctuations) with the size of the computational domain, i.e. we regard this simulation as occurring on the scale of the fluctuations. The large scale, on the other hand, is defined by the projection of a quantity onto the wavenumber $\mathbf{k} = 0$, i.e. by the volume average over the entire computational domain. This definition essentially assumes that the large scale component is infinitely long. Furthermore, α is related to the average value of $\langle \mathbf{u} \times \mathbf{B} \rangle$, where the angle brackets correspond to the projection onto $\mathbf{k} = 0$ and the average must be performed either over realizations or over long times (the procedure adopted in this paper). Either way, the quantity $\langle \mathbf{u} \times \mathbf{B} \rangle$ fluctuates in a way similar to a classical random walk and therefore it approaches its mean value roughly as $1/\sqrt{N}$, where N is either the number of independent realizations or the number of elapsed velocity correlation times. Since the velocity correlation time is not prescribed, but rather emerges naturally from the solution of the momentum equation with the chosen random forcing, the calculation of meaningful average values can be a delicate matter requiring, possibly, very long computations. Explicitly, we calculated the three components of $\mathbf{u} \times \mathbf{B}$ averaged over space and time:

$$\alpha_i = \frac{1}{T} \int_{t_0}^{t_0+T} \langle \mathbf{u} \times \mathbf{B} \rangle \cdot e_i \, dt, \tag{4}$$

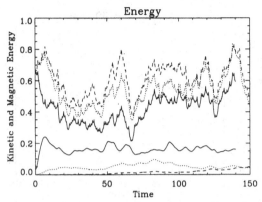

Figure 1. Energy densities vs. time. Solid, $M^2 = 30$; dots, $M^2 = 32^2$; dashes, $M^2 = 256^2$.

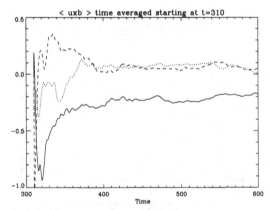

Figure 2. Time-averaged α vs. time. $M^2 = 256^2$. Solid, x-component; dashed, y; dotted, z.

where the angle brackets denote averaging in space (or, alternatively, calculating the $k = 0$ component of $u \times B$) and e_i are the directional unit vectors. Only one of the components – the one parallel to the large scale magnetic field – will have a non-zero time average.

4 RESULTS
We compute four cases with M^2 ranging from 256^2 (weaker field) to 30 (stronger field). In all cases the Reynolds number and the magnetic Reynolds number are of order 130 (the parameter values of ν and η are both 130^{-1}), and we use the same random set forcing coefficients throughout. In the $M^2 = 256^2$ case, the one we take to be our 'kinematic' case, the velocity field remains a stationary random process (see Figure 1). The value of the time-averaged α-coefficient is approximately 0.2, the same order as the root-mean-squared velocity (see Figure 1), which we expect on dimensional grounds (Figure 2). The reason that there are three components is because

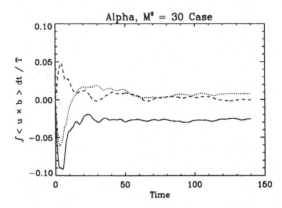

Figure 3. Time-averaged α vs. time. $M^2 = 30$. Solid, x-component; dashed, y; dotted, z.

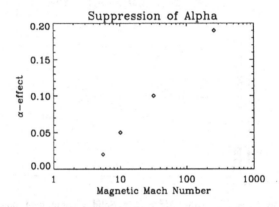

Figure 4. Suppression of α. Summary of calculations.

even though the forcing is statistically isotropic, it is not isotropic at any given instant. Therefore, the space average $\langle \mathbf{u} \times \mathbf{B} \rangle$ in the other two directions are nonzero. However, as we expect, they have zero time averages.

As the initial magnetic field strength increases (as we decrease M^2), the value of the time-averaged α decreases, eventually going to a value of 0.02 at the case $M^2 = 30$ (where the initial magnetic energy density is a mere 3% of the equipartition value) (Figure 3). We summarize our calculations in Figure 4. The interesting result is that even though the time-averaged value of α decreased as we decrease M^2, the amplitude of the time fluctuations of $\langle \mathbf{u} \times \mathbf{B} \rangle$ did not.

The reason for this reduction of the α coefficient is associated with the dynamics of the small-scale magnetic fields. We show in Figure 5 the time history of the Taylor microscale of the magnetic field, $\lambda^2 = \langle \mathbf{B}^2 \rangle / \langle |\nabla \times \mathbf{B}|^2 \rangle$, for the cases $M^2 = 256^2$, 32^2 and 30. In the kinematic case, we see that the

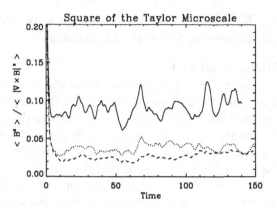

Figure 5. The magnetic Taylor microscale vs. time. Solid, $M^2 = 30$; dots, $M^2 = 32^2$; dashes, $M^2 = 256^2$.

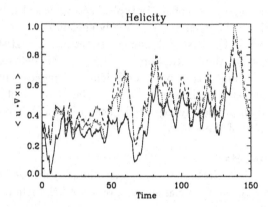

Figure 6. The fluid helicity vs. time. Solid, $M^2 = 30$; dots, $M^2 = 32^2$; dashes, $M^2 = 256^2$.

small diffusive scales are formed right away, whereas in the weak field case, the diffusive scales are never formed. The small scales that are necessary for the α coefficient to be nonzero are not present; the magnetic fields on these scales resist the turbulent motions of the fluid as equipartition fields are created on scales larger than the diffusive scales. The dynamics of these small-scale fields effectively modify the small-scale motions, while leaving the readily observable large-scale behavior unchanged (see Figure 6 – the fluid helicity density of our dynamical calculation approaches that of the 'kinematic' case after the initial transient; the same occurs in the kinetic energy density time history in Figure 1).

This mechanism of the suppression of the α-effect is quite different from α quenching results obtained with first-order smoothing approximation. In the case when $R_m \ll 1$, the quenching occurs when $M^2 \approx 1$, and the reduction of the velocity field is responsible for the quenching of α (see Vainshtein &

Cattaneo, 1992). In the case when $R_m \gg 1$, even a weak field can severely suppress α, while the velocity field is roughly unchanged.

5 CONCLUDING REMARKS

Our calculations suggest that the α-effect can be effectively suppressed by a large-scale field much weaker than equipartition. The suppression occurs through a subtle modification of the flow field and not through an overall reduction of the velocity amplitude – as is normally assumed in the quenching models based on first order smoothing. It is therefore possible to exhibit two flow fields characterised by similar kinetic energies and helicities but radically different α-effects. Furthermore, these calculations support the hypothesis that the critical large-scale magnetic field strength for effective suppression is related to the equipartition value by an inverse power of R_m with an exponent greater than 0.5.

Because of the huge values of R_m in typical astrophysical situations and the near-equipartition field strengths commonly encountered, we must conclude that the standard α-effect based dynamos cannot provide satisfactory models of dynamo action for these situations. In view of this deficiency and of the fact that if the turbulent transport coefficients are indeed suppressed in the manner proposed above, then primordial stellar and galactic fields should survive longer than previously expected. The question then remains: what is the origin and fate of these fields?

Acknowledgements
We would like to thank Robert Rosner for stimulating discussions. L.T. would like to acknowledge the financial support of a GAANN fellowship.

REFERENCES
Batchelor, G.K. 1959 Small-scale variation of convected quantities like temperature in turbulent fluid. *J. Fluid Mech.* **5**, 113–133.

Canuto, C., Hussaini, M.Y., Quarteroni, A. & Zang, T.A. 1988 *Spectral Methods in Fluid Dynamics*. Springer.

Cattaneo, F. & Vainshtein, S.I. 1991 Suppression of turbulent transport by a weak magnetic field. *Astrophys. J.* **376**, L21–L24.

Lesieur, M. 1987 *Turbulence in Fluids*. Martinus Nijhoff Publishers.

Moffatt, H.K. 1978 *Magnetic Field Generation in Electrically Conducting Fluids*. Cambridge University Press.

Vainshtein, S.I. & Cattaneo, F. 1992 Nonlinear restrictions on dynamo action. *Astrophys. J.* **393**, 165–171.

Turbulent Magnetic Transport Effects and their Relation to Magnetic Field Intermittency

S.I. VAINSHTEIN

Department of Astronomy and Astrophysics
University of Chicago, Chicago, IL 60637 USA

L. TAO

Department of Physics
University of Chicago, Chicago, IL 60637 USA

F. CATTANEO & R. ROSNER

Department of Astronomy and Astrophysics
University of Chicago, Chicago, IL 60637 USA

We discuss the consequences of nonlinear effects on the effective magnetic field transport coefficients in a magnetofluid; such transport effects lie at the heart of modern astrophysical dynamo theories. The particular focus of our discussion is on the distinction between fully turbulent and quasi-steady flows; we show that these two types of flows both show suppression of effective magnetic field transport, but are distinguished by the amplitude of the suppression effect: suppression is substantially more profound in a fully turbulent flow.

M.R.E. Proctor, P.C. Matthews & A.M. Rucklidge (eds.)
Theory of Solar and Planetary Dynamos, 311–320
©1993 Cambridge University Press.

1 INTRODUCTION

An essential aspect of virtually all astrophysical magnetic dynamos is the role played by turbulent magnetic field diffusion. From the analytical perspective, discussions of turbulent diffusion have until recently been generally couched in the language of mean field theory, and in particular, within a kinematic context (cf. Moffatt 1978; Krause & Rädler 1980). Indeed, the great theoretical elegance of mean field electrodynamics, together with its attractive intuitiveness, have led to a situation where basic constructs of this theory, such as turbulent diffusion and the 'α-effect', have carried over into domains, such as numerical simulations, where their meaningfulness is not *a priori* obvious (cf. Glatzmaier 1985). In a recent series of papers, we have examined precisely the question of how such notions can be carried over into the nonlinear domain, and further have asked under what circumstances nonlinear effects are likely to matter (Vainshtein & Rosner 1991; Cattaneo & Vainshtein 1991; Vainshtein & Cattaneo 1992; Tao, Cattaneo & Vainshtein 1993). Our results show that, for example, the effective magnetic diffusivity or the effective α-effect can be substantially suppressed below their 'classic' turbulent values when the large-scale magnetic field strength exceeds a certain threshold; the previously unexpected aspect of this result is the very low value of this threshold under astrophysical conditions, i.e., well below the equipartition value.

An obvious question is how general these results are, and in particular, what restrictions exist on the nature of the hydrodynamic flows in order to reach our conclusions. This question is urgent because a large number of both analytical and numerical studies carried out in the context of quasi-steady (or steady) flows produce results in apparent conflict (cf. Proctor & Weiss 1982). The purpose of this paper is to examine this issue, and therefore to focus on the connection between flow properties and the value of the effective transport coefficients.

2 FORMULATION OF THE TWO-DIMENSIONAL PROBLEM

We consider the classical problem of diffusion of a magnetic field over some large scale L due to the effects of a velocity field u. In particular, we wish to study the changes in diffusion time due to changes in the strength of the initial field. Our aim is to show that the diffusion of the magnetic field becomes greatly reduced once the initial field strength exceeds a critical value.

Clearly the process of diffusion depends also on the properties of the velocity field u. We examine two distinct and representative cases: First, the case in which the velocity is turbulent, in the sense that the correlation time is comparable to the turnover time (section 3); second, the case in which the velocity is quasi-steady or steady, e.g., the correlation time is long compared to the turnover time (section 4). As we shall see, the different degree of spatial intermittency of the magnetic fluctuations in the two cases leads to different rates of diffusion and to different rates of suppression of diffusion.

The evolution of a magnetic field $\mathbf{B}$ is governed by the induction equation

$$\partial_t \mathbf{B} = \nabla \times (\mathbf{u} \times \mathbf{B}) + \eta \nabla^2 \mathbf{B}, \tag{1}$$

where η is the molecular diffusivity. In two-dimensional geometry, $\mathbf{B} = \nabla \times (A\mathbf{e}_z)$, where A is the (scalar) flux function and $\mathbf{e}_z$ is a unit vector along the ignorable coordinate. Equation (1) can be rewritten as an advection-diffusion equation for the quantity A, namely

$$\partial_t A + \mathbf{u} \cdot \nabla A = \eta \nabla^2 A. \tag{2}$$

In two dimensions, the process of diffusion or field decay is clearly equivalent to the process of dispersion or homogenization of the quantity A because dynamo action is impossible. By multiplying (2) by A and averaging over the entire volume we have

$$\partial_t \langle A^2 \rangle = -2\eta \langle |\nabla A|^2 \rangle = -2\eta \langle |\mathbf{B}|^2 \rangle. \tag{3}$$

Equation (3) is *exact*, and assumes only that there are no external sources of magnetic field. It also shows that the interaction term $\mathbf{u} \cdot \nabla A$ does not contribute to the destruction of A^2-stuff.

We now assume that the velocity $\mathbf{u}$, characterized by a magnitude $U = \langle |\mathbf{u}|^2 \rangle^{1/2}$, is maintained by some forces chosen so that in the absence of a magnetic field, the energy-containing scale is $\ell \ll L$. In the turbulent case, the forcing is chosen so that the velocity correlation time $\tau \approx \ell/U$ (section 3), and in the quasi-steady case so that $\tau \gg \ell/U$ (section 4). We now examine the process of diffusion in each case in turn.

3 THE CASE $\tau \approx \ell/U$

In order to proceed, we make three assumptions, listed below:

I. The velocity $\mathbf{u}$ is chaotic, i.e., it has positive Liapunov exponents almost everywhere, in the kinematic regime.

II. The diffusion time in the kinematic regime is independent of η, or is at most only weakly (logarithmically) dependent on η for small η.

III. On average, the magnetic energy does not exceed the kinetic energy, i.e.,

$$\langle |\mathbf{B}|^2 \rangle \leq \rho_0 \mu_0 U^2.$$

The first two assumptions characterize the diffusion process in the kinematic regime, while the third is necessary to derive results in the nonlinear regime. Assumption (I) is reasonable since in the absence of a magnetic field the velocity depends on the forcing alone, and we can choose it so that (I) is satisfied. Assumption (II) is phenomenological, and is based on experimental data (e.g., Monin & Yaglom 1971). Assumption (III) cannot be rigorously justified, but is nevertheless generally believed to be valid (and is certainly not contradicted by the observations). It will be recalled that the large scale field is supposed to be very weak in all our considerations.

Let t_T denote the diffusion time for a kinematic field over a scale L. Then (II) implies

$$t_T \approx L^2/U\ell, \qquad (4)$$

which represents classical turbulent diffusion (Monin & Yaglom 1971). Estimating the relative magnitude of each term in (3) and using (4) we obtain

$$\frac{U\ell}{\eta L^2}\langle A^2\rangle \approx \langle|\mathbf{B}|^2\rangle; \qquad (5)$$

furthermore, since $A \approx B_0 L$, where B_0 is the large-scale component of the magnetic field, we have

$$\langle|\mathbf{B}|^2\rangle \approx R_m\langle B_0^2\rangle, \qquad (6)$$

where $R_m = U\ell/\eta$ is the magnetic Reynolds number (Zel'dovich 1957; see also Moffatt 1961, who exploited the analogy between vorticity and the magnetic field). Expression (6) indicates a substantial growth of the magnetic energy due to the turbulence.

We can also obtain a corresponding expression in terms of field strength by noting that in two dimensions the growth in magnetic field due to line stretching is accompanied by a corresponding process of scale decrease. If δ denotes the thickness of the current sheets generated by the turbulence we have

$$B_{max} \approx B_0\ell/\delta, \qquad (7)$$

where B_{max} is a least upper bound on the field strength. If we further note that in the kinematic regime $\delta \approx \delta_d = \ell R_m^{-1/2}$, which corresponds to a balance between advection and diffusion in the induction equation, we obtain

$$B_{max} \approx B_0 R_m^{1/2}. \qquad (8)$$

Clearly in order for expressions (6) and (8) to be consistent, the magnetic field must be close to its maximum value over a substantial fraction of the entire volume. Although this result might appear surprising, it is implicit in our assumption (I), which states that the flow is chaotic and that therefore field lines are stretched at an exponential rate almost everywhere in the fluid.

It is perhaps useful to note that if the effect of the velocity on the individual Fourier components of A on scales smaller than ℓ can be approximated by a uniform shear, or in more abstract language if the first-order velocity structure functions are linear, then it follows from assumptions (I) and (II) that the spectrum of A behaves like k^{-1} in the range $\ell^{-1} < k < \delta^{-1}$. This result is in essence Batchelor's (1959) theory for the spectrum of a passive scalar in the convective (subinertial) range.

Two important points emphasized in Batchelor's original work concerning the behavior of the spectrum as $\eta \to 0$ are relevant here. Clearly a k^{-1} spectrum diverges, albeit weakly, as $\eta \to 0$. This problem is resolved if the time it takes to set up the spectrum over the entire range exceeds the lifetime of the A-stuff; otherwise, one can assume that weak (logarithmic) corrections are present – which we allowed for in assumption (II). Furthermore, a k^{-1} spectrum requires large (divergent) gradients of A over a finite volume of the fluid, and not at a set of isolated points (which would be consistent with a k^{-2} spectrum).

In the case at hand, the spectrum for ∇A, or magnetic field, goes as k, resulting again in expression (6). As noted by Batchelor, this spectrum is self-consistent if the field of the order of (8) fills the entire space.

In the nonlinear regime, the growth of magnetic fluctuations is limited by dynamical processes, and according to (III) we have

$$B_{max} = \frac{B_0 \ell}{\delta_{eff}} \leq U \sqrt{\rho_0 \mu_0}. \tag{9}$$

Expression (9) is written in terms of the peak field value, as opposed to the energy density, to emphasize the difference between the kinematic and nonlinear regimes. In the former regime, a small increase in B_0 leads to a proportional increase in B_{max} while δ remains equal to δ_d; in the latter regime, an increase in B_0 causes an increase in δ while the peak field strength remains the same and close to equipartition. Expressions (3) and (9) can now be used to estimate the (effective) diffusion time; we find

$$\frac{1}{t_{eff}} \langle A^2 \rangle \approx \eta U^2 \rho_0 \mu_0,$$

which gives

$$t_{eff} \approx \frac{B_0^2}{U^2 \rho_0 \mu_0} \frac{L^2}{\eta} = t_T \frac{R_m}{M^2}, \tag{10}$$

where, for convenience we have introduced the magnetic Mach number $M = U/B_0\sqrt{\rho_0\mu_0}$. Formula (10) shows a substantial increase of the effective diffusion time as the initial large-scale magnetic field increases, or equivalently, strong suppression of turbulent diffusion. Expression (10) shows that the effective diffusion time approaches the Ohmic diffusion time as M^2 approaches unity, and the turbulent time for $M^2 \approx R_m$. It is convenient to introduce the ratio $F \equiv R_m^{1/2}/M$ (Vainshtein & Cattaneo 1992), so that (10) is re-written as

$$t_{eff} \approx t_T F^2; \tag{11}$$

$F \leq 1$ defines the kinematic regime.

4 THE CASE $\tau \gg \ell/U$

We now consider the case of steady or quasi-steady flow, characterized by $\tau \gg \ell/U$, for which a different set of assumptions is needed:

I'. The magnetic field is concentrated into current sheets of thickness δ occupying a fraction δ/ℓ of the entire volume. Presumably, the field will be concentrated at stagnation lines (Proctor & Weiss 1982).

II'. The local field strength does not exceed the equipartition value, i.e.,

$$B_{max} \le U\sqrt{\rho_0\mu_0}.$$

Assumption (I') is phenomenological, and is supported by many numerical studies. Assumption (II') – analogous to our previous assumption (III) – is necessary in order to extend the analysis to the nonlinear regime; here we find it more appropriate to express it in terms of the peak field strength rather than the average magnetic energy.

We begin by estimating the diffusion time in the kinematic regime. We consider (3) and assumption (I') to obtain

$$\frac{1}{t_s}\langle A^2 \rangle \approx \eta B_{max}\frac{\delta}{\ell} = \eta B_0^2 R_m^{1/2}, \tag{12}$$

which yields

$$t_s \approx \frac{L^2}{U\ell}\frac{U\ell}{\eta}R_m^{-1/2} = t_T R_m^{1/2}, \tag{13}$$

assuming, as before, that in the kinematic regime a balance between advection and diffusion is achieved when $\delta \approx \delta_d = \ell R_m^{1/2}$. Expression (13) shows a substantial increase in diffusion time due to the effects of spatial intermittency of the magnetic field. Nonlinear effects due to the magnetic back-reaction can be included in an analogous way to the turbulent case. We have

$$\frac{1}{t_{eff}}\langle A^2 \rangle \approx \eta B_{max}^2\frac{\delta_{eff}}{\ell},$$

which gives

$$t_{eff} \approx \frac{L^2}{\eta}\frac{B_0}{U\sqrt{\rho_0\mu_0}} = t_s R_m^{1/2}/M, \tag{14}$$

or, using our previous notation,

$$t_{eff} \approx t_s F. \tag{15}$$

Again, expressions (14) and (15) show that t_{eff} approaches the kinematic value for $M^2 \approx R_m$, and the Ohmic value for M of order unity; however, comparison with the corresponding equations (10) and (11) shows that the rate of suppression is less effective here than in the turbulent (chaotic) case.

5 THE THREE-DIMENSIONAL CASE

In three dimensions, the analogy between magnetic field diffusion and the evolution of the gradient field of a diffusing scalar field no longer holds; furthermore, there exists the possibility of dynamo action, a possibility which is entirely excluded in the two-dimensional case. It is therefore fair to ask whether the effects we have discussed so far are restricted to the two-dimensional case, or can be extended to three dimensions; if not, then the two-dimensional results may well be interesting in their own regard, but would have little if any relevance to the astrophysical case. In a nutshell, what we would like to know, for example, is the corresponding result to equation (6) (if it exists), for three dimensions; thus, for example, Vainshtein & Rosner (1991) have posited the relation

$$\langle |\mathbf{B}|^2 \rangle \approx R_m^n \langle B_0^2 \rangle, \tag{16}$$

where $n = 2$ if the field is compressed into magnetic flux tubes which fill the entire space, and $n < 2$ more generally than this extreme case. We note in this connection that Gilbert & Childress (1990) have shown that the field structure in chaotic regions becomes highly fluctuating, i.e., the net flux generated (corresponding to B_0) is small, while the unsigned flux generated (corresponding to $|\mathbf{B}|$) is very large.

A partial answer to this essential question is given in this volume by Tao, Cattaneo & Vainshtein (1993) on the basis of self-consistent numerical solutions of the coupled Navier–Stokes and induction equations: precisely because of the possibility of dynamo action in three dimensions, they focus on the efficacy of the α-effect (rather than on turbulent diffusion) as the strength of the large-scale magnetic field is varied; that is, they consider helical turbulence, where the helicity of the turbulence is defined as

$$H \equiv \langle \mathbf{u} \cdot \nabla \times \mathbf{u} \rangle. \tag{17}$$

The magnetic field generation coefficient α can then be expressed in terms of the helicity H (e.g., Moffatt 1978),

$$\alpha \approx -H\tau. \tag{18}$$

Note that α is also a *transport* coefficient, as is the turbulent diffusivity η_T; this fact can be easily appreciated from the Lagrangian representation of α, namely when it is expressed in terms of the displacements $\boldsymbol{\xi}$ of fluid particles,

$$\alpha = -\frac{d}{dt} \langle \boldsymbol{\xi} \cdot \nabla \times \boldsymbol{\xi} \rangle \tag{19}$$

(Moffatt 1974), to be compared with the corresponding expression for η_T,

$$\eta_T = \frac{1}{3} \frac{d}{dt} \langle \xi^2 \rangle. \tag{20}$$

The key result of Tao *et al.* (1993) is that the α-effect is suppressed in a fashion similar to the suppression of turbulent diffusion discussed in section 3 above (see also Kulsrud & Anderson 1992 for related discussions of nonlinear effects in dynamos); in particular, if the flow is chaotic (in the sense of section 3 above), then Tao *et al.* demonstrate that the effective α coefficient becomes

$$\alpha_{\textit{eff}} = \frac{\alpha}{F^2}. \tag{21}$$

For extremely weak fields, i.e., in the kinematic regime, $F = 1$. In the other limiting case, namely when turbulent diffusion is totally suppressed ($\eta_{\textit{eff}} = \eta$), the turbulence behaves like a wave ensemble, $M = 1$, and the calculation of α is based on the molecular diffusivity (in this case, $F = R_m^{1/2}$; see Moffatt 1978).

Can this result be understood physically? Vainshtein & Cattaneo (1992) argue on the basis of a phenomenological model that equation (11) may still apply in three dimensions, but with F replaced by $F_{3D} \equiv R_m^{1/2}/M^{1/n}$ (see (16) above), e.g., the effective diffusivity is then very conservatively given by

$$\eta_{\textit{eff}} = \eta_T/F_{3D}^2, \tag{22}$$

where $\eta_T \equiv U\ell$. In that case, we may use (19) to construct an estimate for α,

$$|\alpha| \le \frac{d}{dt}\frac{\langle \xi^2 \rangle}{\ell} \le \frac{\eta_T}{\ell}, \tag{23}$$

which, when combined with (22), yields our desired result.

The above discussion holds for the case of $\tau \approx \ell/U$; what about the case corresponding to our discussion in section 4 (where $\tau \gg \ell/U$)? The essence of that previous argument reduces to the fact that the filling factor of the volume in which field line stretching takes place is vastly different in the chaotic and quasi-steady cases; it is this effect which is largely responsible for the difference between equations (11) and (15). We expect the same argument to apply to the present case.

6 DISCUSSION

In this paper, we have discussed the mechanisms by which turbulent transport effects are suppressed by nonlinear effects in a magnetofluid. These effects are central elements in modern astrophysical dynamo theories, and their efficacy lies at the heart of attempts to understand actual astrophysical dynamos such as are believed to exist in late-type stars such as our Sun.

Our principal interest was in distinguishing between the behavior in a fully chaotic flow (in which the Liapunov exponents are positive, and the velocity correlation time $\tau \approx \ell/U$), and the quasi-steady case in which $\tau \gg \ell/U$. Our essential result is that these two cases differ largely in the volume filling factor in which magnetic field line stretching takes place; as a consequence, the effective diffusion time scales as

$$\tau_{eff} \approx t_T F^2$$

in the fully turbulent case, and as

$$\tau_{eff} \approx t_T F$$

for the quasi-steady (or steady) case (where $F \equiv R_m^{1/2}$).

Now, which case is astrophysically relevant? It strikes us as highly improbable that the dominant flows in astrophysical magnetofluids, such as occur within stars such as the Sun, can be characterized as quasi-steady. Indeed, observations of the Sun's surface layers strongly suggest that $\tau \approx \ell/U$ is an appropriate description, and we therefore contend that the discussions of sections 3 and 5 should apply to such fluids.

It is essential to understand that the effects discussed here, as well as the suppression of the α-effect discussed by Tao, Cattaneo & Vainshtein (1993) in this volume, are entirely different from what has been previously understood by nonlinear suppression of turbulent transport by the magnetic backreaction. In the present approach, the effects of the Lorentz back-reaction do *not* diminish the amplitude of U, that is, the intensity of the turbulence. In other words, the *turbulence itself is not suppressed*, but rather it is the turbulent diffusion that is strongly affected by the magnetic field. Thus, we argue that from an Eulerian point-of-view, it is difficult (if at all possible) to determine from examination of the statistical fluid properties whether suppression has occurred; it is only a Lagrangian analysis which reveals the subtle modifications of the fluid statistical properties which lead to the suppression effects (Cattaneo & Vainshtein 1991; see also Gilbert *et al.* 1993). This is an especially important point from the perspective of astrophysics since the effects we have discussed become operative under much less constrained circumstances than the back-reaction effects previously discussed in the literature (e.g., Stix 1972; Jepps 1975; Ivanova & Ruzmaikin 1977; Krause & Meinel 1988; Brandenburg *et al.* 1989).

REFERENCES

Batchelor, G.K. 1959 Small-scale variations of convective quantities like temperature in turbulent fluid. *J. Fluid Mech.* **5**, 113–133.

Brandenburg, A., Krause, F., Meinel, R., Moss, D. & Tuominen, I. 1989 The stability of nonlinear dynamos and the limited role of kinematic growth rates. *Astron. Astrophys.* **213**, 411–422.

Cattaneo, F. & Vainshtein, S.I. 1991 Suppression of turbulent transport by a weak magnetic field. *Astrophys. J.* **376**, L21–L24.

Gilbert, A.D. & Childress, S. 1990 Evidence for fast dynamo action in a chaotic web. *Phys. Rev. Lett.* **65**, 2133–2136.

Gilbert, A.D., Otani, N.F. & Childress, S. 1993 Simple dynamical fast dynamos. In this volume.

Glatzmaier, G.A. 1985 Numerical simulations of stellar convective dynamos. *Astrophys. J.* **291**, 300–307.

Ivanova, T.S. & Ruzmaikin, A.A. 1977 A nonlinear MHD-model of the dynamo of the Sun. *Astron. Zh. (USSR)* **54**, 846–858.

Jepps, S.A. 1975 Numerical models of hydromagnetic dynamos. *J. Fluid Mech.* **67**, 629–646.

Krause, F. & Meinel, R. 1988 Stability of simple nonlinear α^2-dynamos. *Geophys. Astrophys. Fluid Dynam.* **43**, 95–117.

Krause, F. & Rädler, K.-H. 1980 *Mean-Field Magnetohydrodynamics and Dynamo Theory.* Pergamon Press, Oxford.

Kulsrud, R.M. & Anderson, S.W. 1992 The spectrum of random magnetic fields in the mean field dynamo theory of the galactic magnetic field. *Astrophys. J.* **396**, 606–630.

Moffatt, H.K. 1961 The amplification of a weak applied magnetic field by turbulence in fluids of moderate conductivity. *J. Fluid Mech.* **11**, 625–635.

Moffatt, H.K. 1974 The mean electromotive force generated by turbulence in the limit of perfect conductivity. *J. Fluid Mech.* **65**, 1–10.

Moffatt, H.K. 1978 *Magnetic Field Generation in Electrically Conducting Fluids.* Cambridge University Press.

Monin, A.S. & Yaglom, A.M. 1971 *Statistical Fluid Mechanics.* M.I.T. Press, Cambridge, Massachusetts.

Proctor, M.R.E. & Weiss, N.O. 1982 Magnetoconvection. *Rep. Prog. Phys.* **45**, 1317–1379.

Stix, M. 1972 Non-linear dynamo waves. *Astron. Astrophys.* **20**, 9–12.

Tao, L., Cattaneo, F. & Vainshtein, S.I. 1993 Evidence for the suppression of the α-effect by weak magnetic fields. In this volume.

Vainshtein, S.I. & Cattaneo, F. 1992 Nonlinear restrictions on dynamo action. *Astrophys. J.* **393**, 165–171.

Vainshtein, S.I. & Rosner, R. 1991 On turbulent diffusion of magnetic fields and the loss of magnetic flux from stars. *Astrophys. J.* **376**, 199–203.

Zel'dovich, Ya.B. 1957 Magnetic field in a turbulent conductive fluid in two-dimensional motion. *Soviet Phys.–JETP* **4**, 460–462.

Proving the Existence of Negative Isotropic Eddy Viscosity

M. VERGASSOLA[1,2], S. GAMA[1,3] & U. FRISCH[1,2]

[1]CNRS, Observatoire de Nice
BP 229, 06304 Nice Cedex 4, France

[2]Isaac Newton Institute for Mathematical Sciences
University of Cambridge, 20 Clarkson Rd., Cambridge, CB3 0EH UK

[3]FEUP, Universidade de Porto
R. Bragas, 4099 Porto Codex, Portugal

We demonstrate the existence of a two-dimensional incompressible flow having a negative and isotropic eddy viscosity. Here, we
understand by 'eddy viscosity' the sum of the molecular viscosity
and of the small-scale flow contribution. The flow is deterministic,
time-independent, space-periodic and has $\pi/3$ rotational invariance. The eddy viscosity is calculated by multiscale techniques.
The resulting equations for the transport coefficients are solved
(i) by a Padé-resummed Reynolds number expansion and (ii) by
direct numerical simulation. Results agree completely.

It is known that the action of a small-scale incompressible flow (having suitable symmetries) on a large-scale perturbation of small amplitude is 'formally'
diffusive (Kraichnan 1976; Dubrulle & Frisch 1991). There are two essential
assumptions. The first one is scale-separation: the ratio ϵ between the typical
length-scale of the basic flow and that of the perturbation is small. The second one is the absence of a large-scale AKA effect (Frisch et al. 1987). If the
basic flow is parity-invariant (i.e. has a center of symmetry), this condition
is automatically satisfied. By 'formally' diffusive, we understand that, unlike
the case of the eddy diffusivity for a passive scalar (Frisch 1989), the eddy
viscosity tensor need not be positive definite. There are indeed examples of
strongly anisotropic flows (e.g. the Kolmogorov flow), where some components
of the tensor are negative, resulting in a large-scale instability (Meshalkin &
Sinai 1961; Green 1974; Sivashinsky 1985; Sivashinsky & Yakhot 1985).

When the eddy viscosity tensor is isotropic, the equation for the perturbation reduces to an ordinary diffusion equation, with diffusion coefficient ν_E.

M.R.E. Proctor, P.C. Matthews & A.M. Rucklidge (eds.)
Theory of Solar and Planetary Dynamos, 321–327
©1993 Cambridge University Press.

In order to avoid misunderstandings, we stress that ν_E contains the positive molecular viscosity contribution ν. For time-independent basic flow, Dubrulle & Frisch (1991) proved that, in an expansion in the Reynolds number R, the first correction to the eddy viscosity is positive. Sivashinsky & Frenkel (1992) recently pointed out that this need not be the case for a time-dependent basic flow. Nevertheless, because such a negative correction is small compared to the molecular viscosity when R is small, the eddy viscosity will still be positive. Furthermore, for finite R's knowing only the first two terms of the expansion is not enough. As a matter of fact, not a single flow having $\nu_E < 0$ was known. The aim of this paper is actually to prove the existence of two-dimensional flows having a negative and isotropic eddy viscosity.

Because we will focus on the two-dimensional case, it is convenient to introduce the stream function $\Psi(\mathbf{x}, t)$ and write the Navier–Stokes equations in the form

$$\partial_t \partial^2 \Psi + J(\partial^2 \Psi, \Psi) = \nu \partial^2 \partial^2 \Psi - \varepsilon_{ij} \partial_i f_j. \tag{1}$$

Here, ∂^2 is the Laplacian, J is the Jacobian, the velocity $u_i = \varepsilon_{ij} \partial_j \Psi$, and ε_{ij} is the fundamental antisymmetric tensor having $\varepsilon_{12} = 1$. We will suppose the force $\mathbf{f}$ periodic in x, y and t. The condition $\langle \mathbf{f} \rangle = 0$ (where $\langle \cdot \rangle$ denotes the average over the periodicities) is also imposed to exclude trivial advection effects. The restrictive hypothesis of periodicity is made just for the sake of simplicity. The random case can be handled by the same method with minor modifications.

Let us now consider a small-amplitude large-scale perturbation with stream function $\psi(\mathbf{x}, t)$. As long as the amplitude of ψ is sufficiently small, nonlinear terms are irrelevant. The linearized Navier–Stokes equation for the perturbation is

$$\partial_t \partial^2 \psi + J(\partial^2 \psi, \Psi) - \nu \partial^2 \partial^2 \psi = -J(\partial^2 \Psi, \psi). \tag{2}$$

The equation is here written in such a way as to stress the presence of the stretching term $J(\partial^2 \Psi, \psi)$, which can produce amplification of the perturbation. The presence of the r.h.s. reflects the fact that vorticity is not a passive scalar.

The method used to treat (2) if scale separation holds is multiscale expansion, as discussed in Bensoussan *et al.* (1978) and Dubrulle & Frisch (1991). The expected large-scale diffusive time is order ϵ^{-2}, where $\epsilon \ll 1$ is the parameter controlling the scale separation. We then introduce slow variables, *viz.* $\mathbf{X} = \epsilon \mathbf{x}$ and $T = \epsilon^2 t$. As usual, the multiscale expansion pretends that fast variables and slow variables are independent variables. It follows that

$$\partial_i \rightarrow \partial_i + \epsilon \nabla_i, \qquad \partial_t \rightarrow \partial_t + \epsilon^2 \partial_T, \tag{3}$$

where, for the sake of clarity, we denote the derivatives with respect to fast space variables by the symbol ∂ and those with respect to slow space variables by ∇. The solution $\psi(\mathbf{x}, t; \mathbf{X}, T)$ is sought as a series in ϵ

$$\psi = \psi^{(0)} + \epsilon\psi^{(1)} + \epsilon^2\psi^{(2)} + \dots, \tag{4}$$

where all the functions $\psi^{(n)}$ depend *a priori* on both fast and slow variables. In the following, we shall consider only time-independent basic flows, so that the dependence on fast time is absent. When the derivatives (3) and the series (4) are inserted into the equation (2), and terms having equal powers in ϵ are equated, a hierarchy of equations is generated, which all have the form

$$Af \equiv J(\partial^2 f, \Psi) + J(\partial^2 \Psi, f) - \nu\partial^2\partial^2 f = g. \tag{5}$$

Here, f is the unknown function and g involves the solutions of lower-order equations. Because the operator A has derivatives on the left of all the terms, the condition $\langle g \rangle = 0$ has to be imposed (Fredholm alternative).

The equations needed for the calculation of the eddy viscosity are the following. At order ϵ^0

$$A\psi^{(0)} = 0, \tag{6}$$

where the operator A has been defined above. It can be proved that the only relevant solution of (6) is a constant, i.e. $\psi^{(0)} = \psi^{(0)}(\mathbf{X}, T)$.

The equation at order ϵ is

$$A\psi^{(1)} = (\varepsilon_{\alpha i}\partial_i\partial^2\Psi)(\nabla_\alpha\psi^{(0)}). \tag{7}$$

The operator A being linear, the solution may be written as

$$\psi^{(1)} = \mathbf{Q} \cdot \nabla\psi^{(0)}, \tag{8}$$

where $\mathbf{Q}$ is a suitable vector depending only on the fast space variables. Additional terms in the null-space of A (which in our case are constants) are irrelevant for the calculation of the eddy viscosity and will be omitted.

At order ϵ^2, we obtain

$$\begin{aligned} A\psi^{(2)} = &[-\varepsilon_{i\alpha}(\partial_i\partial^2\Psi)Q_\beta + 2\varepsilon_{ij}(\partial_i\Psi)(\partial_j\partial_\alpha Q_\beta) \\ &+ \varepsilon_{i\alpha}(\partial_i\Psi)\partial^2 Q_\beta + 4\nu\partial_\alpha\partial^2 Q_\beta]\nabla_\alpha\nabla_\beta\psi^{(0)}. \end{aligned} \tag{9}$$

As above, the solution may be written as

$$\psi^{(2)} = S_{\alpha\beta}\nabla_\alpha\nabla_\beta\psi^{(0)}, \tag{10}$$

where the tensor $S_{\alpha\beta}$ depends only on fast variables.

The various solvability conditions can be obtained by standard techniques (see e.g. Dubrulle & Frisch 1991) or, in a somewhat quicker way, by inserting (4) into the equation

$$\epsilon^4 \partial_T\nabla^2\langle\psi\rangle + \epsilon^2\varepsilon_{\alpha i}\langle(\partial_i\Psi)(2\partial_\beta\nabla_\beta + \epsilon\nabla^2)\nabla_\alpha\psi\rangle = \epsilon^4 \nu\nabla^2\nabla^2\langle\psi\rangle. \tag{11}$$

This equation is obtained from (2) by using (3), averaging and observing that

Figure 1. Streamlines of the basic flow.

a number of terms vanish or cancel. The non-trivial solvability conditions are the following. At order ϵ^3

$$\varepsilon_{\alpha i}\langle(\partial_\beta\psi^{(1)})(\partial_i\Psi)\rangle = 0 \quad \forall\,\alpha,\beta. \tag{12}$$

This condition corresponds to the absence of the AKA effect. In the case of a parity-invariant basic flow Ψ, the condition (12) is automatically satisfied.

At order ϵ^4, we finally obtain the equation describing the evolution of the large-scale perturbation $\psi^{(0)} = \langle\psi^{(0)}\rangle$

$$\partial_T\nabla^2\psi^{(0)} = \nu_{\alpha\beta\gamma\eta}\nabla_\alpha\nabla_\beta\nabla_\gamma\nabla_\eta\psi^{(0)}. \tag{13}$$

The expression of the eddy-viscosity tensor is

$$\nu_{\alpha\beta\gamma\eta} = \nu\delta_{\alpha\beta}\delta_{\gamma\eta} - \varepsilon_{\alpha i}\delta_{\beta\gamma}\langle Q_\eta\partial_i\Psi\rangle - 2\varepsilon_{\alpha i}\langle(\partial_\beta S_{\gamma\eta})(\partial_i\Psi)\rangle. \tag{14}$$

The functions $\mathbf{Q}$ and $S_{\alpha\beta}$ have been defined in (8) and (10). The calculation of these functions involves the solution of linear inhomogeneous partial differential equations with non-constant coefficients. In general the solution cannot be found analytically.

For reasons which will become clear in the sequel, we have concentrated our study on the following flow:

$$\begin{aligned}
\Psi(x,y) = &-\tfrac{1}{2}[\cos(2x) + \cos(x + \sqrt{3}y) + \cos(x - \sqrt{3}y)] \\
&+ \tfrac{1}{2}[\cos(4x + 2\sqrt{3}y) + \cos(5x - \sqrt{3}y) + \cos(x - 3\sqrt{3}y)] \\
&- \tfrac{1}{2}[\cos(4x) + \cos(2x + 2\sqrt{3}y) + \cos(2x - 2\sqrt{3}y)] \\
&+ \tfrac{1}{2}[\cos(4x - 2\sqrt{3}y) + \cos(5x + \sqrt{3}y) + \cos(x + 3\sqrt{3}y)].
\end{aligned} \tag{15}$$

The streamlines are shown in Figure 1.

	Resolution 64^2	Resolution 256^2	Padé Prediction
ν	ν_E	ν_E	ν_E
1	1.348	1.348	1.348
0.8	0.961	0.961	0.961
0.7	0.642	0.642	0.642
0.6	0.134	0.134	0.134
0.59	0.066	0.066	0.066
0.58	−0.007	−0.007	−0.007
0.57	−0.085	−0.085	−0.085
0.56	−0.169	−0.169	−0.169
0.55	−0.259	−0.259	−0.260

Table 1. Eddy-viscosity ν_E for the flow given by (15).

It is clear that the flow is invariant under rotations of $\pi/3$ (and so, parity-invariant). This choice is motivated by the fact that in a triangular lattice, fourth-order tensors are isotropic (see p. 40 of Landau & Lifshitz 1970). In particular, the eddy viscosity tensor will then be isotropic. Observe that our flow does not possess any mirror symmetry compatible with the Navier–Stokes equations such as $x \mapsto -x$, $y \mapsto y$, $\Psi \mapsto -\Psi$. This observation has consequences for the nonlinear terms discussed elsewhere (Vergassola 1992).

Equations (7) and (9) corresponding to the flow (15) have been solved using two different methods.

The first one is the numerical integration of the equations. The prototype equation to be solved is

$$J(\partial^2 f, \Psi) + J(\partial^2 \Psi, f) - \nu \partial^2 \partial^2 f = g, \qquad (16)$$

where f is the unknown function and g is a given periodic function having zero mean. The equation is solved by adding a term $\partial_t \partial^2 f$ and letting the system relax to equilibrium. The integration is performed using a spectral method. Since the Reynolds numbers considered are order one, the method is not particularly demanding as far as spatial resolution is concerned (this has been checked also by analysing the energy spectrum). The values of the eddy viscosity obtained for two different resolutions (64^2 and 256^2) and various molecular viscosity values are shown in Table 1. The eddy viscosity changes sign in the neighbourhood of $\nu = 0.58$, then becoming conspicuously negative, actually achieving negative values with absolute value well above the round-off errors (which are about 10^{-4} in absolute terms). The values are stable with respect to modifications of the grid size, confirming the absence of spurious resolution problems. The isotropy of the eddy viscosity tensor has been verified by calculating all the components of the tensor.

We have also verified numerically the stability of the basic flow with respect to small-scale perturbations: no eigenvalue of the operator A has crossed the imaginary axis within the range of viscosities studied. This ensures that the equations for $\mathbf{Q}$ and $S_{\alpha\beta}$ are mathematically well-posed.

The second independent method used to solve the equations for $\mathbf{Q}$ and $S_{\alpha\beta}$ is a perturbation expansion. The solution is constructed as a power-series in R. A small number of terms can be calculated by hand. Actually, the basic flow was chosen so as to have as many negative coefficients as possible in a Reynolds number expansion, in order to help achieving negative eddy viscosities. It is well known that in the case of nonlinear Navier–Stokes or Euler equations, clustering of singularities in the complex plane makes the use of such series very dangerous. In our case, the situation is much better because a compactness argument indicates that the solutions will be meromorphic functions in R. In general, because of the presence of poles in the complex R-plane, the Taylor series in R for the eddy viscosity (more precisely, for ν_E/ν) will have a finite radius of convergence and may well not (and actually does not in our case) become negative within its disc of convergence. Still, the solution can be reliably continued by Padé approximants beyond the disc of convergence. When the eddy viscosity of the flow (15) is expressed as a power series in the Reynolds number R, odd terms are absent, so that our expansion is actually in terms of R^2, i.e. $\nu_E/\nu = 1 + R^2 F(R^2)$. Actually, one can prove that this is a general property of basic flows having streamlines (not velocities) which are mirror symmetrical. The results for the diagonal Padé [5, 5] for the function $F(\cdot)$ (which involves terms up to R^{22} in the series for ν_E/ν) are also shown in Table 1. The agreement with the values obtained by numerical simulations is clearly not accidental.

For a Reynolds number exceeding the value where the eddy viscosity changes sign, large-scale perturbations which are initially small grow exponentially, until higher-order dissipative terms and nonlinearities become relevant. The growth rate is isotropic and proportional to the square of the wavenumber. For a marginally negative isotropic eddy viscosity, the ensuing nonlinear dynamics can be studied in the same spirit as Nepomnyachtchyi (1976) and Sivashinsky (1985) did for the Kolmogorov flow. Results will be published elsewhere.

We have then shown that two-dimensional flows having an isotropic and negative eddy viscosity indeed exist. The result is obtained by using two independent strategies which lead to the same result. Problems similar to those discussed here can be posed also in the case of magnetic eddy diffusivity. In two dimensions, instabilities are ruled out, since the induction equation is equivalent to the passive advection of the magnetic potential. In three dimensions, the presence of a stretching term in the equations suggests that negative magnetic eddy diffusivities are indeed possible.

Acknowledgements

We have benefited from extensive discussions with R. Benzi, P. Collet, B. Dubrulle and A. Vespignani. This work was supported by grants from DRET (91/112), NATO Scientific Studies Program under grant 11/A/89/PO and from the European Community (Human Capital and Mobility 2/ERB-4050PL920014). The numerical calculations were done on the CM-2/CM-200 of the 'Centre Régional de Calcul PACA, antenne INRIA-Sophia-Antipolis' through the R3T2 network.

REFERENCES

Bensoussan, A., Lions, J.-L. & Papanicolaou, G. 1978 *Asymptotic Analysis for Periodic Structures.* North-Holland, Amsterdam.

Dubrulle, B. & Frisch, U. 1991 The eddy-viscosity of parity-invariant flow. *Phys. Rev.* A **43**, 5355–5364.

Frisch, U., She, Z.S. & Sulem, P.L. 1987 Large-scale flow driven by the anisotropic kinetic alpha effect. *Physica* **28D**, 382–392.

Frisch, U. 1989. In *Lecture Notes in Turbulence* (ed. J.R. Herring & J.C. McWilliams) pp. 219–371. World Scientific, Singapore.

Green, J.S.A. 1974 Two-dimensional turbulence near the viscous limit. *J. Fluid Mech.* **62**, 273–287.

Kraichnan, R.H. 1976 Eddy-viscosity on two and three dimensions. *J. Atm. Sci.* **33**, 1521–1536.

Landau, L.D. & Lifshitz, E.M. 1970 *Theory of Elasticity* (revised edition). Pergamon Press, Oxford.

Meshalkin, L.D. & Sinai, Ya.G. 1961 Investigation of the stability of a stationary solution of a system of equations for the plane movement of an incompressible viscous liquid. *J. Appl. Math. Mech.* (PMM), **25**, 1700–1705.

Nepomnyachtchyi, A.A. 1976 On the stability of the secondary flow of a viscous fluid in an infinite domain (in Russian). *Prikl. Mat. Mekh.* **40** (5), 886–891.

Sivashinsky, G.I. 1985 Weak turbulence in periodic flows. *Physica* **17D**, 243–255.

Sivashinsky, G.I. & Yakhot, V. 1985 Negative viscosity effect in large-scale flows. *Phys. Fluids* **28**, 1040–1042.

Sivashinsky, G.I. & Frenkel, A.L. 1992 On negative eddy viscosity under conditions of isotropy. *Phys. Fluids* A **4**, 1608–1610.

Vergassola, M. 1992 Chiral nonlinearities in forced 2-D Navier–Stokes flows. *Phys. Rev. Lett.* (submitted).

Dynamo Action Induced by Lateral Variation of Electrical Conductivity

J. WICHT & F.H. BUSSE

Institute of Physics
University of Bayreuth
D-8580 Bayreuth, Germany

The recent evidence for the possibility of laterally varying electrical conductivity in the lowermost mantle of the Earth has motivated us to consider in more detail the problem of dynamo action induced by this kind of inhomogeneity. An earlier model (Busse & Wicht 1992) has been extended in that the assumption of a thin layer of sinusoidal varying conductivity is replaced by the assumption of a thick layer. In the new formulation the toroidal field as well as the poloidal field are determined explicitly in the domain of varying conductivity. The results support the conclusion based on the earlier thin layer assumption that the dynamo action is too weak to be of geophysical importance.

1 INTRODUCTION

The influence of varying conductivity on the dynamo process has been investigated for example for galaxies (Donner & Brandenburg 1990) and accretion disks (Stepinski & Levy 1991) and found to be negligible there. Jeanloz's (1990) interpretation of the D'' layer as a laterally inhomogeneous distribution of conducting and insulating alloys, resulting from chemical reactions at the core–mantle boundary and the percolation of iron into the mantle, has motivated us to consider the possibility of a dynamo induced by varying conductivity on the Earth's dynamo. Two questions arise in this context. Firstly, one may ask how a lower mantle with laterally varying conductivity will affect the extrapolation of magnetic fields from the Earth surface to the core. Poirier & le Mouël (1992) have investigated this question in detail and found the effect to be negligible. Jeanloz's (1990) view of pinned fieldlines is too dramatic. We consider a second question, namely whether a laterally varying conductivity in the D'' layer induces a dynamo by itself independently of the

329

M.R.E. Proctor, P.C. Matthews & A.M. Rucklidge (eds.)
Theory of Solar and Planetary Dynamos, 329–337
©1993 Cambridge University Press.

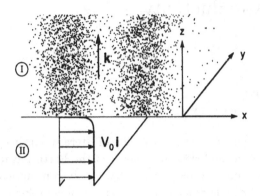

Figure 1. Model of two halfspaces, region I with varying conductivity and region II with constant conductivity. Region II is moving relative to region I with $\mathbf{v} = V_0\mathbf{i}$.

dynamo process in the core. In our first paper (Busse & Wicht 1992) we have investigated the problem in a thin layer approximation. In this paper we consider a two layer model which is less relevant in the geophysical point of view but shows that the results of Busse & Wicht (1992) hold without the thin layer approximation. The model is presented in section 2 and the results in section 3. We draw our final conclusions in section 4.

2 MATHEMATICAL FORMULATION

The lateral variation of conductivity enables us to choose a very simple one-dimensional and purely toroidal velocity field. We would like to point out that Elsasser's theorem does not apply in this case and we can obtain dynamo action without any α-effect. Using the length scale d which will be defined later and the timescale d^2/λ_0 where λ_0 is a typical magnetic diffusivity we can write the dimensionless dynamo equation in the form

$$\partial_t\mathbf{B} = \nabla \times (\mathbf{v} \times \mathbf{B}) + \xi\nabla^2\mathbf{B} - (\nabla\xi) \times (\nabla \times \mathbf{B}) . \qquad (1)$$

Here ξ is a normalised magnetic diffusivity

$$\xi = \frac{\lambda}{\lambda_0} = \frac{\sigma_0}{\sigma}$$

and the dimensionless velocity vector $\mathbf{v}$ has the form

$$\mathbf{v} = \frac{\mathbf{V}d}{\lambda_0} ,$$

where $\mathbf{V}$ denotes the dimensional velocity vector. A typical magnitude of $\mathbf{v}$ can therefore be identified with the magnetic Reynolds number. With σ we denote the electrical conductivity.

Our model consists of two halfspaces separated by the $z = 0$ plane. A Cartesian coordinate system is chosen as shown in Figure 1. Region I has the constant magnetic diffusivity λ_0 and moves relative to region II with a uniform velocity $\mathbf{v} = V_0 \mathbf{i}$ in the x-direction. The magnetic diffusivity in region II varies like $\xi = \xi_0 + \kappa_0 \cos(\alpha x)$ such that ξ_0 denotes the average magnetic diffusivity, κ_0 is the amplitude of the variation and $1/\alpha$ is the length scale in the x-direction.

Without losing generality we may write the solenoidal vector $\mathbf{B}$ as a sum of a poloidal and a toroidal field

$$\mathbf{B} = \nabla \times (\nabla \times \mathbf{k}h) + \nabla \times \mathbf{k}g,$$

where $\mathbf{k}$ is the normal vector of the boundary at $z = 0$.

Looking for a stationary onset of dynamo action, we set $\partial_t \mathbf{B} = 0$ and can now identify V_0 with a critical magnetic Reynolds number

$$R_m^c = V_0.$$

The onset of oscillatory solutions will be explored in future work. Taking the z-component of equation (1) and the z-component of the curl of equation (1), we get two equations for each region:
Region I: $z \geq 0$, $\mathbf{v} = 0$, $\xi = \xi_0 + \kappa_0 \cos(\alpha x)$,

$$(\partial_x \xi)(\partial_x \nabla^2 h) + \xi \nabla^2 \Delta_2 h + (\partial_x \xi) \partial_y \partial_z g = 0, \tag{2}$$

$$\partial_x [(\Delta_2 g) \frac{1}{\xi}(\partial_x \xi)] + \nabla^2 \Delta_2 g = 0; \tag{3}$$

Region II: $z < 0$, $\mathbf{v} = V_0 \mathbf{i}$, $\xi = 1$,

$$\xi \Delta_2 \nabla^2 h^{II} = V_0 \partial_x \Delta_2 h^{II}, \tag{4}$$

$$\xi \Delta_2 \nabla^2 g^{II} = V_0 \partial_x \Delta_2 g^{II}. \tag{5}$$

The upper index II denotes the solution in region II; g and h without an index are the solutions in region I. Δ_2 is the two-dimensional Laplacian

$$\Delta_2 = \partial_{xx}^2 + \partial_{yy}^2.$$

The form of the magnetic diffusivity variation in region I suggests a Fourier expansion in x with wavenumber α. A more general Floquet *ansatz* will be considered in following publications. Since there is translational symmetry in y, we can choose a trigonometric dependence with wavenumber β. We further demand that the solution should decay exponentially for large values of $|z|$ with a decay rate λ. Region I: $z \geq 0$, $\mathbf{v} = 0$, $\xi = \xi_0 + \kappa_0 \cos(\alpha x)$,

$$g = \sum_n G_n e^{in\alpha x} e^{-\lambda z} \sin(\beta y), \tag{6}$$

$$\nabla^2 h = \sum_n H_n e^{in\alpha x} e^{-\lambda z} \cos(\beta y) \,, \tag{7}$$

$$\Rightarrow h = \sum_n H_n e^{in\alpha x} e^{-\lambda z} \cos(\beta y)(1 + A_n e^{(\lambda - \gamma_n)z})/(\lambda^2 - \gamma_n^2) \,, \tag{8}$$

$$\gamma_n^2 = n^2 \alpha^2 + \beta^2 \,.$$

Region II: $z < 0$, $\mathbf{v} = V_0 \mathbf{i}$, $\xi = 1$,

$$g^{II} = \sum_n G_n^{II} e^{in\alpha x} e^{\epsilon_n z} \sin(\beta y) \,, \tag{9}$$

$$h^{II} = \sum_n H_n^{II} e^{in\alpha x} e^{\epsilon_n z} \cos(\beta y) \,. \tag{10}$$

Because only $\nabla^2 h$ is needed in equation (2), we choose the simple form (7) for $\nabla^2 h$ and obtain (8) by integration. The decay rates λ and the amplitudes A_n must be determined through the boundary conditions. With

$$\epsilon_n^2 = \gamma_n^2 + in\alpha V_0 \,,$$

the equations (4) and (5) are readily fulfilled.

With *ansatz* (6), equation (3) becomes an eigenvalue problem with the eigenvalues λ_l^2 and the eigenvectors $\tilde{G}_{n,l}$. The tilde denotes that the eigenvalue problem has a free amplitude Γ_l that still must be determined

$$G_{n,l} = \Gamma_l \tilde{G}_{n,l} \,.$$

The number L of eigenvalues depends on the dimension of the eigensystem, i.e. on the truncation level N of n. Since the solution must decay for large values of $|z|$, we choose the square root with positive real part of λ_l^2. The general solution in region I is the sum over all possible eigensolutions:

$$g = \sum_{n,l} G_{n,l} e^{in\alpha x} e^{-\lambda_l z} \sin(\beta y) \,, \tag{11}$$

$$h = \sum_{n,l} H_{n,l} e^{in\alpha x} e^{-\lambda_l z} \cos(\beta y)(1 + A_{n,l} e^{(\lambda_l - \gamma_n)z})/(\lambda_l^2 - \gamma_n^2) \,. \tag{12}$$

Instead of the continuity of the horizontal components of the electric field $\mathbf{E}$ we demand the equivalent continuity of the x and y derivatives of $\mathbf{E}$. Because the continuity of $\mathbf{k} \cdot \nabla \times \mathbf{E}$ is fulfilled with that of $\mathbf{B}$, only the continuity of $\nabla_2 \cdot \mathbf{E}$ remains to be satisfied. ∇_2 is the two-dimensional divergence $(\partial_x, \partial_y, 0)$. Via Ohm's law this condition leads to the equation at $z = 0$

$$\xi \Delta_2 \partial_z g - (\partial_x \xi) \nabla^2 \partial_y h + (\partial_x \xi) \partial_z \partial_x g = \Delta_2 \partial_z g^{II} - V_0 \partial_y \Delta_2 h^{II} \,. \tag{13}$$

To satisfy this matching condition the general solutions for g and h are needed.

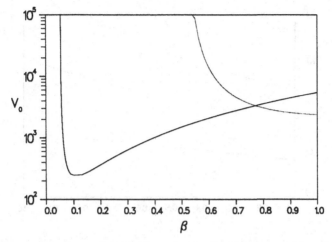

Figure 2. V_0 as a function of β for fixed $\alpha = 0.1$ (solid) and $\alpha = 1.0$ (dotted). $\xi_0 = 1.0$, $\kappa_0 = 0.9$ and $N = 10$ for both.

The boundary conditions for the magnetic field have the conventional form of the continuity of g, h and $\partial_z h$ at $z = 0$, which yield

$$A_{n,l} = -\frac{\lambda_l + \epsilon_n}{\gamma_n + \epsilon_n}\,, \qquad G_n^{II} = \sum_l G_{n,l}\,, \qquad H_n^{II} = \sum_l \left(\frac{1 + A_{n,l}}{\lambda_l^2 - \gamma_n^2}\right) H_{n,l}\,.$$

Equations (2) and (13) now become a system of homogeneous linear equations which we solve numerically for a truncation N. We use the solvability condition that the determinant of a linear homogeneous system must vanish to determine the critical magnetic Reynolds number $V_0 = R_m^c$ and then calculate the magnetic field at this onset value.

3 RESULTS

3.1 Parameter dependence of R_m^c

The length scale of the problem is given by $1/\alpha$. If the ratio α/β is kept fixed, the critical magnetic Reynolds number scales with α and β. Therefore it is sufficient to determine the β-dependence for one fixed α or vice versa. The two curves for $\alpha = 0.1$ and $\alpha = 1.0$ in Figure 2 illustrate the scaling of V_0. As mentioned above, V_0 can be identified with the critical magnetic Reynolds number and is therefore plotted in all the results. As in Busse & Wicht (1992) the system prefers states with $\alpha \approx \beta$.

The dependence of V_0 on the normalised average diffusivity ξ_0 in region I is shown in Figure 3 for various ratios κ_0/ξ_0. The efficiency of the dynamo increases with decreasing ξ_0, i.e. with increasing average conductivity in region I.

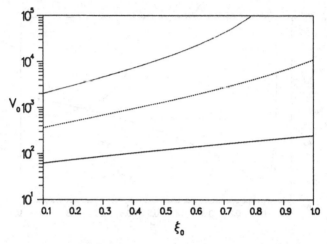

Figure 3. V_0 as a function of ξ_0 for fixed ratios $\kappa_0/\xi_0 = 0.5$ (dotted), $\kappa_0/\xi_0 = 0.7$ (dashed) and $\kappa_0/\xi_0 = 0.9$ (solid). $\alpha = \beta = 0.1$ and $N = 10$ for all.

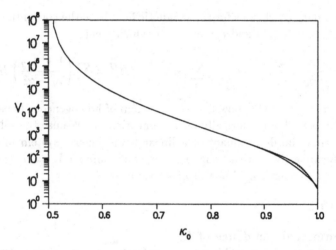

Figure 4. V_0 as a function of κ_0 for different truncation levels $N = 10$ (solid), $N = 15$ (dashed) and $N = 17$ (dotted). $\alpha = \beta = 0.1$ and $\xi_0 = 1.0$ for all.

Figure 4 indicates that the dynamo effect vanishes for small κ_0, i.e. for small variation of the conductivity in region I. For $\xi_0 = 1.0$ no solution could be found for $\kappa_0 \leq 0.5$. The three curves for different truncation levels indicate the good convergence of the numerical solutions.

The parameter dependence of the critical magnetic Reynolds number exhibits basically the same behaviour as shown in Busse & Wicht (1992). A new feature is the vanishing dynamo action for small amplitudes of the conductivity variation.

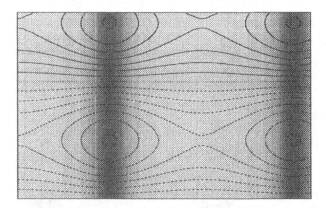

Figure 5. Contour lines of the toroidal scalar field g in the xy-plane at $z = 0$ for $\alpha = \beta = 0.1$ and $N = 12$. Solid lines for $g > 0$, dashed lines for $g < 0$ and dotted lines for $g = 0$. The degree of shading is a measure of the conductivity for $\xi_0 = 1.0$ and $\kappa_0 = 0.9$.

3.2 The magnetic field

An interesting question connected with this dynamo is the form of the produced magnetic field, especially of the poloidal field that would reach the Earth's surface. In Figure 5 contour lines of the toroidal scalar field g are plotted in the xy-plane at $z = 0$. These lines can be identified with fieldlines of the toroidal magnetic field. As expected the field is periodic in the x and y directions and the maxima are concentrated at the extrema of the conductivity.

The magnetic fieldlines of the poloidal part in the same plane are plotted in Figure 6. They are periodic as well but shifted in the direction of the flow. The field has a strong gradient at the stripes of high conductivity and its strength is correlated with the conductivity gradient. In Busse & Wicht (1992) we did not see the shift of the toroidal field in the x-direction owing to the thin layer approximation.

Figure 7 presents the magnetic fieldlines in the xz-plane at $y = 0$. They are plotted as contour lines of $\partial_x h$. According to the ideas of the frozen flux approximation the fieldlines are transported with the conductor in region II. The differential motion at $z = 0$ leads to a stretching of the fieldlines that are pinned at the regions of higher conductivity in halfspace I. This effect can produce a toroidal field out of a poloidal field and enters the system in the boundary condition (13). It is equivalent to the so called ω-effect in spherical systems. To gain a working dynamo we need a second effect that produces a poloidal field out of a toroidal field. Looking at equation (2) we notice that the toroidal field is coupled with the poloidal field via the diffusivity gradient $\partial_x \xi$. Thus the second effect is caused by the varying diffusion of the toroidal field in the z-direction.

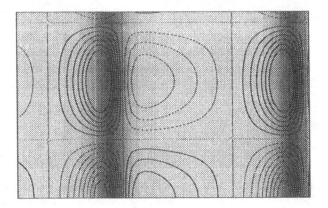

Figure 6. Contour lines of B_z in the xy-plane at $z = 0$ for $\alpha = \beta = 0.1$ and $N = 12$. Solid lines for $B_z > 0$, dashed lines for $B_z < 0$ and dotted lines for $B_z = 0$. The degree of shading is a measure of the conductivity for $\xi_0 = 1.0$ and $\kappa_0 = 0.9$.

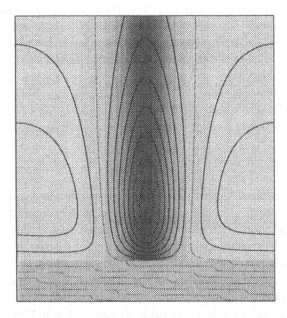

Figure 7. Contour lines of $\partial_x h$ in the xz-plane at $y = 0$ for $\alpha = \beta = 0.1$ and $N = 12$. Solid lines for $\partial_x h > 0$, dashed lines for $\partial_x h < 0$ and dotted lines for $\partial_x h = 0$. The degree of shading is a measure of the conductivity for $\xi_0 = 1.0$ and $\kappa_0 = 0.9$. Fieldlines in the interior of the loops have not been plotted.

4 CONCLUSIONS

As pointed out in section 3 the dynamo effect depends on the gradient of the conductivity in region I. If we identify this conductivity variation with that in the Earth's D'' layer it becomes clear that the effect is very small. We have calculated the critical magnetic Reynolds number for the more realistic model of a thin D'' layer in Busse & Wicht (1992) and found it to be of order 10^4 if we assume that the average conductivity of the thin layer is the same as the core conductivity and its maximum conductivity is twice that of the core. This assumption is likely seriously to overestimate the D''-conductivity, but it still requires unrealistic velocities of the order $10\,\mathrm{cm\,s^{-1}}$ if a tenth of the core radius is assumed for the thickness of the D''-layer. Similar estimates can be obtained from the present model. Although the dynamo effect is not geophysically relevant, the model demonstrates that conductivity variations near the core–mantle boundary may have a considerable effect on the observed geomagnetic field.

REFERENCES

Busse, F.H. & Wicht, J. 1992 A simple dynamo caused by conductivity variations. *Geophys. Astrophys. Fluid Dynam.* **64**, 135–144.

Donner, K.J. & Brandenburg, A. 1990 Magnetic field structure in differentially rotating discs. *Geophys. Astrophys. Fluid Dynam.* **50**, 121–129.

Jeanloz, R. 1990 The nature of the Earth's core. *Annual Rev. Earth Planet. Sci.* **18**, 357–386.

Poirier, J.-P. & le Mouël, J.-L. 1992 Does infiltration of core material into the lower mantle affect the observed geomagnetic field? *Phys. Earth and Planet. Inter.* **73**, 29–37.

Stepinski, T.F. & Levy, E.H. 1991 Dynamo magnetic field modes in thin astrophysical disks: an adiabatic computational approximation. *Astrophys. J.* **379**, 343–355.

Spherical Inertial Oscillation and Convection

KEKE ZHANG

Department of Mathematics
University of Exeter
Exeter, EX4 4QJ UK

Inertial oscillation is coupled with convection in rapidly rotating
spherical fluid systems. It is shown that the combined effects of
Coriolis forces and spherical curvature enable the equatorial region
to form an equatorial waveguide tube. Two new convection modes
which correspond to the inertial waves described by the Poincaré
equation with the simplest structure along the axis of rotation and
equatorial symmetry are then identified. On the basis of solutions
of the Poincaré equation and taking into account the effects of the
Ekman boundary layer, we establish a perturbation theory so that
analytical convection solutions in rotating fluid spherical systems
are obtained.

---

1 INTRODUCTION

Rotating fluid dynamics is of primary importance in the understanding of the
origin of planetary magnetic fields which are generated by dynamo processes
in the rotating fluid interiors of planets. There are two important but traditionally separate branches in the subject of rotating fluid dynamics: inertial
oscillation and convection. Both have been extensively investigated. Inertial oscillation in rotating systems is governed by the Poincaré equation; it
was also shown by Malkus (1967) that the problem of hydromagnetic inertial
oscillation can be changed to the Poincaré problem with a special form of
the basic field. A classic introduction and most of the earlier research results
concerning this problem can be found in Greenspan's monograph (1969). The
important application to the dynamics of the Earth's fluid core was discussed
by Aldridge & Lumb (1987). Recent studies on the form of the inertial waves
(Zhang 1993a) revealed that fluid motions with sufficiently simple structure
along the axis of rotation and small azimuthal scale must be trapped in an

M.R.E. Proctor, P.C. Matthews & A.M. Rucklidge (eds.)
Theory of Solar and Planetary Dynamos, 339–346
©1993 Cambridge University Press.

equatorial waveguide tube with characteristic latitudinal radius $(2/m)^{1/2}$ and radial radius $(1/m)$, m being the azimuthal wavenumber of a wave. A classic introduction to convection and the earlier research results were contained in Chandrasekhar's monograph (1961). The asymptotic theories of convection in rapidly rotating spherical systems were given by Roberts (1968), Busse (1970) and Soward (1977). Recent investigations (Zhang 1992) suggest, however, that the form of the convection pattern is dependent on the size of the Prandtl number, which is a property of the fluid. A comprehensive review of convection and the influences of the magnetic field on convection can be found, for example, in Fearn *et al.* (1988).

This paper reviews the recent finding of a link between the two important but traditionally separate branches in rotating fluid dynamics: inertial oscillation and convection. A perturbation theory is set up on the basis of the link and analytical convection solutions can therefore be obtained in rotating spherical fluid systems with small Prandtl numbers. Numerical solutions of the full equations are also calculated and the results of the numerical analysis are compared with those obtained from the perturbation theory. A satisfactory quantitative agreement between the analytical results in a full sphere and the numerical results in a thick spherical shell is reached.

2 GOVERNING EQUATIONS

The model considered in this paper is the same as that discussed by Chandrasekhar (1961). With the thickness of the shell, $d = r_o - r_i$, as the length scale, ν/d as the unit of speed, and $\beta d^2 \nu/\kappa$ as the unit of temperature fluctuation, the problem of convective instability is governed by the following equations

$$i\omega\mathbf{u} + 2\mathbf{k} \times \mathbf{u} = -\nabla P + R(1-\eta)^4\Theta\mathbf{r} + (E/(1-\eta)^2)\nabla^2\mathbf{u}, \qquad (1)$$

$$\nabla \cdot \mathbf{u} = 0, \qquad (2)$$

$$[\nabla^2 - iP_r(1-\eta)^2\omega/E]\Theta = -\mathbf{r} \cdot \mathbf{u}, \qquad (3)$$

where $\mathbf{k}$ is a unit vector parallel to the axis of rotation, η being r_i/r_o, and $\mathbf{u}$ is the three-dimensional velocity field, (u_r, u_θ, u_ϕ), in spherical polar coordinates. For a convenient comparison with the standard form of the Poincaré equation, the frequency ω in equations (1–3) is re-scaled by the rotation rate Ω. The non-dimensional parameters, the Rayleigh number R, the Prandtl number P_r and the Ekman number E are defined as

$$R = \alpha\beta\gamma r_o^4/(\Omega\kappa), \quad P_r = \nu/\kappa, \quad E = \nu/(\Omega r_o^2).$$

The velocity boundary conditions assumed in this paper are stress-free, perfectly thermally conducting and impenetrable,

$$\partial(u_\phi/r)/\partial r = \partial(u_\theta/r)/\partial r = \Theta = u_r = 0. \qquad (4)$$

3 CLASSIFICATION OF CONVECTION MODE

Consider a rapidly rotating spherical system and imagine that the thermal diffusivity, κ, is increased from a small value compared to ν ($\kappa \ll \nu$, $P_r \gg 1$) to an asymptotically large value ($\kappa \gg \nu$, $P_r \ll 1$) while keeping all other parameters of the system unchanged. We observe four different forms of convection in the following sequence: (I) a columnar convection mode with rolls aligned with the axis of rotation and intercepting the outer spherical surface at middle latitudes (Busse 1970; see also Zhang 1991); (II) a spiralling columnar convection mode in the form of prograde spiralling drifting columnar rolls (Zhang 1992); (III) an equatorially trapped Poincaré convection mode travelling in the eastward direction (Zhang 1993b), where the velocity **U** of convection is associated with the Poincaré equation for the pressure field P

$$\frac{1}{s}\frac{\partial}{\partial s}s\frac{\partial P}{\partial s} - \frac{m^2 P}{s^2} + (1 - \frac{4}{\omega^2})\frac{\partial^2 P}{\partial z^2} = 0 \qquad (5)$$

with the boundary condition at $z^2 + s^2 = 1$,

$$s\frac{\partial P}{\partial s} + \frac{2m}{\omega}P + (1 - \frac{4}{\omega^2})z\frac{\partial P}{\partial z} = 0, \qquad (6)$$

where m is the azimuthal wavenumber, and cylindrical coordinates, (s, ϕ, z), with the axis of rotation at $s = 0$ are used. **U** is approximately represented by a particular class of solutions of equations (5) and (6), namely,

$$U_s = -i[a_s s^{m+1} + bz^2 s^{m-1} + cs^{m-1}]\exp[i(m\phi + \omega t)], \qquad (7)$$

$$U_\phi = [a_\phi s^{m+1} + bz^2 s^{m-1} + cs^{m-1}]\exp[i(m\phi + \omega t)], \qquad (8)$$

$$U_z = -idzs^m \exp[i(m\phi + \omega t)], \qquad (9)$$

which have the following symmetry

$$(U_s, U_z, U_\phi)(s, z, \phi) = (U_s, -U_z, U_\phi)(s, -z, \phi + \Phi), \qquad (10)$$

where Φ is a constant, and the coefficients a_ϕ, a_s, b, c are functions of m and ω. At leading order, the frequency of convection is the negative root (ω^-) of the following equation

$$2m P^m_{m+2}(\omega/2) - (4 - \omega^2)\frac{d}{d\omega}P^m_{m+2}(\omega) = 0; \qquad (11)$$

and (IV) an equatorially trapped Poincaré mode but travelling in the westward direction. The frequency is the positive root (ω^+) of equation (11) and the corresponding velocity of convection can be also obtained from equations (7–9) by replacing ω^- with ω^+.

4 A PERTURBATION THEORY

The fact that the relevant solutions of the Poincaré equation are relatively simple and coupled with convection suggests an analytical perturbation approach. We can assume the following expansion

$$\mathbf{u} = \mathbf{U} + \mathbf{u}_1, \quad P = P_0 + P_1, \quad \omega = \omega_0 + \omega_1,$$

where $\mathbf{u}_1, P_1$ and ω_1 represent small perturbations from the solutions of the Poincaré equation. The requirement for the validity of the above expansion is $Em^2 \ll \omega_0$, that is, the perturbation expansion is not valid for slow columnar convection, where $\omega \sim E^{1/3}$ and $m \sim E^{-1/3}$. Substituting the expansion into equations (1–3), the zero order of the perturbation problem yields

$$i\omega_o \mathbf{U} + 2\mathbf{k} \times \mathbf{U} + \nabla P_0 = 0; \quad \nabla \cdot \mathbf{U} = 0, \tag{12}$$

where $\mathbf{U}$ satisfies only the inviscid boundary condition $U_r = 0$. Equation (12) can be combined to form equation (5) (Greenspan 1969). The next order of the perturbation gives rise to

$$i\omega_o \mathbf{u}_1 + 2\mathbf{k} \times \mathbf{u}_1 + \nabla P_1 = R\mathbf{r}\Theta + E\nabla^2(\mathbf{U} + \mathbf{u}_1) - i\omega_1 \mathbf{U}, \tag{13}$$

$$(\nabla^2 - i\omega_o E^{-1} P_r)\Theta = -\mathbf{r} \cdot \mathbf{U}, \tag{14}$$

$$\nabla \cdot \mathbf{u}_1 = 0. \tag{15}$$

Here $\mathbf{u}_1 = \mathbf{u}_i + \mathbf{u}_b$, where $\mathbf{u}_i$ is the perturbation of the interior flow and $\mathbf{u}_b$ is the boundary flow associated with the Ekman layer in the vicinity of the outer spherical boundary surface. The corresponding solvability conditions can be expressed as

$$Re[R \int_v \mathbf{U}^* \cdot \mathbf{r}\Theta dV] = -Re[E \int_v \mathbf{U}^* \cdot \nabla^2(\mathbf{U} + \mathbf{u}_b)dV], \tag{16}$$

$$Im[R \int_v \mathbf{U}^* \cdot \mathbf{r}\Theta dV] = -Im[E \int_v \mathbf{U}^* \cdot \nabla^2(\mathbf{U} + \mathbf{u}_b)dV] + \omega_1 \int_v |\mathbf{U}|^2 dV. \tag{17}$$

A complete solution of convection with given E and P_r may be represented by $(R_c, m_c, \mathbf{U}, \omega_o + \omega_1, \Theta)$, where $\mathbf{U}$ and ω_o are given by equations (7–9) and (11), and R_c, m_c, ω_1 can be determined once the temperature Θ is obtained.

In the limit $P_r m^{-5/2} \ll E$, the temperature Θ may be decomposed into two parts, where $\Theta = \Theta_o(r, \theta, \phi, t) + T(r, \theta, \phi, t)$. Θ_o satisfies an inhomogeneous differential equation but with a homogeneous boundary condition,

$$\Theta_o(s, \phi, z, t) = i(As^{m+4} + Bs^{m+2} + Cz^2 s^{m+2}) \exp[i(m\phi + \omega t)];$$

here the constants A, B, C are function of m. The quantity T satisfies $\nabla^2 T = 0$ but with an inhomogeneous boundary condition

$$T(r, \theta, \phi, t) = -i \sin^m \theta \sum_{k=m}^{k=m+4} T_k r^k \frac{\partial^m P_k(\cos\theta)}{\partial(\cos\theta)^m} \exp[i(m\phi + \omega t)], \tag{18}$$

where T_k are coefficients and $P_k(\cos\theta)$ is the Legendre function. The argument of equatorial symmetry $T(r,\theta,\phi) = T(r,\pi - \theta,\phi)$ leads to $T_k = 0$ if $k = m+1, m+3$. The stability analysis is then connected with the evaluation of the three complex integrals in equations (16) and (17). From equation (14) it can be shown that

$$H_\Theta = \int_v \mathbf{U}^* \cdot \mathbf{r}\Theta dV = \int_v |\nabla(T + \Theta_0)|^2 dV = H_\Theta(m),\qquad (19)$$

where H_Θ is an analytical function of m. However, the integral related to the viscous dissipation is

$$\int_v \mathbf{U}^* \cdot \nabla^2 \mathbf{U}\, dV = 0$$

for either the ω^+ or ω^- modes. This implies that convection in this case is driven by the boundary stresses resulting from the unrealistic inviscid boundary condition. The Ekman boundary layers are therefore of essential importance in determining the stability properties of the problem. The boundary flow **u** satisfies the following condition

$$\frac{\partial(u_\theta/r)}{\partial r} = -\frac{\partial(U_\theta/r)}{\partial r}, \quad \frac{\partial(u_\phi/r)}{\partial r} = -\frac{\partial(U_\phi/r)}{\partial r},$$

on the outer spherical surface. The real part of the solvability condition (16) in the limit $P_r m^{-5/2} \ll E$ becomes

$$\frac{R}{E}\int_v |\nabla(T + \Theta_0)|^2 dV = \int_v \mathbf{U}^* \cdot \nabla^2(\mathbf{U} + \mathbf{u})dV,$$

the right-hand side of which at leading order is

$$H_b = \int_s 2(|U_\phi|^2 + |U_\theta|^2)dS - \int_v |\nabla \times \mathbf{U}|^2 dV.\qquad (20)$$

The expression for the integral H_b in equation (20) can be further simplified and the Rayleigh number R can be expressed as

$$R = E\int_s [U_\phi^* \frac{\partial(U_\phi/r)}{\partial r} + U_\theta^* \frac{\partial(U_\theta/r)}{\partial r}]dS \Big/ \int_v |\nabla\Theta|^2 dV = EH_b/H_\Theta,\qquad (21)$$

where H_Θ and H_b are analytical functions of m in closed form. Furthermore, H_Θ and H_b are pure real numbers, equation (17) indicating that

$$\omega_1 \int_v |\mathbf{U}|^2 dV = 0.$$

For studying the dependence of the character of the system on the Prandtl number without the limit $P_r m^{-5/2} \ll E$, we expand the total temperature Θ in spherical Bessel functions,

$$\Theta = \sum_{l,n}^{N} \Theta_{lmn} Y_l^m(\theta,\phi) j_l(B_{ln}r)\exp(i\omega t)$$

where Θ_{lmn} are complex constants and B_{ln} are chosen such that $j_l(B_{ln}) = 0, n = 1, 2, 3 \ldots$. It can be shown that the analytical expression for the Rayleigh number R is

$$R = \frac{E}{H_b} \sum_{l,n}^{N} \frac{[a_s Z_{ln}^1 + (d+b) Z_{ln}^2 + c B_{ln}^2 Z_{ln}^3]^2}{B_{ln}^{2m+8} j_{l+1}^2 (B_{l,n}) [P_r^2 E^{-2} \omega^2 + B_{ln}^4]}, \tag{22}$$

where H_b is given in equation (21), and $l = m, m+2$, $n = 1, 2..N$; Z_{ln}^1, Z_{ln}^2 and Z_{ln}^3 are associated with the following integrals:

$$\int_{-1}^{1} (1-x^2)^{m/2} P_l^m(x) dx, \quad \int_{0}^{B_{ln}} y^{m+2} j_m(y) dy, \quad \int_{0}^{B_{ln}} y^{m+4} j_m(y) dy,$$

which have simple analytical expressions (Zhang 1993a). A similar expression for ω can be also found with the form,

$$\omega = \omega_0 (1-h) \tag{23}$$

where h is a small positive constant: the effect of the Prandtl number is always to reduce the speed of the travelling wave.

5 DISCUSSION OF THE RESULTS AND REMARKS

The perturbation analysis on the basis of the Poincaré modes accurately predicts the critical Rayleigh number R_c, frequency ω_c and the azimuthal wavenumber m_c for convection in a thick spherical shell. For example, numerical analysis in a thick spherical shell with $\eta = r_i/r_o = 0.2$ for $E = 10^{-5}$ and $P_r = 0.02$ gives rise to the most unstable mode characterized by ($R_c = 23.9$, $\omega_c = -0.2537$, $m_c = 7$), while the minimization of R by (22) and the frequency given by (23) produce ($R_c = 25.68$, $\omega_c = -0.2548$, $m_c = 7$). A typical convection pattern at $E = 10^{-5}$, $m_c = 7$ and $P_r = 0.02$ is also illustrated in Figure 1, where a moving frame with the phase speed of azimuthally travelling waves is used to illustrate the stationary pattern. There are very few noticeable differences between the patterns of convection obtained from the numerical and analytical methods. Convective motions are trapped in the equatorial waveguide tube with characteristic latitudinal radius $(2/m)^{1/2}$ and radial radius $(1/m)$. Very little influence can be thus exerted on this type of convection by the presence of a small inner sphere. The agreement between the analytical and numerical analysis is still quite satisfactory for global scale convection, owing largely to the rapid decay of the convection from the outer spherical surface (Zhang 1993a).

In this paper, we have briefly reviewed the link between convective instability and inertial waves in rotating spherical fluid systems. It should be pointed out that this link has an important mutual benefit for both the problems. On the one hand, relatively simple solutions of the Poincaré equation shed light

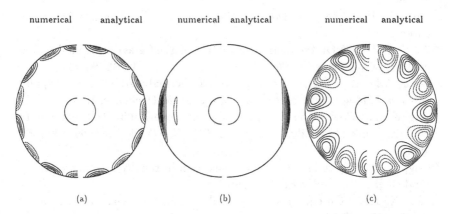

numerical   analytical   numerical   analytical   numerical   analytical

(a)   (b)   (c)

Figure 1. (a) Contours of u_ϕ at the equatorial plane, (b) contours of u_ϕ in a meridional plane and (c) contours of Θ at the equatorial plane.

on the understanding of rotating spherical convection, which not only enables us to obtain analytical solutions but also further studies of weakly nonlinear convection become feasible because of the availability of analytical linear solutions. On the other hand, thermal instability provides a mechanism by which inertial oscillation can be excited and sustained and a way to select an inertial mode by the mechanism of convective instability. In short, this link opens up a new line of research for fluid dynamics in rotating spherical systems.

Acknowledgements

This work is supported by the SERC GR/H55437 and partially supported by the Nuffield foundation. I would also like to take this opportunity to thank Dr. N. Barber for many useful suggestions and comments on various papers.

REFERENCES

Aldridge, K.D. and Lumb, L.I. 1987 Inertial waves identified in the Earth's fluid outer core. *Nature* **325**, 421–423.

Busse, F.H. 1970 Thermal instabilities in rapidly rotating systems. *J. Fluid Mech.* **44**, 441–460.

Chandrasekhar, S. 1961 *Hydrodynamic and Hydromagnetic Stability.* Clarendon Press, Oxford.

Fearn, D.R., Roberts, P.H. and Soward, A.M. 1988 Convection, stability and the dynamo. In *Energy, Stability and Convection* (ed. B. Straughan and P. Galdi), pp. 60–324. Longman.

Malkus, W.V.R. 1967 Hydromagnetic planetary waves. *J. Fluid Mech.* **28**, 793–802.

Greenspan, H.P. 1969 *The Theory of Rotating Fluids*. Cambridge University Press.

Roberts, P.H. 1968 On the thermal instability of a self-gravitating fluid sphere containing heat sources. *Phil. Trans. R. Soc. Lond.* A **263**, 93–117.

Soward, A.M. 1977 On the finite amplitude thermal instability of a rapidly rotating fluid sphere. *Geophys. Astrophys. Fluid Dynam.* **9**, 19–74.

Zhang, K. 1991 Convection in a rapidly rotating spherical fluid shell at infinite Prandtl number: steadily drifting rolls. *Phys. Earth Planet. Inter.* **68**, 156–169.

Zhang, K. 1992 Spiralling columnar convection in rapidly rotating spherical fluid shells. *J. Fluid Mech.* **236**, 535–556.

Zhang, K. 1993a On equatorially trapped boundary inertial waves. *J. Fluid Mech.* **248**, 203–217.

Zhang, K. 1993b On coupling between the Poincaré equation and the heat equation. *J. Fluid Mech.* (submitted).

Hydrodynamic Stability of the ABC Flow

O. ZHELIGOVSKY

International Institute of Earthquake Prediction Theory and
Mathematical Geophysics
Academy of Sciences USSR, 79 k.2, Warshavskoe Avenue
Moscow, 113556 Russia

A. POUQUET

Observatoire de la Côte d'Azur
CNRS URA 1362
BP 229, 06304 Nice Cedex 4, France

ABC flows, which may be good candidates for fast dynamos, are hydrodynamically unstable at Reynolds numbers R greater than a critical value R_C. We report here numerical results concerning the nonlinear stability of one such flow with $A = B = C = 1$ denoted $\mathcal{A}_1$, for which $R_C \approx 13.05$ when 2π-periodicity is assumed. When the bifurcation parameter $\epsilon = R/R_C - 1$ varies in the window $0.004 \leq \epsilon \leq 0.92$, a series of bifurcations towards progressively more complex flows takes place, with a relaminarization window and with modifications of symmetries. Temporal plateaus are observed for two different states – the initial $\mathcal{A}_1$ flow and a steady state $\mathcal{A}_2(\epsilon)$, probably revealing an underlying heteroclinic structure; for $\epsilon \geq 0.126$, they are separated by intermittent bursts involving all Fourier modes. Once $\mathcal{A}_2(\epsilon)$ emerges, it remains close to the *same* ABC flow $\mathcal{A}_2 \neq \mathcal{A}_1$ with $A \neq B = C$ for the rest of the ϵ-window. The intensified chaotic character of the flow as the Reynolds number increases occurs *without* significant amplification of the noise level in the non-ABC Fourier modes: spatially, the flow remains somewhat coherent; it stems from stronger and more numerous oscillations between these two ABC states. For $\epsilon = 0.92$, the bursts become contiguous and spatio-temporal turbulence appears to be nascent.

M.R.E. Proctor, P.C. Matthews & A.M. Rucklidge (eds.)
Theory of Solar and Planetary Dynamos, 347–354
©1993 Cambridge University Press.

1 INTRODUCTION

We give in this paper preliminary results on the nonlinear stability of the ABC flow whose velocity, with A, B and C constant coefficients, is given by

$$\mathbf{v}_{ABC} = (A \sin k_0 z + C \cos k_0 y, B \sin k_0 x + A \cos k_0 z, C \sin k_0 y + B \cos k_0 x)$$

where k_0 is its characteristic wavenumber and $E = \frac{1}{2}(A^2 + B^2 + C^2)$ is its energy. By *nonlinear stability* we mean the following: the flow $\mathbf{v} = \mathbf{v}_{ABC}$, which is a Beltrami flow $\nabla \times \mathbf{v} = \boldsymbol{\omega} = k_0 \mathbf{v}$, is a particular steady solution of the Navier–Stokes equations:

$$\frac{\partial \mathbf{v}}{\partial t} + \mathbf{v} \cdot \nabla \mathbf{v} = -\nabla p + \nu \nabla^2 \mathbf{v} + \nu \mathbf{F} , \qquad (1)$$

when the force is $\mathbf{F} = k_0^2 \, \mathbf{v}_{ABC}$. This solution, unique and stable for sufficiently large ν, bifurcates towards increasingly complex flows as ν decreases. Six Fourier modes are present in ABC flows: $(0, 0, \pm k_0)$ and cyclic permutations. The present study is restricted to the case of the ABC flow

$$\mathcal{A}_1 : \ A = B = C = 1; \qquad k_0 = 1, \qquad (2)$$

subject to perturbations with the same 2π-periodicity in x, y and z, thus ruling out any large-scale instabilities, although they are known to occur in the dynamo problem (Galanti *et al.* 1992).

The motivation for this work is dual. On the one hand, the ABC flows are often invoked in the context of fast dynamos. However, dynamos take place in both astrophysics and geophysics at high Reynolds numbers, for which the ABC flow is known to be linearly unstable. It may thus be of interest to investigate towards what states the ABC flows evolve in the nonlinear case, possibly to shed some light on what might be better candidates for fast dynamos. On the other hand, as a pure hydrodynamical problem, ABC flows may be viewed as 3D prototypes for the study of the route to spatio-temporal turbulence. The two problems are close but not identical (Friedlander & Vishik 1991).

2 NUMERICAL PROCEDURE

The emergence of spatio-temporal turbulence is a problem for which numerical data in 3D is scarce. Kida *et al.* (1989) have performed a study up to Reynolds numbers of 10^3 of the route to turbulence in a highly symmetric flow. Here, we consider the ABC flow (2) on whose underlying chaotic structure there is a wealth of data (Dombre *et al.* 1986; Galanti *et al.* 1992). We integrate in time equations (1) with $\nabla \cdot \mathbf{v} = 0$. Denote by C_K the spectral shell of unit width constructed around the wavenumber K, and by E_K the energy density in C_K. All runs are performed with the same initial perturbation $\tilde{\mathbf{v}}$ with equal energy in each shell $\tilde{E}_K = 10^{-6}$. The bifurcation parameter

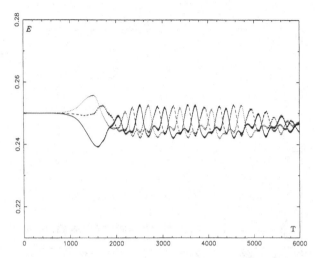

Figure 1. Evolution of the energy of the basic modes of the $\mathcal{A}_1$ flow for $\epsilon = 0.115$ ($R = 13.2$) in the doubly periodic regime. Solid line: mode (100); dashed line: (010); dotted line: (001).

is $\epsilon = R/R_C - 1$ with $R = \nu^{-1}$ the Reynolds number and R_C the critical value for the onset of linear instability; $R_C \approx 13.05$ for $\mathcal{A}_1$ (Galloway & Frisch 1987). For the runs reported here $0.004 \leq \epsilon \leq 0.92$ or $13.1 \leq R \leq 25$ (see also Galanti *et al.* 1992). Runs at higher Reynolds numbers are planned.

We use a pseudo-spectral code and periodic boundary conditions. The grid is uniform with resolutions from 16^3 to 64^3 points. The time-stepping scheme is either Adams–Bashforth or Runge–Kutta–Gill. The time-steps range from 0.05 to 0.005. Although the notion of eddy turn-over time is ill-defined for Beltrami flows, we choose it as a unit time for our computations, which are performed for up to $t = 2 \cdot 10^4$. The code runs at approximately 250 MFlops on the 8K-sequencer of the CM–2. Tests of both temporal and spatial precisions have been performed, as well as recovering the linear growth rate of the fastest growing mode of perturbation of $\mathcal{A}_1$.

3 THE THREE REGIMES

3.1 The weakly perturbed case: temporal chaos
For $0.004 \leq \epsilon \leq 0.027$, the perturbation stays at a low level $\tilde{E} < 10^{-2}$; it is found to have at first two main frequencies and then becomes chaotic. The global energy level is slightly lower (by 5%) than for $\mathcal{A}_1$. Once the perturbation has set in, after times ranging from 2000 to 500 for increasing ϵ, its temporal behavior consists of exchanges of energy between the three basic modes of $\mathcal{A}_1$, with all symmetries lost (see Figure 1). Other Fourier modes are

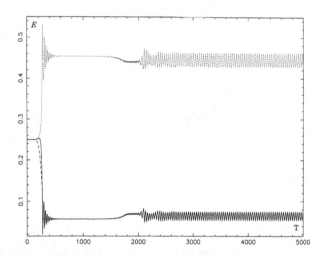

Figure 2. Evolution of the energy of the basic modes of the $\mathcal{A}_1$ flow for $\epsilon = 0.119\,(R = 14.3)$. Note the emergence of the $\mathcal{A}_2$ flow.

excited at a much lower level and all follow the same pattern (slave modes).

3.2 The persistence of a secondary ABC flow

After a relaminarization, the flow evolves for $0.034 \leq \epsilon \leq 0.53$ from a stable solution to a simply periodic one, to a doubly periodic one and finally to a chaotic regime. All have in common the following properties: (i) the energy of the perturbation $\tilde{E}$ is now significant, between 0.3 and 1.2; (ii) the flow $\mathcal{A}_1$ destabilizes sooner than in the weak regime; (iii) the dominant structure corresponds to a new steady flow $\mathcal{A}_2(\epsilon)$, close to the ABC flow $\mathcal{A}_2$

$$\mathcal{A}_2 : \quad A \approx 1.34; \quad B = C \approx 0.48; \qquad k_0 = 1 \tag{3}$$

without stagnation points and with an energy 75% lower than for $\mathcal{A}_1$; (iv) the energy of all but the six modes with unit wave-number remains low ($\approx 10^{-3}$) and remains on average *quasi-constant* with ϵ, as the system evolves from periodic to chaotic. After the flow has settled on $\mathcal{A}_2(\epsilon)$, oscillations occur (see Figure 2 for a run at $\epsilon = 0.119$) – sooner and with larger amplitudes as ϵ grows – up to $\epsilon \approx 0.12$.

The flow is chaotic at $\epsilon \geq 0.126$. Intermittency is present, with strong bursts occurring on periods ≈ 700; their average amplitude does not increase with ϵ either. During these bursts, $\tilde{E} \to 0$ sporadically; the flow has a behavior consistent with an underlying homoclinic ($\epsilon = 0.126$) structure (Šilnikov 1965) or heteroclinic ($\epsilon \geq 0.15$). At $\epsilon \geq 0.38$, the flow $\mathcal{A}_2(\epsilon)$ appears together with the other two steady flows symmetric to $\mathcal{A}_2(\epsilon)$, with the coefficients of

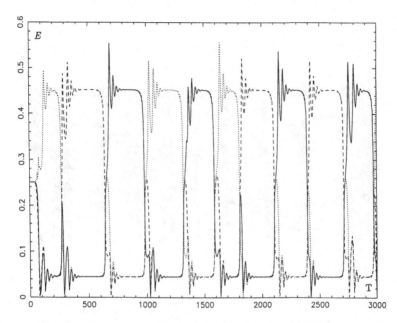

Figure 3. Evolution of the energy of the basic modes of $\mathcal{A}_1$ for $\epsilon = 0.53$ ($R = 20$). Note the emergence of two other flows, symmetric to $\mathcal{A}_2(\epsilon)$.

$\mathcal{A}_2$ permuted (Figure 3). Now, $\tilde{E} \geq 0.1$ in the bursts; they are more frequent but somewhat less intense. The relative helicity $\rho = \langle \mathbf{v} \cdot \boldsymbol{\omega}\rangle / \sqrt{\langle \mathbf{v}^2\rangle \langle \boldsymbol{\omega}^2\rangle}$ remains most of the time close to its maximal value (unity), indicative of a flow close to a Beltrami field. Only during the intermittent bursts does it depart strongly from unity, the smaller-scale modes participating then in the dynamics of the flow. Note again the coherence of the modes in all shells, and the increased perturbation levels during the intermittent bursts (see Figure 4).

3.3 The emergence of the spatio-temporal regime

In the last regime, for $\epsilon = 0.92$, the bursts corresponding to jumps between the steady flows are occurring too often for any structure to be recognizable, the $\mathcal{A}_2$-plateaus having almost totally disappeared although the $\mathcal{A}_2$ modes of lowest energy are still present. The flow is strongly chaotic, the energy of the perturbation is equal to that of the basic flow, the Beltrami property is partially lost (see Figure 5), characteristic times are shorter and, in space, smaller eddies become apparent. The energy in the shells E_K, $K \neq 1$ is rising, more so for higher K, and remain coherent in time; $E_2 \approx 0.02$ is now a factor of twenty above that of the previous regime but still somewhat weak for a fully turbulent flow.

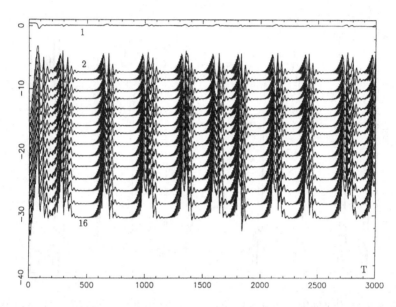

Figure 4. Temporal evolution of $\ln(E_K)$ for $\epsilon = 0.53$ ($R = 20$) in the spectral shells C_K, $1 \leq K \leq 16$ (a few are labelled).

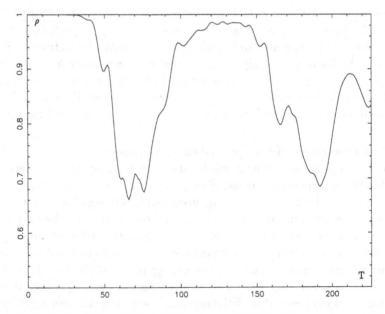

Figure 5. Evolution of the relative helicity ρ for $\epsilon = 0.92$ ($R = 25$).

4 CONCLUSION

We report here preliminary results on the nonlinear evolution of a perturbed ABC flow when the Reynolds number R is varied; a more detailed description will be given elsewhere (Zheligovsky & Pouquet, in preparation). The main result concerns the emergence of a heteroclinic structure with two steady states, the original $\mathcal{A}_1$ flow and $\mathcal{A}_2(\epsilon)$, remaining close to the ABC flows $\mathcal{A}_2$ given in (3); the increased level of turbulence stems from the increased temporal complexity of the transitions between these ABC states. The spatial structure, on the other hand, does not seem to rise in complexity as R progressively increases – all Fourier modes remaining at a constant level – except in the last regime: indeed the temporal complexity has become too high, and the intermittent bursts are now contiguous, calling for the quasi-permanent increased level of non-ABC modes, which can in turn interact nonlinearly. The same behaviour has been obtained with different initial perturbations, including when $\tilde{E}|_{t=0} \approx 1$.

In the domain of Reynolds number covered here, the solution is structurally dominated by a secondary steady flow $\mathcal{A}_2(\epsilon)$. Several questions arise: how to determine $\mathcal{A}_2$ analytically? What of the instability of $\mathcal{A}_2$ itself? At what ϵ does $\mathcal{A}_2(\epsilon)$ appear? Does $\mathcal{A}_2(\epsilon)$ exist for all $\epsilon > 0.27$, and how does it asymptotically approach $\mathcal{A}_2$, as $\epsilon \to \infty$? Does the system possess other steady states, rather than $\mathcal{A}_1$ and $\mathcal{A}_2(\epsilon)$, and other attractors? What is the spatial structure of $\mathcal{A}_2$ and how does it intertwine with $\mathcal{A}_1$? Is the instability for $\mathcal{A}_1$ (and possibly $\mathcal{A}_2$) fast, or does the growth rate depend on R? Also, other relaminarization windows may occur, and other chaotic states appear as the Reynolds number is increased further. These questions clearly need to be addressed but most of them require huge amounts of resources, given the long times of integration involved.

Furthermore, it is known that the flow $\mathcal{A}_1$ may not be the best candidate for a fast dynamo, in terms of magnitude of the growth rate. Indeed, there are other ABC flows with substantially larger growth rates of the magnetic field, for example (i) $A = B = C = 1$ and $k_0 \neq 1$; (ii) a two-dimensional ABC flow with $C = 0$ but with periodic in time polarization that reintroduces chaotic behavior (Galloway & Proctor 1992). In the former case, the fastest growing mode has a scale slightly larger than the basic ABC scale. In fact, among all possible ABC flows with the constraints $B = C$ and $k_0 = 1$, the flow that has both maximal magnetic field growth rate – a fact known only for moderate magnetic Reynolds number, however – and maximal Lyapunov exponent (Galanti *et al.* 1992) is 1:0.38:0.38, close to the $\mathcal{A}_2$ parameters. In the latter case, the role of chaos can be investigated on the growth rate of the instability, both for MHD (Ponty *et al.* 1993) and in hydrodynamics, using standard tools of chaotic dynamics in Hamiltonian systems. The linear and nonlinear stability of these flows should also be investigated.

We acknowledge fruitful discussions with V.I. Arnold, U. Frisch, D. Galloway and P.L. Sulem. O. Zheligovsky is grateful to Observatoire de la Côte d'Azur, where she has worked as a traveling scientist. The computations were performed on the CM-2 at INRIA, Sophia-Antipolis, France.

REFERENCES

Dombre, T., Frisch, U., Greene, J.M., Hénon, M., Mehr, A. & Soward, A. 1986 Chaotic streamlines in the ABC flows. *J. Fluid Mech.* **167**, 353–391.

Friedlander, S. & Vishik, M.M. 1991 Dynamo theory, vorticity generation and exponential stretching. *Chaos* **1**, 198–205.

Galanti, B., Sulem, P.L. & Pouquet, A. 1992 Linear and non-linear dynamos associated with ABC flows. *Geophys. Astrophys. Fluid Dynam.* **66**, 183–208.

Galloway, D.J. & Frisch, U. 1987 A note on the stability of a family of space-periodic Beltrami flows. *J. Fluid Mech.* **180**, 557–564.

Galloway, D.J. & Proctor, M. 1992 Numerical calculations of fast dynamos for smooth velocity field with realistic diffusion. *Nature* **356**, 691–693.

Kida, S., Yamada, M. & Ohkitani, K. 1989 A route to chaos and turbulence. *Physica* **37D**, 116–125.

Ponty, Y., Pouquet, A., Rom-Kedar, V. & Sulem, P.L. 1993 Dynamo in a nearly integrable chaotic flow. In this volume.

Šilnikov, L.P. 1965 A case of the existence of a denumerable set of periodic motions. *Sov. Math. Dokl.* **6**, 163–166.

Dynamos with Ambipolar Diffusion

ELLEN G. ZWEIBEL

Dept. of Astrophysical, Planetary, and Atmospheric Sciences
University of Colorado
Boulder, CO 80309-0391 USA

MICHAEL R.E. PROCTOR

Department of Applied Mathematics and Theoretical Physics
University of Cambridge, Silver St., Cambridge, CB3 9EW UK

Ambipolar diffusion, or ion-neutral drift, has important effects on the transport of magnetic fields in weakly ionized media such as the galactic interstellar medium. Ambipolar diffusion can inhibit the development of small scale magnetic structure because the field ceases to be kinematic with respect to the ions at strengths well below equipartition with the neutrals. On the other hand, magnetic nulls are characterized by steep profiles in which the current density diverges. The addition of ambipolar diffusion to mean field α-ω dynamos makes the equations nonlinear and can lead to steady states or traveling waves.

1 INTRODUCTION

The theory of linear, kinematic, mean field dynamos has been studied extensively since the pioneering paper by Parker (1955). In such dynamos, the mean magnetic field grows despite the action of resistivity through the combined action of small-scale, helical motions (α effect) and large-scale shear flows (ω effect). If the background state is time independent, the mean field evolves exponentially in time, and saturation of the field amplitude must occur through effects not included in the model.

Astrophysical systems typically have very low resistivities and correspondingly high magnetic Reynolds numbers R_m (of order 10^8–10^{10} in the Solar convection zone and 10^{18}–10^{20} in the galactic disk). This raises a problem for dynamo theory: if the resistivity is assumed to be molecular, the fastest growing wavelengths are extremely short and it is difficult to see how large scale fields could be generated. Moreover, the resistivity plays a central role

M.R.E. Proctor, P.C. Matthews & A.M. Rucklidge (eds.)
Theory of Solar and Planetary Dynamos, 355–362

in the calculation of the α effect (e.g., Moffatt 1978). Most workers therefore assume that turbulent resistivity is present.

A further complication in the application of dynamo theory to the galactic magnetic field arises because most of the mass of the interstellar medium is only weakly ionized. The magnetic field is therefore coupled directly to the charged component by the usual electrodynamic effects and only indirectly to the neutral component, through friction. An ion-neutral relative drift, or ambipolar drift, is inherent in this frictional coupling. Thus, if resistivity were completely absent, the magnetic field would remain frozen to the ions, but the magnetic flux through a surface comoving with the *neutrals* would vary over time.

Ambipolar diffusion affects dynamo activity within the galactic disk in at least three ways. First, and most obviously, the mean field is transported relative to the bulk fluid; this is probably most significant in the direction perpendicular to the galactic plane because the Galaxy is highly flattened. Second, the slight phase lag between magnetic and bulk fluid velocity fluctuations caused by ambipolar drift leads to an α effect if one sign of helicity predominates, much as occurs in a resistive medium (Zweibel 1988). Finally, the properties of fluctuations on small scales are affected substantially by ambipolar diffusion, and it appears that turbulent resistivity is therefore suppressed under certain conditions.

In this paper we discuss the third and then the first of these issues. Some of the work described here has been reported in greater detail earlier (Proctor & Zweibel 1992).

2 BASIC EQUATIONS

We take the medium to consist of two fluids, one charged and one neutral. For simplicity we suppose the charged fluid is constituted by electrons and one species of ion of mass m_i and particle density n_i and that a single neutral species of mass m_n and number density n_n is present. We make the following assumptions: (i) the ionized mass fraction is much less than unity, (ii) the ion-neutral collision frequency ν_{in} and ion mean free path are much less than any other timescale or lengthscale in the problem, and (iii) Reynolds stress terms in the momentum equation are negligible. Under these conditions the inertia of the ionized fluid is negligible and the Lorentz force on the ions balances the frictional force arising from collisions with neutrals (see e.g., Zweibel 1987). Equating these two forces yields the relative drift velocity $\mathbf{v}_D \equiv \mathbf{v}_i - \mathbf{v}_n$ between ions and neutrals:

$$\mathbf{v}_D = \frac{\mathbf{J} \times \mathbf{B}}{\rho_i \nu_{in} c}, \tag{1}$$

where $\rho_i \nu_{in} = m_i m_n n_i n_n \langle \sigma v \rangle / (m_i + m_n)$ and $\langle \sigma v \rangle$ is the collision cross-section averaged over the distribution of relative velocities. Now consider the magnetic induction equation

$$\frac{\partial \mathbf{B}}{\partial t} = \nabla \times (\mathbf{v}_i \times \mathbf{B}) + \lambda \nabla^2 \mathbf{B}, \qquad (2)$$

where λ is the resistive diffusion coefficient. Using the definition of $\mathbf{v}_D$, assumption (i), and Ampere's law, (2) can be approximated as

$$\frac{\partial \mathbf{B}}{\partial t} = \nabla \times (\mathbf{v} \times \mathbf{B}) + \nabla \times \left\{ \left[(\nabla \times \mathbf{B}) \times \mathbf{B} \right] \times \frac{\mathbf{B}}{4\pi \rho_i \nu_{\text{in}}} \right\} + \lambda \nabla^2 \mathbf{B}, \qquad (3)$$

where $\mathbf{v}$ is the center of mass velocity of the fluid. The second term on the right-hand side of (3) represents ambipolar diffusion, but it is only a diffusion-like operator for certain geometries. We now simplify (3) by considering a two-dimensional magnetic field in a slab, i.e., $\mathbf{B} = \hat{x} B_x(z,t) + \hat{y} B_y(z,t)$. With this form for $\mathbf{B}$, (3) reduces to

$$\frac{\partial \mathbf{B}}{\partial t} = \nabla \times (\mathbf{v} \times \mathbf{B}) + \frac{\partial}{\partial z} \frac{\mathbf{B}}{8\pi \rho_i \nu_{\text{in}}} \frac{\partial B^2}{\partial z} + \lambda \nabla^2 \mathbf{B}. \qquad (4)$$

Equation (4) is the basic equation for the remainder of this paper.

3 AMBIPOLAR DRIFT AND TURBULENT RESISTIVITY

Standard calculations have shown that a magnetic field can develop small-scale structure if it is passively advected by a flow. If the gradients in the magnetic field become sharp enough, resistivity becomes important. Although the flow need not be turbulent (see e.g., Moffatt & Kamkar 1983), when resistive effects arise in this way the resistivity is said to be turbulent. In a medium with large R_m, any resistivity must be turbulent resistivity. Recently it has been argued that Lorentz forces quench the buildup of structure on small scales and suppress turbulent resistivity (Cattaneo & Vainshtein 1991).

What is the effect of ambipolar diffusion on turbulent resistivity? We do not have a complete answer to this question, but we note two effects, which oppose each other. The first effect is due to the low inertia of the charged fluid component. Essentially, the field ceases to behave kinematically at strengths well below equipartition with the *neutrals*, slipping through the bulk fluid and therefore not becoming too strong or too tangled. On the other hand, if $\mathbf{B}$ vanishes, so does the ambipolar drift. This means that steep gradients can persist in the presence of a magnetic null, much as a fluid with temperature-dependent thermal conductivity can support sharply bounded thermal fronts (Zel'dovich & Raizer 1967). We give a simple example which illustrates both effects.

Consider a steady shear flow $\mathbf{v} = \hat{x} U(y)$ and a magnetic field of the form $\mathbf{B} = B_o(\hat{x} f(y) + \hat{y})$. We can integrate (4) once in y and obtain

$$U + (\lambda + \lambda_A f^2) \frac{df}{dy} = K, \qquad (5)$$

where K is a constant and λ_A, the ambipolar diffusion coefficient, is defined as $B_o^2/4\pi\rho_i\nu_{in}$. Consider the two limiting cases $\lambda_A = 0$ and $\lambda = 0$. In the first case,

$$f_R = \int_{y_o}^{y} \frac{(K - U(y'))}{\lambda} dy' \tag{6}$$

and in the second case

$$f_A = \left(\int_{y_o}^{y} \frac{3(K - U(y'))}{\lambda_A} dy' \right)^{1/3}, \tag{7}$$

where the subscripts R and A denote resistive and ambipolar diffusion, respectively. Consider the specific example $U(y) = U_o/(1 + (y^2/L^2))$. Then $f_R = (U_oL/\lambda)\arctan(y/rL)$ while $f_A = ((3U_oL/\lambda_A)\arctan y/L)^{1/3}$. Two points are noteworthy about this example. The first is that as $y \to \pm\infty$,

$$\frac{f_R}{f_A} \to \left(\frac{\pi^2 U_o L}{12\lambda} \right)^{2/3} \left(\frac{\lambda_A}{\lambda} \right)^{1/3}.$$

We evaluate this ratio for the interstellar medium, measuring B in μ-gauss, L in parsecs, n in cm^{-3}, temperature T in 100 K, U_o in km s^{-1} and let $x \equiv n_i/n_n$, $m_i/m_n = 10$. Then we find

$$\frac{f_R}{f_A} \sim 3 \times 10^{14} \left(\frac{BU_oL}{n_n x^{1/2}} \right)^{2/3} T^{3/2}.$$

Evidently ambipolar diffusion allows much more slip of the magnetic field with respect to the flow than does Ohmic diffusion, so the field remains much weaker.

The second point to note is the behavior of f_A as $y \to 0$: $f_A \sim y^{1/3}$. Thus $df_A/dy \sim y^{-2/3}$; there is a current sheet in the plane $y = 0$. This divergence is just what is needed to keep the diffusive flux constant as $f \to 0$. Of course, resistivity would resolve this singular layer, but for the parameters of the interstellar medium the current would still become extremely large.

Therefore, the role of ambipolar diffusion in the formation of current sheets is complex. The low inertia of the ionized component suggests that ambipolar diffusion inhibits the development of structure on small scales, a point also made recently by Kulsrud & Anderson (1992). On the other hand, singularities may occur at magnetic nulls – which, however, are probably rare.

4 DYNAMOS WITH NONLINEAR DIFFUSION

If we regard (4) as an equation for the mean field, assume that $\mathbf{v} = \hat{x}\Omega y$, and add an α effect, we have the standard model for the α-Ω dynamo in a slab, with the addition of nonlinear diffusion. We consider an interval $(0, L)$ in z

with spatially periodic boundary conditions on **B** and are interested only in states for which the azimuthal and radial fluxes are normalized:

$$\int_0^L B_x \, dz = LB_o \quad \text{and} \quad \int_0^L B_y \, dz = 0.$$

We then choose a unit of time $\tau = 8\pi \rho_i \nu_{\text{in}}/B_o^2$ and rescale the mean field equations so that they take the form

$$\dot{B}_x = \Omega B_y + \alpha B_y' + \lambda B_x'' + \left(B_x B^{2'}\right)', \tag{8a}$$

$$\dot{B}_y = -\alpha B_x + \lambda B_y'' + \left(B_y B^{2'}\right)', \tag{8b}$$

on the interval $(0,1)$ in z with dot and prime denoting time and space derivatives respectively, and $\lambda_A = 2$ in our units.

We first consider small perturbations about the state $B_x = 1$, $B_y = 0$ and assume the perturbations go as $\exp(pt + ikz)$. The linear dispersion relation corresponding to (8) is

$$p^2 + 2p(1 + \lambda) + \lambda(2 + \lambda)k^4 + ik\alpha(\Omega + ik\alpha) = 0. \tag{9}$$

If $\Omega = 0$, dynamo waves are amplified if

$$\alpha > \alpha_c \equiv 2\pi \left(\lambda(2 + \lambda)\right)^{1/2}, \tag{10}$$

where we have set $k = 2\pi$ to satisfy periodic boundary conditions. Note that if $\lambda = 0$ the waves are amplified; ambipolar diffusion alone does not amplify dynamo waves. When $\Omega > 0$, the critical α for instability is reduced below α_c by the factor $(1 + \Omega/(1 + \lambda)^2)^{1/2}$.

We now consider the full nonlinear equations. When $\Omega = 0$ we can qualitatively understand the solutions using analytical methods, although we also integrated the equations numerically. When $\Omega > 0$, we made little progress analytically beyond the linear theory, but found numerical solutions.

In the case $\Omega = 0$ the linear modes have zero phase velocity, so we look for steady state solutions of (8). Integrating once, and using the flux constraint to eliminate one constant of integration, we find

$$\alpha B_y + \lambda B_x' + 2B_x(B_x B_x' + B_y B_y') = 0, \tag{11a}$$

$$-\alpha B_x + \lambda B_y' + 2B_y(B_x B_x' + B_y B_y') = 0. \tag{11b}$$

Adopting a polar representation $B_x = \rho \cos \phi$, $B_y = \rho \sin \phi$ and manipulating (11), we arrive at

$$(\lambda + 2\rho^2)\rho' = -\alpha\beta \sin \phi, \tag{12a}$$

$$\lambda\rho\phi' = \alpha\rho - \alpha\beta \cos \phi, \tag{12b}$$

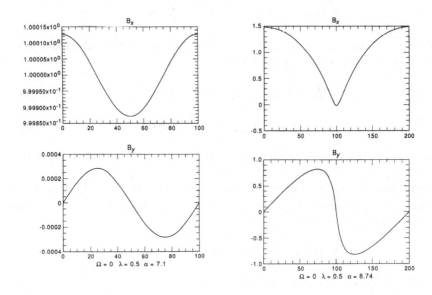

Figure 1. Magnetic profiles for two steady states with $\Omega = 0$. In the second pair of profiles, α is nearly equal to $\alpha_{\max}$.

with $\rho(0) = \rho(1)$, $\phi(1) = \phi(0) + 2n\pi$. We have considered only the cases $n = 0, 1$, corresponding respectively to the cases that B_x does not or does change sign within the interval.

Equations (12) possess a first integral; it may be verified that all solutions satisfy the relationship

$$\beta \cos\phi = \rho - \frac{\lambda}{2\rho}\left[1 - F\exp\left(\frac{-\rho^2}{\lambda}\right)\right].$$

The requirements $\phi(1) = \phi(0)$ or $\phi(1) = \phi(0) + 2\pi$ together with the flux constraint on B_x determine the integration constants F and β (Proctor & Zweibel 1992). It turns out that for any given λ, steady solutions exist only within a finite range for α: $2\pi\lambda \leq \alpha \leq \alpha_{\max}$, where $\alpha_{\max}$ must be determined numerically. For $\alpha_c < \alpha < \alpha_{\max}$ there are two solutions: a lower amplitude, apparently stable solution in which B_x does not change sign, and a larger amplitude, unstable solution in which B_x does change sign. For $\alpha > \alpha_{\max}$ there are no solutions, and numerical integration of the mean field equations leads to blowup in a finite time. For $2\pi\lambda < \alpha < \alpha_c$ there is only one (unstable) solution, which diverges in amplitude as $\alpha \to 2\pi\lambda$ from above. Plots of B_x and

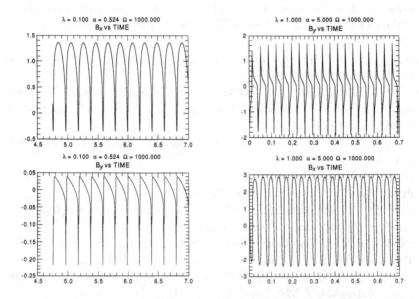

Figure 2. Magnetic profiles for two steady traveling waves with Ω large.

B_y versus z are shown in Figure 1. These plots were obtained by integrating the mean field equations until the solutions converged to a steady state. In the first pair of plots α is well below α_{max}; in the second pair α is nearly α_{max}. The steepening of the profiles and near-zero minimum of B_x is apparent in the second solution.

Next consider the case $\Omega > 0$. We expect traveling wave rather than stationary solutions. Lacking an analytical theory, we present results of some numerical integrations of initial value problems. We found that small perturbations are damped if linear theory predicts stability. If the perturbations are linearly unstable the solution either blows up or settles down to a traveling wave. As expected from linear theory, the phase velocity of the wave scales as $(\alpha\Omega)^{1/2}$ when Ω is large. Some examples of the waveforms are shown in Figure 2.

Acknowledgements

We are happy to acknowledge the hospitality of Princeton University and the Isaac Newton Institute, where part of this work was carried out. Additional support was provided by NSF Grant ATM 9012517 and by a University of Colorado Faculty Fellowship. A. Ruzmaikin provided a useful reference.

REFERENCES

Cattaneo, F. & Vainshtein, S. 1991 Suppression of turbulent transport by a weak magnetic field. *Astrophys. J.* **376**, L21–L24.

Kulsrud, R.M. & Anderson, S. 1992 The spectrum of random magnetic fields in the mean field dynamo theory of the galactic magnetic field. *Astrophys. J.* **396**, 606–630.

Moffatt, H.K. 1978 *Magnetic Field Generation in Electrically Conducting Fluids.* Cambridge University Press.

Moffatt, H.K. & Kamkar, H. 1983 The time-scale associated with flux expulsion. In *Stellar and Planetary Magnetism* (ed. A.M. Soward), pp. 91–111. Gordon and Breach, London.

Parker, E.N. 1955 Hydromagnetic dynamo models. *Astrophys. J.* **122**, 293–314.

Proctor, M.R.E. & Zweibel, E.G. 1992 Dynamos with ambipolar diffusion drifts. *Geophys. Astrophys. Fluid Dynam.* **64**, 145–161.

Zel'dovich, Y.B. & Raizer, Y.P. 1967 *Physics of Shock Waves and High Temperature Hydrodynamic Phenomena.* Academic Press.

Zweibel, E.G. 1987 The theory of the galactic magnetic field. In *Interstellar Processes.* (ed. D. Hollenbach & H. Thronson), pp. 195–221. Reidel.

Zweibel, E.G. 1988 Ambipolar diffusion drifts and dynamics in turbulent gases. *Astrophys. J.* **329**, 384–391.

Subject Index

Solar and Stellar Magnetic Fields

Galactic Magnetic Fields